Classical and Modern Mechanics

To my friend

ANDERS REIZ

(Director, Copenhagen Observatory)

James H. Bartlett

Classical and Modern Mechanics

The University of Alabama Press
University, Alabama

Contents

Contents

Contents

Preface

The present book is the outgrowth of three stages in the
author's experience: (1) as a graduate student, (2) as a
teacher, and (3) as a research worker. As progress was made
from one stage to the next, the emphasis changed from one
aspect of mechanics to another, with a resulting broadening
of viewpoint.

In the first stage, the material barely extended to the
Hamilton-Jacobi equation, where no clearcut presentation of
this equation, or of its utility, was given. There was the
traditional material on the calculus of variations, but it
all seemed to be mathematical jargon with no real motivation.
This tradition persists today, and mechanics courses rarely
go beyond the work of Hamilton, if indeed they get that far.

During the second stage, various excrescences were dropped,
and it gradually became evident, especially with the appearance of Minorsky's Non-Linear Mechanics, that mechanics has
progressed since the time of Hamilton and that there is a
"Modern Mechanics," which was essentially founded by
Poincaré. This subject is no more complicated mathematically
than the earlier "Classical Mechanics"; it can be presented
simply, and a student of today is really not well informed
unless he has become acquainted with it. Furthermore, its
great practical significance becomes apparent when one
attempts to design accelerators with beams which remain stable
during many revolutions, or when one wishes to learn about the
stability of various astronomical systems such as the solar
system or a galaxy.

The third stage has entailed (1) applying the Hamiltonian
methods to the restricted 3-body problem and (2) studying the
stability of this and of simpler systems. Since there has
not existed any one book which brought all the material
together, the author embarked on the task of writing one,
fully aware that much surgery and reorganization was necessary in order to keep the treatment within a reasonable length.

The primary aim of the book is to present, in as elementary
a way as possible, the main lines of development of classical
and modern mechanics. To this end, all excess baggage in the
form of subscripts, superscripts, and summation signs, and
footnotes, has been avoided where possible. If an idea can
be illustrated by using just two variables, this is regarded
as preferable to a general formulation with n variables. The
author believes that the story is truly fascinating, especially when it is told simply and directly.

The theoretical treatment is divided into four parts:
(1) elementary theory of Hamiltonian systems, in Chapters 1-2;
(2) advanced theory of Hamiltonian systems, in Chapter 8;
(3) elementary theory of autonomous systems, in Chapter 9;
and (4) the theory of non-autonomous, but periodic, systems,
in Chapter 10. When the reader has mastered the material in
these chapters, he should have enough of a feel for the
subject so that he can read further literature on specialized
topics.

Applications of the theory are given in other chapters of
the book. Chapter 3 introduces elliptic functions, which are
then used in detailed discussions of the top and pendulum,
and also later on (Chapter 7) in connection with experimental
tests of general relativity. Chapter 4 treats the Kepler
2-body problem, which is a limiting case of the circular
restricted 3-body problem, to which Chapter 12 is devoted.
Chapters 5 and 6 contain discussions of elementary electromagnetic theory and of the dynamics of charged particles, and
are a necessary introduction to part of Chapter 7, to part of
Chapter 10, and to Chapter 11. The latter chapter is on the
motion of charged particles in the field of a magnetic dipole,
and this material has been useful in the analysis of cosmic
rays. (This is admittedly an idealized case, to be modified
for the earth and for pulsars.)

Chapter 7, on the relativistic dynamics of particles, has
been included at the suggestion of an anonymous reviewer.
It discusses in some detail the foundations of the theory of
special relativity, and then later shows how to treat the
motion of a particle in a spherically symmetrical gravitational field, with Einstein's equations presumed as given.
This presentation is the outcome of a semester course on
special and general relativity, given by the author at the
University of Alabama. Experience has shown that the average
physics graduate student is not overly interested in how
Einstein arrived at his generalization of Poisson's equation,
and hence such a student is only too happy if the differential
geometry is dispensed with and Einstein's equations are
written down directly as a good heuristic guess. (The

student who thirsts for the differential geometry can easily
find it in other texts, to some of which reference is made
at the end of this book.) The solution of Einstein's
equations gives the "metric," which is essentially another
word for the (gravitational) kinetic term in the Lagrangian
function. With this, one can then proceed in a straight-
forward manner to find the motion of a particle in a gravitational field. As long as the student keeps firmly in mind
that this is what is going on, he will have his feet firmly
on the ground (in a world with three space dimensions and one
time dimension) and should not get lost. (At least, no more
than the rest of us, with our anthropomorphic cosmology
extrapolated to the ends of space and of time.)

Chapter 8 is mainly concerned with the way in which a
collection of particles, in the form of a beam, spreads out
or is focused in space. It starts with a discussion of how
geometrical optics is the limiting case of wave optics when
the wave length goes to zero, and shows that the same
mathematics is applicable for geometrical mechanics as a
limiting case of wave mechanics. The concept of a contact
transformation is introduced and applied (via the Huygens'
construction) to the determination of the spreading of a wave
front or surface of constant action. Variations in initial
conditions give rise to variations in the subsequent motion,
and it is shown how area in the phase plane is preserved for a
Hamiltonian system. (This method more or less replaces the
standard perturbation theory, which is somewhat superfluous
when a computer is available.) Transformation of the equations
of motion to a different coordinate system is accomplished
neatly and with minimum effort by the use of an invariantly
connected differential form. (This is useful for reducing
the order of a system which has an energy integral.) Thus,
the entire theory is presented without reference to minimum
(or extremum) principles, which may have intrinsic mathematical interest but which are not useful for the further work
in this book.

Chapter 9, although limited to autonomous systems (where
the time does not enter explicitly in the equations of motion),
introduces the methods of Poincaré. There is a discussion of
singular (or critical) points, characteristic exponents,
stability, and limit cycles (periodic orbits). This uses as
simple examples the damped harmonic oscillator and the Van
der Pol oscillator. For the latter, Lefschetz's topological
display of the results is quite illuminating.

Chapter 10 concerns those non-autonomous systems for which
the coefficients in the equations of motion are periodic.
The question of primary interest here is whether or not

such a system will be stable over a long period of time. To
resolve this, it suffices to sample the motion at intervals
of one period, or in other words to see how a representative
point in phase space is transformed or mapped when multiples
of one period has elapsed. This method of approach was
initiated by Poincaré, and continued at first by Birkoff, and
later by Arnol'd and Moser, and still later by the author and
his collaborators. The problem of determining the boundary
of eternal stability to a high degree of precision is one
which has so far not been solved, but it does seem probable,
on the basis of computer experiments, that such a boundary
exists, i.e., that there is a stable region. The chapter
begins with the linear case (Mathieu-Hill equation) and its
applications, moves on to a simple non-linear equation, and
concludes with a treatment of a simple area-preserving mapping
(since area preservation is the one feature that Hamiltonian
systems, even though periodic, have in common).

   Chapter 12 brings together the main features of the restricted (circular) 3-body problem, including the topology of the
periodic orbits and a short discussion of the stability in
general. This problem, which up until 1960 was still somewhat
of a mystery, has now, thanks to computer work by Bartlett,
Hénon and others, been essentially solved in all but minor
details.

   The epilogue, plus the bibliography, should give the reader
an idea of what has been accomplished and where to turn for
further specialized reading.

   Various anonymous reviewers are to be thanked for sugges-
tions on how to improve the original manuscript. The final
version has been read in its entirety by Dr. W. C. Saslaw
and by Dr. Joseph Ford. Dr. Saslaw suggested the inclusion
of methods of moments, adiabatic invariants, and of the
epilogue. Dr. M. Hénon has read and constructively criticized
Chapter 10. In addition, Mr. J. C. Clayton and Dr. M. Cahill
have edited Chapters 1, 2, 7 and 8. The author is deeply
indebted to all the above scientists for their efforts, and
to Mrs. Helen Snider for her excellent work in preparing the
typed manuscript. Finally, acknowledgment must be made to
Mr. James Travis of The University of Alabama Press for his
constant encouragement.

Classical and Modern Mechanics

Introduction

Ever since the time of NEWTON, the subject of dynamics, i.e., how objects move under the action of forces, has received much attention from both physicists and mathematicians. For simplicity, we shall study mainly the motion of groups of objects (particles), each of which is concentrated at a point.

If a particle is constrained to move in the x-direction, then Newton's second law states that the motion is governed by the second order differential equation

$$m\ddot{x} \equiv m(d^2x/dt^2) = X(x,t) \qquad (1)$$

where m is a constant representing the mass, and X is some given function of x and the time t. The <u>dynamical problem</u> <u>consists in the integration of this differential equation</u>, i.e., to find a function $x(t)$ which satisfies the equation. If $\partial X/\partial x$ is bounded (see Appendix A), we can find many such functions, but only one for which $x(t_o) = x_o$ and $\frac{dx}{dt}(t_o) = \dot{x}_o$, where x_o and $\dot{x}_o$ are preassigned values of position and velocity at time $t = t_o$.

One method of integration, provided X is finite, is the graphical one. One can, on a plot of x vs. t, start the solution off with a short arc which goes through x_o, with the slope of its tangent $\dot{x}_o$, and with curvature $d^2x/dt^2 = X(x_o\ t_o)/m \equiv X_o/m$. Then at time $t_o + \Delta t$, where Δt is small, repeat the process with an arc through $x = x_o + \dot{x}_o \Delta t$, $\dot{x} = \dot{x}_o$ + $X_o \Delta t/m$ and $\ddot{x} = X(x,t)/m$. In this way, we can rather laboriously construct an approximate solution. The accuracy can be improved considerably by the use of local power series expansions, and thus a single trajectory can be found.

For some dynamical problems, it is possible to perform one integration analytically, which results in a first integral. In the present instance, this happens if X does not depend explicitly on t, i.e., if X is a function of x alone. Equation (1), when multiplied on both sides by $\dot{x}dt = dx$, yields $m \dot{x} \, d\dot{x} = X \, dx$, which can be integrated immediately to give

$$\frac{1}{2} m \dot{x}^2 - \int_a^x X \, dx = h \ (= \text{const.}) \tag{2}$$

or $$T + U = h \tag{3}$$

where $T = m\dot{x}^2/2$ depends only on the velocity $\dot{x}$ and is called the kinetic energy, $U = -\int_h^x X \, dx$ depends on the position x and is termed the potential energy, while h is the energy integral, which is here the sum of the kinetic and potential energies.

Naturally, if a first integral does exist, this makes the solution of the problem that much easier. Equation (3) can be rewritten as

$$dt = (m/2)^{1/2} \ [h - U(x)]^{-1/2} \ dx$$

and the further integration may be carried out, analytically or numerically, giving t as a function of x. Accordingly, it is always well to find out if the problem is of a type such that a first integral exists, and then to make maximum use of this fact. Equation (3) is said to express conservation of energy, while other first integrals express conservation, or constancy, of linear momentum and angular momentum.

When the motion takes place with more than one degree of freedom, or when more than one particle is involved, or when it is desired to refer to a rotating coordinate system, then it is of interest to ascertain whether or not first integrals can be found. The systematic treatment of such systems is due to LAGRANGE, who showed in particular how one can exploit any symmetry of the force field and write down equations equivalent to Newton's equations in the appropriate coordinate system, and thus have a better insight into the dynamical relations. If one is given three particles, for instance, whose initial positions and velocities are known, and if the interparticle forces are assumed to be known, then a straight-forward integration may be performed by a computer and the subsequent motion determined as a function of time. The methods of Lagrange are thus quite appropriate for finding the motion of one particle, or one system of particles, when the initial conditions are given.

Such was the state of dynamics prior to the work of HAMILTON. The problem to be solved was a rather narrow one, for one chose a particular set of initial conditions and inquired into the details of the motion as a function of time. For each set, one must make the appropriate calculation, and that can be done. However, this process can be quite tedious if many trajectories are to be studied.

Hamilton asked the questions: (1) How does the motion depend upon the initial conditions? (2) What happens to a beam of particles when it encounters a force field? (3) Can a beam be focussed, and what are the imaging relations? (4) If particles of a given energy leave a certain point, where can they get to? The emphasis here is not on the time

relations between the trajectories, but upon their space relations. This part of the dynamical problem is one of geometrical optics, and we shall see later how the mathematics for
beams of massive particles is equivalent to that for beams of
photons. Hamilton's great contribution was to stress that one
must study both the time and the space relations of a trajectory if a comprehensive attack on the problem is to be made.
Even though Hamilton died over a hundred years ago (Sept. 2,
1865), the usual course on dynamics and the associated text-
books do not cover adequately these methods of Hamilton, and
the very important subsequent researches in dynamics by
POINCARÉ, LIAPUNOV, and BIRKHOFF are completely ignored. We
shall try to remedy this in the present treatment.

The equations of dynamics are in general not easy to solve
because of their non-linearity, i.e., because the function X
in Equation (1) is seldom proportional to the coördinate x.
For this reason, physicists (who would like to explain all
natural phenomena in terms of linear equations) have tended
to avoid the subject. On the other hand, mathematicians (for
whom the infinite holds no terror) have regarded dynamical
problems as a challenge, and have had some success in predict-
ing what may be expected. The discipline is neither dead nor
static, and much theoretical progress has been accomplished
in recent years. However, since the subject is vast and complex, there is much room for future development.

From the beginning, much time and effort has been spent in
studying periodic motion, for there are many dynamical
systems, as for example the planets moving around the sun,
where the motion is periodic and for which the theoretical
predictions can be compared with the observations. Periodic
motion can occur for a wide variety of cases, namely for
systems (a) where an energy integral exists, (b) where the

differential equation is of the <u>autonomous</u> form

$$\ddot{x} + f(x, \dot{x})\dot{x} + g(x, \dot{x}) = 0 \qquad (4)$$

i.e., where the time is not involved explicitly but where
energy is not conserved, and (c) where there is explicit
periodic dependence on time, such as $X(x,t) = p(t)f(x)$ with
$p(t)$ periodic. The associated mathematics is more complicated
than previously, because the simple dynamical problem involved
solving a differential system subject to certain initial conditions, while here we must find <u>solutions of a non-linear</u>
<u>differential equation</u> (system) <u>which satisfy given initial</u>
<u>and final conditions</u>. (The technique of obtaining the solu-
tions is no more difficult than when the equation is linear).

Where the time does not enter explicitly, as in the <u>autono-</u>
<u>mous</u> equation (4), it may be eliminated by the substitution
$y = \dot{x}$, the integration may be partly accomplished, and the
integral curves in the <u>phase plane</u> (x,y) may be examined.
Equation (4), which becomes $y \, dy/dx = - fy - g$, was replaced
by

$$dy/dx = Y(x,y)/X(x,y) \qquad (5)$$

where X and Y are well-behaved functions of x and y, and
studied first by BRIOT and BOUQUET and then later by
POINCARÉ. The former concerned themselves with the behavior
of the solutions of Equation (5) near <u>singular points</u>, i.e.
points where $X = 0$, $Y = 0$. POINCARÉ did not limit himself
to this <u>local</u> behavior, but obtained qualitative (topological) results about how the integral curves looked in the
whole plane in relation to the singular points. In consequence of his work, it became easier to look for periodic
solutions of (5). One can therefore discuss with intelli-
gence some of the basic properties of autonomous systems,
which are encountered very commonly even when energy is not

conserved, as with various electronic devices and with diffusion-controlled reactions.

Non-autonomous systems are less familiar and are more difficult in general to deal with, but the case where the explicit dependence on the time is periodic is amenable to treatment. Perhaps the most simple instance is where the right-hand side of Equation (1) is $X = - p(t)x$, with $p(t)$ periodic in time. The theory is due to FLOQUET, who showed that the essential thing is a knowledge of <u>how the $(x, \dot{x})$</u> <u>plane maps into itself</u> when the time has increased by exactly one period τ, or how any point $P(x, \dot{x})$ is transformed into another point $T(P) = (x', \dot{x}')$. This theory was later applied by MATHIEU and HILL to special cases, and has become of importance in discussions of stability, energy bands in solids, and in the design of particle accelerators. It is somewhat limited because X is proportional to x, but this restriction can be removed (the present author has taken $X = p(t) x^3$) and one can find the mapping of the $(x, \dot{x})$ plane for a general $X = p(t)f(x)$.

Finally, even though one has located a periodic orbit, it is of very practical importance to determine whether or not it is <u>stable</u>, and it is this question which has occupied the minds of many mathematicians (LIAPUNOV, POINCARÉ, BIRKHOFF, ARNOL'D and MOSER) for most of the last century. The periodic orbit is defined to be <u>stable</u> if a nearby motion stays in the neighborhood as $\tau \to \pm \infty$, but unstable otherwise. For instance, suppose that a body of negligible mass, executes a periodic motion in the plane of the motion of the earth and moon, which might be considered to rotate in circles about their center of mass. This is known as the <u>restricted 3-body</u> <u>problem</u>, and has many types of periodic motion. An astronomer would like to know

(1) If a small mass originally moves in a periodic orbit
     and is then displaced slightly, will it move away from
     the vicinity of this orbit?

(2) Will the small mass then seek out the vicinity of some
     other periodic orbit and stay there?

(3) Where are the regions in which a small mass will stay,
     and where are the regions it will shun?

Some of these questions concern local stability and some
concern stability in the large. Whereas some progress has
been made with the theory of local stability, the overall
picture is still obscure and much work remains to be done on
it before we can say the theory is adequate. There is some
prospect that motion with one degree of freedom will be
understood in the not too distant future, while that for two
degrees of freedom is considerably more complicated and
because of that all the more difficult to handle.

I

Simple Motions

Newton's second law of motion for one degree of freedom, namely

$$m\ddot{x} = X(x,t) \qquad (1.1)$$

applies to a particle moving at slow speeds in the x-direction and presupposes (1) that there is a constant m, called the mass, which is a property of the particle and does not depend upon its position or velocity, and (2) that there is a function X of x and t, called the force, which determines the acceleration through Equation (1.1).

If we know the velocity of the particle when it is at position x at time t, then the trajectory can be found graphically as outlined in the introduction, regardless of whether the force X depends on t or not. If it does not, then multiplication of (1.1) by $\dot{x}\,dt$ on both sides results in $m\dot{x}\,d\dot{x} = X\,dx$ which upon integration yields

$$\Delta T = (m\dot{x}^2/2) - (m\dot{x}_o^2/2) = \int_{x_o}^{x} X\,dx \qquad (1.2)$$

where we define $T = m\dot{x}^2/2$ as the _kinetic energy_, a quantity associated with motion, and ΔT denotes its change during the motion from x_o to x.

The integral on the right is defined to be the work done on the particle, by the force X, during the displacement from x_o to x. Equation (1.2) states that <u>the change of kinetic energy of a particle equals the work done by the force acting on it</u>. This integral depends on the initial coordinate x_o and the final coordinate x, and not on the time. If we want to have a conservation law, it is natural to say that an increase of kinetic energy would be due to a decrease of another kind of energy called <u>potential energy</u>, and to define the potential energy U(x) as a function of x such that its change is

$$\Delta U = U(x) - U(x_o) = -\int_{x_o}^{x} X\ dx \qquad (1.3)$$

which with (1.2) gives

$$\Delta T + \Delta U = 0$$

and the <u>total</u> energy, defined as T + U, is constant, i.e.

$$T + U = h \qquad (1.4)$$

This is the expression for the <u>conservation of energy</u>, in perhaps its simplest form. Another formulation is that (1.4) states that a certain function of the velocity plus another function of position (but not of time) is a <u>constant of the motion</u>. (Later on we shall have occasion to replace (1.4) by other equations which hold when the situation is more complicated, e.g. when the mass varies with the velocity.)

From (1.3), the force is related to the potential energy by the equation

$$X = -\ dU/dx$$

where the potential energy is assumed to be a smooth function of distance. It may become infinite at certain points, and hence the above equation is only to hold away from these points.

Rectilinear Motion for a General U(x).

Inspection of the general nature of the potential energy
curve can tell us much about the motion, so that we can say
qualitatively what happens, without necessarily integrating
the equations in detail. Inserting $m\dot{x}^2/2$ for T in (1.4) and
solving for $\dot{x}^2$,

$$\dot{x}^2 = (2/m)\ (h - U(x)) \equiv f(x) \qquad (1.5)$$

The square of the velocity is a definite function of position, denoted by $f(x)$, for a given value of the total energy
h. [For problems involving more than one dimension, one may
encounter situations where $f(x) = a(b - U^*(x))$, with a>0 and
b some constants and $U^*(x)$ an effective potential energy.
The following discussion will be applicable to this case as
well]. The motion is confined to the region where $f(x)$ is
positive or zero, and the velocity will be positive or negative for any $f(x) > 0$.

The acceleration is given by

$$\ddot{x} \equiv \dot{x}\,\frac{d\dot{x}}{dx} = \frac{1}{2}\frac{df}{dx} = -\frac{1}{m}\frac{dU}{dx} \qquad (1.6)$$

where (1.5) has been differentiated re x.

If the slope of the potential energy is negative, the force
and acceleration are positive and will cause a particle initially at rest to move toward greater values of x. If the
slope is positive, the particle would move toward smaller x,
while if the slope is zero, the particle will stay put indefinitely at its initial rest position $x = x_o$, which is then
called a position of equilibrium.

For the problems ordinarily encountered, U(x) is a well-
behaved function, having at most a singularity such as $1/x$ or
$1/x^2$ at x = 0. Apart from this, U(x) will have a finite non-

zero slope except at a maximum, minimum or inflection point.
Figure 1.1 shows a plot of such a function $U(x)$. For $h = h_1$
there are non-zero slopes of $f(x)$ at x_1, x_2, and x_3, while for
$h = h_4$, there is one at x_6. At each of these points, from
(1.6), the acceleration is finite and will cause a particle
initially at rest there to move. Motion from rest at x_6, x_1,
and x_3 will be toward more positive x, and from x_2 toward more
negative x.

For a short time, the motion will be that in a uniform field,
e.g. near x_1

$$x = x_1 + \frac{1}{2} a\, t^2 \qquad\qquad (1.7)$$

where $a = -(1/m)\ [\partial U/\partial x]_x$. The velocity is $\dot{x} = a\, t$, so that

$$x = x_1 + (\dot{x}^2/2a) \qquad\qquad (1.8)$$

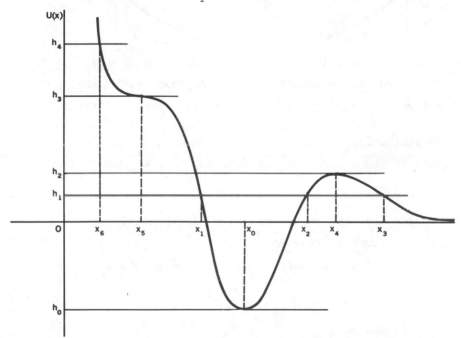

Figure 1.1.   Potential Energy vs. Distance

Figures 1.2 and 1.3 show the parabolas which
are just approximate representations of the motion

near the ends of its path. For $h = h_1$, the motion goes back
and forth between x_1 and x_2, with the velocity, from (1.5),
being (1.9): $\dot{x} = \pm (f(x))^{1/2}$. It thus describes a closed
oval in the <u>phase plane</u> $(x,\dot{x})$, symmetric about the x-axis.
Such a motion is periodic and is called a <u>libration</u>.

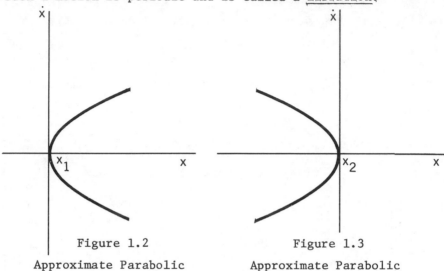

Figure 1.2	Figure 1.3
Approximate Parabolic	Approximate Parabolic
Motion near x_1 (Fig. 1.1)	Motion near x_2 (Fig. 1.1)

The period is

$$\tau = 2 \int_{x_1}^{x_2} \frac{dx}{\{f(x)\}^{1/2}} \qquad (1.10)$$

Motion near the points x_3 and x_6 is obviously of the same
general nature as that near x_1.

Special attention must be paid to the points x_0, x_4 and x_5
where $\partial U/\partial x = 0$.

<u>Harmonic Oscillator</u>

Near x_0 the potential energy $U(x)$ can usually be approximated by a parabola $U = h_0 + \frac{1}{2} U_2 (x-x_0)^2$ $\qquad$ (1.11)

The force is then $F = - U_2 (x-x_o)$, a restoring one proportional to the displacement from equilibrium.

The equation of motion is, from (1.6),

$$\ddot{x} = - \frac{U_2}{m} (x - x_o) \qquad (1.12)$$

and has for a solution, if $\omega^2 = U_2/m$,

$$x - x_o = (x_2 - x_o) \sin \omega t \qquad (1.13)$$

Because this is sinusoidal, the particle is said to execute <u>harmonic oscillations</u>. The total energy is, with x_2 one limit of the motion,

$$h = U(x_2) = h_o + \frac{1}{2} U_2 (x_2 - x_o)^2 \qquad (1.14)$$

Thus, if we fix h arbitrarily and find x_2 graphically, the quantity U_2 can be calculated and then the frequency $\omega = (U_2/m)^{1/2}$.

The motion in the phase plane is, from (1.5), along the ellipse

$$\frac{m}{2} \dot{x}^2 + \frac{U_2}{2} (x - x_o)^2 = h - h_o \qquad (1.15)$$

It always stays near the equilibrium point x_o, which is then defined to be a <u>point of stable equilibrium.</u>

Limitation Motion

Near x_4, let the curve $U(x)$ be approximately a parabola with negative curvature.

$$U = h_2 - \frac{1}{2} U_2 (x - x_4)^2 \qquad (1.16)$$

with the force always away from x_4,

$$F = U_2 (x - x_4)$$

and the equation of motion

$$\ddot{x} = \frac{U_2}{m} (x - x_4)$$

The solution is

$$x - x_4 = Ae^{\alpha t} + Be^{-\alpha t} \qquad (1.17)$$

with $\alpha^2 = U_2/m$.

At time $t = 0$,

$$x^o - x_4 = (A + B), \quad \dot{x}^o = \alpha(A - B) \qquad (1.18)$$

If the total energy is h_2, then the motion in the phase plane obeys the equation

$$\frac{m}{2} \dot{x}^2 - \frac{1}{2} U_2 (x - x_4)^2 = 0 \qquad (1.19)$$

Substituting (1.18) into this, $(A - B)^2 = (A + B)^2$, which means that either $A = 0$ or $B = 0$.

If $A = 0$ and $x^o < x_4$, then $B < 0$ and $\dot{x}^o > 0$. The motion obeys the equation

$$x - x_4 = (x^o - x_4) e^{-\alpha t} \qquad (1.20)$$

The particle moves constantly toward more positive x, but is slowed because of the acceleration in the opposite direction, and takes an infinite time to reach the point of equilibrium $x = x_4$. This is responsible for the term limitation motion.

If $B = 0$ and $x^o < x_4$, then $A < 0$ and $\dot{x}^o < 0$. The motion is then according to

$$x - x_4 = (x^o - x_4) e^{\alpha t} \qquad (1.21)$$

Now the particle moves steadily toward more negative x, accelerating all the time. For this reason, the point $x = x_4$ is called a point of unstable equilibrium. Even though there is one motion which will eventually get close to it, there is another motion which flies away. This is in contrast to the motion near the minimum of U, where the particle continually stays in the vicinity of the minimum at $x = x_o$, which is therefore called a point of stable equilibrium.

Motion Near Inflection Point

Let us suppose we have an inflection point such as at x_5, but let it be at the origin, for simplicity. Then the motion can be essentially as

$$\dot{x}^2 = ax^3 \qquad\qquad (1.22)$$

The acceleration is

$$\ddot{x} = \frac{3}{2} ax^2$$

This is always toward more positive x, so that the point of equilibrium is unstable.

Integrating (1.22)

$$a\, t = -\, 2(x^{-1/2} - (x^\circ)^{-1/2})$$

A particle which starts at some initial $x^o > 0$ and has negative velocity will reach the origin in infinite time. This is again a limitation motion.

Motion Near Singularity

Suppose that $U(x) = -\kappa/x$, the potential for an attractive Coulomb field, and let a particle go toward $x = 0$. Equation (1.5) now becomes

$$\dot{x}^2 = (2/m)\ [h + (\kappa/x)]$$

Then $\qquad\qquad dt = -\ (m/2)^{1/2}\ dx/[h + (\kappa/x)]^{1/2}$

If x is small, κ/x is the dominant term, so let us neglect h. This gives

$$t = \frac{2}{3}\,(m/2\kappa)^{1/2}\,x^{3/2}$$

as the finite time for a particle to travel from x to the origin.

Motion Toward Infinity (Repulsive Field)

Let now $U = -x^n$ be the dominant term in the potential energy as x becomes large. If $n = 2$, this is the case already treated, see (1.16) et seq. The motion toward infinity is governed by (1.21), namely,

$$x - x_4 = (x^o - x_4)\,e^{\alpha t}$$

(where x^o is the value of x at $t = 0$) and the particle will take an infinite time to reach infinity.

If $n = 3$, then, if κ is a constant

$$\frac{dt}{dx} = \kappa\,x^{-3/2} \text{ and } t = 2\kappa\,x^{-1/2}$$

The <u>particle requires only a finite time to reach infinity</u>, which is the general behavior when $n > 2$. This is because of the way in which the acceleration increases during the motion.

If $n = 1$, the force is constant and $\dot{x}^2 = (2/m)\,(h + x)$. As $x \to \infty$, h can be neglected. Then

$$x^{-1/2}dx \cong (2/m)^{1/2}dt$$

which integrated to $t \propto x^{1/2}$. The particle will take an infinite time to reach infinity. This holds for $n \leq 2$.

Motion in General U(x) - Summary

We have now shown several properties of the motion

in a general conservative field $U(x)$.

1. When, at $x = x_0$, $U(x)$ has a minimum, this point is one of stable equilibrium. In the immediate neighborhood, a particle can execute periodic motion back and forth between $x_1 < x_0$ and $x_2 > x_0$, where x_1 and x_2 are two roots of $U(x) = h_1$ ($h_1 < h_2$).

2. When $U(x)$ has a finite slope, the force is finite and non-zero and will cause a particle to move from rest.

3. When $U(x)$ has a maximum, at $x = x_4$, this point is one of unstable equilibrium. A particle can start out with just the right velocity to arrive at x_4 with zero velocity, but an infinite time will be required to get there. (The same is true if $U(x)$ has an inflection point.)

4. A particle in a Coulomb field will move to the attracting centre in a finite time.

5. For a repulsive field $U(x) = - x^n$, a particle can move to infinity in a finite time if $n > 2$. If $n = 2$, $t = \ln x$, so that the particle goes to infinity in infinite time. If $n = 1$ (constant force) t varies as $x^{1/2}$, and is likewise infinite for x infinite, as it is for $n < 2$ generally.

Conservation of Linear Momentum

Newton's law that action and reaction are equal and opposite leads to an alternative formulation of (1.1). If two particles of unequal mass interact, so that the particle of mass m_2 exerts a force X on the particle of mass m_1, then

$$m_1 \ddot{x}_1 = X \text{ and } m_2 \ddot{x}_2 = - X$$

Consequently

$$m_1 \ddot{x}_1 + m_2 \ddot{x}_2 = 0$$

and $m_1\ddot{x}_1 + m_2\ddot{x}_2 = $ const.

The quantity $p_x = m\ddot{x}$ is defined to be the momentum associated with mass m. The last equation states that <u>if two</u>
<u>particles interact, then the increase of momentum of one</u>
<u>particle equals the decrease of momentum of the other</u>. This
is known as the <u>Law of Conservation of Linear Momentum</u>. Equation (1.1) can be rewritten as

$$\dot{p}_x = \frac{d}{dt}(m\dot{x}) = X \qquad (1.1a)$$

The time rate of change of momentum equals the applied
force. Since the kinetic energy $T = m\dot{x}^2/2$, the momentum

$$p_x = m\dot{x} = \partial T/\partial \dot{x}$$

and Newton's Equation becomes

$$\frac{d}{dt}\left(\frac{\partial T}{\partial \dot{x}}\right) = X \qquad (1.23)$$

Time - Dependent Forces

When the force X does involve the time explicitly, then
(1.2) takes the form

$$\Delta T = \int_{t_o}^{t} X \dot{x} \, dt$$

To evaluate the integral now, we have to know how x depends
upon t, which means that we must have solved (1.1). No conservation theorems are now evident, which means that such
devices cannot be used as shortcuts to the integration of
(1.1).

Example 1. Let $m\ddot{x} = t$. Then $m\dot{x} = \frac{1}{2}t^2$ and $mx = \frac{1}{6}t^3$
represents the motion with initial position and velocity

equal to zero. The kinetic energy is $T = \frac{1}{2} m\dot{x}^2 = t^4/8m$ and the potential energy is $U = -xt = -t^4/6m$. The sum $T + U = -t^4/24m$, which is not constant.

Example 2. Let $m\ddot{x} = A \sin \omega t$, $\dot{x} = -(A/m\omega) \cos \omega t$, and $x = -(A/m\omega^2) \sin \omega t$. The kinetic energy is $T = (m/2)\dot{x}^2 = (A^2/2m\omega^2) \cos^2 \omega t$ and the potential energy is $U = -Ax \sin \omega t = (A^2/m\omega^2) \sin^2 \omega t$. The sum $T + U$ fluctuates periodically about its average value $(3A^2/4m\omega^2)$.

Example 3. Suppose that there is a light wave incident on an oscillator which has a natural frequency $\omega_o/2\pi$, charge e, and mass m and that we wish to find the average energy transferred to the oscillator.

Let the wave have amplitude E and resolve this in terms of a Fourier distribution

$$E = \int_{-\infty}^{\infty} E(\omega) \, e^{i\omega t} \, d\omega$$

For a single Fourier component ω, the motion of the oscillator is described by

$$\ddot{x} + \omega_o^2 \, x = (e/m) \, E(\omega)\cos (\omega t + \delta_\omega)$$

where δ_ω is the phase of a single component, and the phases are assumed to be distributed at random.

Assume that at time $t = 0$ only the free vibration is excited, i.e. $x = A \sin (\omega_o t + \theta)$. The general solution of the equation of motion will be the sum of a particular solution of the inhomogeneous equation and a solution of the homogeneous equation (free vibration). This may be verified to be

$$x = (e/m) \, E(\omega) \, [1/\omega_o^2 - \omega^2)] \, [\cos (\omega t + \delta_\omega)$$

$$- \cos (\omega_o t + \delta_\omega)] + A \sin (\omega_o t + \theta)$$

The energy transfer per unit time and per unit angular frequency interval equals the work done by the light wave, which is $e\dot{x}E(\omega)\cos(\omega t + \delta_\omega)$. This we integrate over an integral number of periods $n\tau = 2\pi n/\omega$ to give

$$[e^2 \, E^2(\omega) \, \omega_o/m(\omega_o^2 - \omega^2)] \int_0^{n\tau} dt \, \sin(\omega_o t + \delta_\omega) \, \cos(\omega t + \delta_\omega)$$

$$+ e \, E(\omega) \, A \, \omega_o \int_0^{n\tau} dt \, \cos(\omega_o t + \theta) \, \cos(\omega t + \delta_\omega)$$

The second integral depends upon the relation between the phase θ and the phase δ_ω. If $\omega = \omega_o$ and $\theta = \delta_\omega + \pi$, the integral will be negative, and radiation will be emitted. This is called stimulated emission, and is present whenever excited atoms are irradiated by light with frequency near to a natural frequency of the atoms.

If the phases δ_ω are distributed at random, the average contribution of the second integral will be zero. The first integral, if one neglects a very high frequency oscillatory term, is then

$$e^2 \, E^2(\omega) \, \omega_o \, /2m \, (\omega_o^2 - \omega^2)] \, (1 - \cos(\omega_o - \omega) \, n\tau)/(\omega_o - \omega)$$

This is large only near the resonance frequency $\omega = \omega_o$, and the maximum is very strong there. We can integrate over all frequencies from zero to infinity, and obtain the energy absorbed per unit time

$$(e^2 E^2(\omega)/4mn\tau) \int_{-\infty}^{\infty} n\tau \, (1 - \cos u) \, du/u^2 =$$

$$e^2 E^2(\omega)\pi/4m; \quad u = (\omega_o - \omega) \, n \, \tau.$$

The average energy transferred to the oscillator is proportional to the time and to the intensity of the incident radiation at the resonant frequency (since the intensity is itself
proportional to E^2).

Velocity – Dependent Forces

Equation (1.1) can be generalized to the case where the
force depends on the velocity. A simple example is that of
the damped oscillator for which

$$m\ddot{x} = -f(x, \dot{x})\,\dot{x} - g(x, \dot{x})x$$

This equation with variable f and g will be discussed at
length in Chapter 9, where it will be shown how periodic,
non-sinusiodal, oscillations can build up or die away. When
the coefficients are constant, say $f(x, \dot{x}) = p$, $g(x, \dot{x}) = 1$,
this represents the familiar damped harmonic oscillator. We
know that the solution is of the form $x = A \exp (\alpha + i\beta)t$,
where α and β are real numbers. If $\alpha = 0$, the motion is
oscillatory. If $\beta = 0$, $\alpha > 0$, the value of x goes to infinity
as $t \to \infty$; if $\beta = 0$, $\alpha < 0$, then $x \to 0$. If $\beta \neq 0$, then the
amplitude of the oscillation continually builds up if $\alpha > 0$,
but dies down if $\alpha < 0$.

Another example of dependence of force on velocity is that
of a particle moving in an electromagnetic field, where the
force on the particle is

$$\vec{F} = q\,\{\vec{E} + \frac{1}{c}(\vec{v} \times \vec{B})\}$$

with q the electric charge, $\vec{E}$ the electric field intensity.
and $\vec{B}$ the magnetic induction intensity. Treatment of this
case will be deferred until after we have introduced the
Lagrangian methods (see Chapters 6 and 11).

Motion of Fast Particles - One Dimension

When a particle moves at a speed which is comparable to the velocity of light c, then the evidence from magnetic deflection experiments shows that its mass m is a function of velocity and may be written

$$m = m_o/(1 - \frac{v^2}{c^2})^{1/2} \quad ,$$

where m_o is a constant called the rest mass, because m reduces to it when v = 0, and v is the velocity of the particle as observed in the laboratory. Newton's second law (1.1) must be revised, with the new version

$$\frac{d}{dt} \vec{p} = \vec{F}, \quad \vec{p} = m\vec{v} \tag{1.1a}$$

The time rate of change of momentum equals the applied force $\vec{F}$. The ratio of momentum to velocity is a constant m for a given speed, but this constant m, called the mass, is the above function of velocity.

Restricting ourselves to one dimension, we shall search for a function T*, in extension of (1.23), such that the momentum component

$$m\dot{x} = \partial T^*/\partial \dot{x} = m_o \dot{x}/(1 - \frac{\dot{x}^2}{c^2})^{1/2}$$

A simple integration gives

$$T^* = - m_o c^2 (1 - \frac{v^2}{c^2})^{1/2} \tag{1.24}$$

This is a function of v which goes to zero as $v \to c$ and is not to be confused with the kinetic energy (see below).

If the force $X = -\partial U/\partial x$, then we may find an energy integral, as will now be shown. (In more than one dimension, this still holds if the force field is conservative.)

The time rate of change of T^* is

$$\frac{dT^*}{dt} = \frac{dT^*}{d\dot{x}} \frac{d\dot{x}}{dt} = p_x \frac{d\dot{x}}{dt} = \frac{d}{dt}(p_x\dot{x}) - \dot{x}\frac{d}{dt}p_x$$

so that

$$\frac{d}{dt}(p_x\dot{x} - T^*) = \dot{x} X$$

$$= -\dot{x}\frac{dU}{dx} = -\frac{dU}{dt} \qquad (1.25)$$

The rate at which work is done is, by definition, the time derivative of the kinetic energy, so that the quantity $p_x\dot{x} - T^*$ represents up to a constant the kinetic energy. Integrating (1.25),

$$p_x\dot{x} - T^* + U = h, \text{ a constant} \qquad (1.26)$$

This expression for <u>conservation of energy</u> replaces the earlier one

$$(T + U = h).$$

If we now substitute the value of T^* from (1.24) and set $p_x = m_o\dot{x}/(1 - \frac{v^2}{c^2})^{1/2}$, we obtain

$$p_x\dot{x} - T^* = mc^2 \qquad (1.27)$$

When $v \to 0$, this becomes $m_o c^2$, so that the <u>kinetic energy</u>, the excess due to motion, will be (1.27) minus $m_o c^2$, i.e.

$$p_x\dot{x} - T^* - m_o c^2 = (m - m_o)c^2$$

For low velocities, this reduces to $mv^2/2$, as it should.

Squaring the expression for $p = p_x$ and adding $m_o^2 c^2$,

$$p^2 + m_o^2 c^2 = m_o^2 c^2 / (1 - \frac{\dot{x}^2}{c^2}) = m^2 c^2$$

This gives m in terms of p, so we can now substitute in (1.26) and (1.27) to find

$$c (p^2 + m_o^2 c^2)^{1/2} - m_o c^2 + U = h \qquad (1.28)$$

The left-hand side is a function of momentum and position, known as the Hamiltonian function H (p,x). This equation says that H is constant, or that an energy integral exists.

Equation (1.28) holds for three degrees of freedom, if we put

$$p^2 = p_x^2 + p_y^2 + p_z^2.$$

Then

$$H = \sum_{xyz} p_x \dot{x} - T^* + U.$$

The keys here to the existence of an energy integral have been that (1) the momentum is a function of the velocity only and (2) the potential function depends only on the space coordinates and not explicitly on the time.

More general kinetic energies and potential energies, which still allow an energy integral, can be devised, but their discussion will be deferred until after the treatment of the Lagrangian.

MOTION IN TWO DIMENSIONS

The study of motion of a particle with two degrees of freedom reveals features which can be extended to higher dimensionality. The equations of motion are

$$m\ddot{x} = X(x,y), \quad m\ddot{y} = Y(x,y) \qquad (1.29)$$

where the force components X and Y will be assumed not to involve the time explicitly. If we now multiply the first equation by $\dot{x}$ dt and the second by $\dot{y}$ dt, integrate and add, we obtain the generalization of (1.2)

$$\Delta T = \int_{x_o,y_o}^{x,y} (X \, dx + Y \, dy) \qquad (1.30)$$

where

$$T = \frac{1}{2}m \, (\dot{x}^2 + \dot{y}^2) = \frac{1}{2} \, mv^2$$

The solution of (1.29) is given by a pair of functions $x(t)$, $y(t)$. As the time increases, the motion is described by a curve traced out in the x,y plane. The line integral (1.30) will in general depend on the particular curve C traced out. However, some functions X and Y are such that the integral is independent of the path. In this case, we may write the generalization of (1.3)

$$\Delta T = \int_C (Xdx + Ydy) = U(x_o y_o) - U(x,y) \qquad (1.31)$$

with $U(x,y)$ the potential energy. The force components are

$$X = -\partial U/\partial x, \quad Y = -\partial U/\partial y$$

from which

$$\frac{\partial X}{\partial y} = \frac{\partial Y}{\partial x} \qquad (1.32)$$

When there is a potential energy function, then (1.32) must hold, i.e., is a necessary condition. But it may also be proved that, if (1.32) holds and the region is simply-connected, this is sufficient for the existence of a potential energy function U and the validity of (1.31). The force field X, Y is then said to be conservative.

Equation (1.31) states that an increase in kinetic energy equals a corresponding decrease in potential energy. Alternatively, the total energy $T + U = h$ where h is a constant.

To demonstrate the nature of functions X and Y which satisfy (1.32), let us first consider the motion of a liquid rotating uniformly about the origin with angular velocity ω. The components of velocity are

$$v_x = \dot{x} = -y\omega$$

$$v_y = \dot{y} = x\omega$$

Then

$$-\frac{\partial v_x}{\partial y} + \frac{\partial v_y}{\partial x} = 2\omega \qquad (1.33)$$

or, in vector notation

$$\text{curl } \vec{v} = \vec{2\omega}$$

Alternatively, one can obtain (1.33) by calculating the curl as defined from Stokes' theorem for an infinitesimal closed loop with area ΔS. The component of the curl normal to this loop is

$$\text{curl}_n v = \lim_{\Delta S \to 0} \oint \vec{v} \cdot \vec{ds}/\Delta S = \frac{2\pi r v}{\pi r^2} = 2\omega$$

Thus, if (1.32) is satisfied, the curl of the force $\vec{F}$ (components x,y) is zero and the field is said to be irrotational.

When the field is conservative, the dynamical problem is enormously simplified, for one need deal with only one function U(xyzt) instead of with the force components X, Y, Z. When, on the other hand, vortex motion is the rule, as with phenomena of turbulence, great difficulties are present, which is evidenced by the fact that satisfactory theories of turbulence are still to be evolved. We have been lucky to find vast areas of physics, connected with the motion of par- ticles, where the fields are irrotational, but must remember that the domains with rotational fields do exist and present problems for the future.

If we do have a simple vortex motion, the force components can be taken to be

$$X = -y \, f(r), \quad Y = x \, f(r)$$

by analogy with the above hydrodynamic example.

The change in kinetic energy around a circle is then, from (1.31)

$$\Delta T = \oint (X dx + Y dy)$$
$$= \oint f(r)(-y dx + x dy) = \int_0^{2\pi} f(r) r^2 d\,\theta$$
$$= 2\pi \, r^2 \, f(r)$$

The applied force is tangential, and would act to increase the kinetic energy continually, if $f(r)$ were positive and the motion were in a circle about the origin.

In order to investigate the motion in more detail, it is useful to transform to polar coordinates, with $x = r \cos\theta$, $y = r \sin\theta$. Differentiating twice re t,

$$\ddot{x} = \ddot{r} \cos\theta - 2\dot{r}\dot{\theta} \sin\theta - r(\sin\theta \, \ddot{\theta} + \cos\theta \, \dot{\theta}^2) = X/m$$

$$\ddot{y} = \ddot{r} \sin\theta + 2\dot{r}\dot{\theta} \cos\theta + r(\cos\theta \, \ddot{\theta} - \sin\theta \, \dot{\theta}^2) = Y/m$$

The components of force in the r- and θ- directions are then

$$F_r = X \cos\theta + Y \sin\theta = m(\ddot{r} - r\dot{\theta}^2) \quad (1.34)$$

$$F_\theta = -X \sin\theta + Y \cos\theta = m(r\ddot{\theta} + 2\dot{r}\dot{\theta}) \quad (1.35)$$

Equation (1.34) may now be displayed in a form similar to
(1.29), setting $v_\theta = r\dot\theta$, as follows:

$$m\ddot{r} = F_r + (mv_\theta^2/r)$$

The product of mass and radial acceleration equals the
radial component of external force plus a term involving the
angular velocity, which is called the centrifugal force. Our
everyday experience teaches us that a particle under no force
will move in a straight line, and that a force must be applied
if the particle is to be constrained, say by a string, to move
in a circle.

Equation (1.35) may be simplified in form by multiplying
both sides by the radius, to yield

$$\frac{d}{dt}(mr^2\dot\theta) = rF_\theta$$

The quantity $mr^2\dot\theta$ takes the place of the momentum $m\dot{x}$ in
(1.25) and is called a generalized momentum, in this case an
angular momentum, or moment of momentum ($r \cdot mr\dot\theta$). The
quantity rF_θ substitutes for X, and is called a generalized
force, in this case a torque. Equation (1.36) then states
that the time rate of change of the angular momentum is equal
to the applied torque.

If the radial external force F_r is zero, then we have to
solve the two equations

$$\ddot{r} - r\dot\theta^2 = 0 \qquad\qquad (1.37)$$

$$\frac{d}{dt}(mr^2\dot\theta) = rF_\theta \qquad\qquad (1.36)$$

For simplicity, take $rF_\theta = m$. Then $r^2\dot\theta = t$ if we start from
rest at $t = 0$, and $r^3\ddot{r} = t^2$. The angular momentum increases

steadily with the time and the radial acceleration is always
outward. No conservation theorems are evident.

   If, on the other hand, the applied torque is zero, then the
angular momentum is constant, from (1.36). Calling this constant J, and substituting $\dot\theta = J/mr^2$ in (1.34), we have

$$m\ddot{r} - (J^2/mr^3) = F_r \qquad (1.38)$$

If F_r should depend on the angle θ, then

$$\text{curl } F = \lim_{\Delta S \to 0} \oint F \cdot ds/\Delta S = \frac{(\partial F_r/\partial\theta)\,\Delta r\,\Delta\theta}{r\,\Delta r\,\Delta\theta} = \frac{1}{r}\frac{\partial F_r}{\partial\theta}$$

and the field is not conservative. Therefore, with applied
torque zero and a conservative field, F_r must depend only on
the radius. Also, the force must be the gradient of a potential function, and so

$$F_r = -\partial U/\partial r, \quad F_\theta = -\frac{1}{r}\frac{\partial U}{\partial\theta} = 0$$

   This potential function U will depend only on the radius r.
We can then integrate (1.38) by multiplying both sides by
$\dot{r}\,dt$, to obtain the energy integral

$$\frac{m\dot{r}^2}{2} + \frac{J^2}{2mr^2} + U = h \qquad (1.39)$$

   The three terms on the left-hand side are, respectively,
the kinetic energy, the centrifugal potential energy, and
the intrinsic potential energy. Their sum is the total
energy h, which is a constant. It is useful to introduce

$$U^* = U + \frac{J^2}{2mr^2}$$

as an effective potential energy, and then to discuss librations in such a field.

GENERAL REMARKS

When a particle moves in a static electromagnetic field,
there is still conservation of energy. The electric field is
derivable from a potential, and the magnetic field merely
deflects the particle but does not change its speed (or
kinetic energy). Thus, even though the forces depend upon
the velocity here, an energy integral still exists. Conservative forces and conservation of energy are then basic for
electromagnetic and gravitational fields which do not vary
with the time.

When, however, one turns to dynamical problems associated
with electrical circuits and control systems, the picture
changes quite radically. The forces are no longer conservative, as one sees easily in the case of the damped harmonic
oscillator, and the theory is so complicated for one degree
of freedom that it has not been extended much farther. The
basic problem is to investigate the solutions of a modified
Equation (1.1), namely

$$\ddot{x} = X (x, \dot{x}, t) \qquad\qquad (1.40)$$

(The mass constant has been incorporated in the right-hand
side, which now represents force per unit mass.)

Whereas in potential problems one allows the force to be
singular at various points, here this is usually ignored and
one presumes that X may be represented by a convergent ascending power series in x, $\dot{x}$, and t. Essentially, one is considering a system which may oscillate with variable damping
and period. Explicit dependence on t occurs usually because
of the presence of a forcing term f(t) or because the
coefficients of x and $\dot{x}$ (and of higher powers) are periodic
in t. Such systems may be treated with profit but more

general dependence on t seems both difficult in theory and
uninteresting in practice.

The second order differential equation (1.40) can be
replaced by the system.

$$\dot{x} = y, \; \dot{y} = X(x, \; y, \; t) \tag{1.41}$$

When the force X does not involve the time explicitly, the
system is _autonomous_. It will be convenient to generalize
(1.41) and to change notation as follows:

$$x_1 = x, \; x_2 = y$$

$$\dot{x}_1 = X_1(x_1 x_2), \; \dot{x}_2 = X_2(x_1 x_2) \tag{1.42}$$

where X_1 and X_2 are now convergent ascending power series in
x_1 and x_2. For this autonomous system, it may be advantageous to divide the second equation by the first, thus eliminating the time, and to find out qualitatively and quantitatively the nature of the integrated paths in the phase plane
$(x_1 x_2)$. We have

$$\frac{dx_2}{dx_1} = \frac{X_2(x_1 x_2)}{X_1(x_1 x_2)} = f(x_1 x_2)$$

This equation may be examined for singular points, and the
qualitative nature of the trajectories in the neighborhood
of such points can be determined. The integral curves
through all ordinary points can be found by numerical integration.

Once these have been constructed, then a further integration of (1.42) may be made, giving both x_1 and x_2 as
functions of the time.

It is usually easiest to refer the trajectories to the
points where $\dot{x}_1 = 0$ and $\dot{x}_2 = 0$, i.e., the equilibrium points.

Let such a point be x_{10}, x_{20} and make the substitution $\Delta x_1 = x_1 - x_{10}$, $\Delta x_2 = x_2 - x_{20}$. The quantities Δx_1 and Δx_2 are termed <u>variations</u> of x_1 and x_2, and satisfy the <u>variational equations</u>.

$$\frac{d}{dt} \Delta x_1 = X_1(x_1 x_2) - X_1(x_{10} x_{20})$$

$$(1.43)$$

$$\frac{d}{dt} \Delta x_2 = X_2(x_1 x_2) - X_2(x_{10} x_{20})$$

The right-hand sides will now be ascending power series in x_1 and x_2, without constant terms. Integration is easy if the terms of second and higher degrees may be neglected, and frequently this is sufficient to determine the qualitative nature of the solutions near the points of equilibrium. The detailed treatment is given in Chapter 9.

The general theory of autonomous systems, as represented by Equations (1.42), has received much attention in recent years. Complications may arise when the linear terms on the right-hand sides of (1.43) are absent, or when the presence of a parameter may result in the character of the motion near a singular point varying as the parameter does.

We shall study in turn Hamiltonian systems, autonomous systems not in this category, and non-autonomous but periodic systems, this being roughly the historical order of development of interest. As already mentioned, Hamiltonian systems are of central importance for gravitational and electromagnetic fields, or alternatively for celestial mechanics and for design of high-energy accelerators. One begins advantageously by becoming acquainted with the methods of Lagrange, and so now we proceed with this.

Exercises

1. The acceleration of a particle is given by $\ddot{x} = 3 x^2$.
Find (a) the expression for the conservation of energy, and
(b) the explicit expression for x as a function of time. If
the particle is initially at a point $x = x_o > 0$ and is given
a negative velocity, what will happen to it? Can it reach
the origin $x = 0$? Plot the motion in the $(x,\dot{x})$ phase plane
for various values of the total energy. For each such value,
draw the corresponding curve of x vs t.

2. Let the acceleration of a particle be $\ddot{x} = -\sin x$. To
what physical system does this correspond? What is the
expression for conservation of energy? Show by a drawing
what the limits on the motion are, for various values of the
total energy, and show how the motion looks in the phase
plane for these values of the energy.

3. Devise physical examples of conservative force fields
for which energy is not conserved.

4. When a charged particle moves in a magnetic field which
does not vary with time, its energy stays constant. Show
this.

5. Given a particle with non-zero rest mass. How does its
kinetic energy behave as its speed approaches the velocity of
light? Can such particles travel faster than light?

6. How does a photon behave dynamically?

II

General Equations of Motion

When a particle is acted on by a central force, the motion could be found by writing down Newton's equations

$$m\ddot{x} = X, \quad m\ddot{y} = Y, \quad \text{and} \quad m\ddot{z} = Z \qquad (2.1)$$

(where X, Y, Z are functions of x, y, z and t) and solving them. This method will be found to be awkward and laborious, chiefly because maximum use has not been made of the spatial symmetry of the force. If one could only write down the equations of motion directly in spherical coordinates (r, θ, ϕ) much time and trouble would be saved. In this special case, we could make the substitution $x = r\sin\theta\cos\phi$, $y = r\sin\theta\sin\phi$, $z = r\cos\theta$ in Equations (2.1) and obtain equations for $\ddot{r}$, $\ddot{\theta}$, and $\ddot{\phi}$, which are the <u>Lagrange equations</u>. However, rather than blindly go through such a procedure in each case, a general method of obtaining the Lagrange equations has been evolved, and this will now be demonstrated.

Generalized Coordinates

The position of a particle free to move in space can be specified by a set of three numbers, according to which coordinate system is most suitable. The <u>Cartesian coordinates</u> (x,y,z) will be appropriate if the potential energy U depends

only on one such coordinate, i.e., $U = U(x)$. However, if U
depends only on the distance r from a fixed point, i.e.,
$U = U(r)$, then <u>spherical coordinates</u> (r, θ, ϕ) should be used.
If U depends only on the distance ρ from a fixed axis, one
will naturally choose <u>cylindrical coordinates</u> (ρ, ϕ, z). Such
a number triple will be denoted in general by (q_1, q_2, q_3)
where q_1, q_2, and q_3 are called <u>generalized coordinates</u> and
specify the space position uniquely.

  <u>Lagrange's Equations</u>, to be <u>derived</u> immediately, are
expressed in terms of q_1, q_2, and q_3 (for each particle) and
<u>are not at all special to the case of spherical coordinates</u>.
They are equivalent to Newton's Equations, but have two
advantages, namely (1) they allow maximum exploitation of
whatever symmetry the system has, so that constants of the
motion may be readily discovered, and (2) they permit one in
certain important cases to ignore forces of constraint, and
to regard the system as subject only to the external forces,
insofar as finding the motion is concerned.

  We can confine the discussion to one-dimensional motion at
first, and then make an immediate generalization to more than
one dimension. Newton's equations have been written in the
form

$$\dot{P}_x = \frac{d}{dt} \frac{\partial T}{\partial \dot{x}} = X \qquad (2.2)$$

If we transform to a new variable q by setting $x = x(q,t)$
then

$$dx = adq + bdt$$

and $$\dot{x} = a\dot{q} + b \qquad (2.3)$$

where $$a = \partial x/\partial q \text{ and } b = \partial x/\partial t$$

From (2.3), we have $\left.\frac{\partial \dot{x}}{\partial \dot{q}}\right|_{q,t} = \left.\frac{\partial x}{\partial q}\right|_t = a \qquad (2.4)$

Now the kinetic energy $T = (m/2)(a\dot{q} + b)^2$ is a quadratic function of $\dot{q}$, and using (2.4)

$$\frac{\partial T}{\partial \dot{q}} = \frac{\partial T}{\partial \dot{x}} \frac{\partial \dot{x}}{\partial \dot{q}} = \frac{\partial T}{\partial \dot{x}} \frac{\partial x}{\partial q} \qquad (2.5)$$

Let us calculate the time derivative of this, in the hope of finding something similar to (2.2). The result is

$$\frac{d}{dt} \frac{\partial T}{\partial \dot{q}} = \frac{\partial x}{\partial q} \frac{d}{dt} \frac{\partial T}{\partial \dot{x}} + \frac{\partial T}{\partial \dot{x}} \frac{d}{dt} \frac{\partial x}{\partial q}$$

$$= X \frac{\partial x}{\partial q} + \frac{\partial T}{\partial \dot{x}} \frac{\partial \dot{x}}{\partial q} \qquad \text{by (2.2)}$$

$$= X \frac{\partial x}{\partial q} + \frac{\partial T}{\partial q}$$

Rearranging,

$$\frac{d}{dt} \frac{\partial T}{\partial \dot{q}} - \frac{\partial T}{\partial q} = X \frac{\partial x}{\partial q} \qquad (2.6)$$

This is _Lagrange's Equation_ in one dimension. From its method of derivation, it is seen that it is just a transformed variety of the original Newton's Equation. The usefulness of this formulation is demonstrated when we pass to three dimensions, where (2.6) is to be replaced by

$$\frac{d}{dt} \frac{\partial T}{\partial \dot{q}_r} - \frac{\partial T}{\partial q_r} = X \frac{\partial x}{\partial q_r} + Y \frac{\partial y}{\partial q_r} + Z \frac{dz}{dq_r}$$

$$= Q_r, \text{ say} \qquad (2.7)$$

It is relatively easy to express the kinetic energy in terms of the $\dot{q}$'s, and then to differentiate and obtain the equations of motion directly instead of transforming the original equations (expressed in Cartesian Coordinates). If

a potential function U exists, i.e., $X = -\partial U/\partial x$, etc., the
right-hand side of (2.7) becomes $-\partial U/\partial q_r$.

Forces of Constraint

Up until now, we have tacitly assumed that the forces
acting on a particle are all those applied externally and are
known beforehand. However, there do arise situations where
the particle cannot move freely in all directions, such as
when a ball is on one end of a string, the other end of which
is fixed, and the force of gravity is the external force.
The ball is now constrained by the string, provided it is
lower than the point of attachment, to move on the surface of
a sphere (hemisphere). In addition to the force of gravity,
there is now an additional force of constraint, which acts
along the string.

In general, a condition of constraint is a relation between
small increments of the coordinates, such as

$$adx + bdy + cdz + g\,dt = 0 \qquad (2.8)$$

where the coefficients may be functions of x, y, z and t.

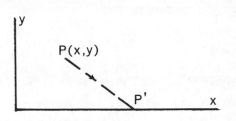

As a simple example, let a
particle at P(x,y) (see
Figure 2.1) move in the
x – y plane toward a point
P' on the x-axis, with
abscissa x'. If the point
P' is stationary, then

Figure 2.1
Pursuit of P', A Point
Moving Along The x-axis

$$\frac{dy}{dx} = \frac{y}{x' - x}$$

can be integrated to

$$y = k\,(x' - x), \qquad (2.9)$$

so the particle is _constrained_ to move along the straight line
from the initial position to P'. The condition of constraint
is _integrable_, i.e., it yields a relation f(x,y) = c, where
c is a constant. The system is then called _holonomic_.

A general integrated condition of constraint will be of the
form

$$f(x, y, z, t) = \text{const.}\qquad\qquad(2.10)$$

The differentials are related by the equation

$$\frac{\partial f}{\partial x}\, dx + \frac{\partial f}{\partial y}\, dy + \frac{\partial f}{\partial z}\, dz + \frac{\partial f}{\partial t}\, dt = 0 \quad (2.11)$$

In the x,y plane, an integrated condition of constraint
would be

$$f(x, y) = c \qquad\qquad (2.12)$$

which would mean that just one curve shall pass through a
given point.

A Non-Integrable Constraint

If, in the above example, the point P' should move along
the axis with constant velocity v, and the particle P is to
move toward it, following a _curve of pursuit_, then

$$\frac{dy}{dx} = \frac{y}{vt - x} \qquad\qquad (2.13)$$

and the slope would depend on where P' is. There is thus an
infinity of possible curves through the point P, and hence
no relation of type (2.12). The slope does not depend on
just x and y, but on t as well. There is no _integrating
factor for (2.13)_ and the system is _non-holonomic_.

Smooth Constraints

We shall from now on assume that the system is <u>holonomic</u>, i.e. that the particle moves on a surface

$$f(x, y, z, t) = \text{const.} \qquad (2.10)$$

and that the surface is smooth (no friction) so that the force exerted on the particle by the surface is normal to the surface.

From (2.10)

$$\frac{\partial f}{\partial x} dx + \frac{\partial f}{\partial y} dy + \frac{\partial f}{\partial z} dz + \frac{\partial f}{\partial t} dt = 0 \qquad (2.11)$$

A <u>possible</u> displacement dx, dy, dz, dt is one which satisfies this equation. If $\partial f/\partial t = 0$, then the constraint is <u>fixed</u>, otherwise it is a moving one. We shall deal mainly with fixed constraints.

A <u>virtual</u> displacement δx, δy, δz is one which satisfies the equation

$$\frac{\partial f}{\partial x} \delta x + \frac{\partial f}{\partial y} \delta y + \frac{\partial f}{\partial z} \delta z = 0 \qquad (2.11a)$$

Now $\partial f/\partial x$, $\partial f/\partial y$, and $\partial f/\partial z$ are the components of a vector $\vec{\nabla} f$, the gradient of f at a particular time. Equation (2.11a) may be written vectorially as

$$(\vec{\nabla} f \cdot \vec{\delta r}) = 0 \qquad (2.11b)$$

Since the gradient of f is normal to the surface f, (2.11b) states that the vector $\vec{\delta r}$, with components δx, δy, δz, must be tangent to the surface.

But the force of constraint is normal to the surface, and hence its components X', Y', Z' have a zero scalar product with δx, δy, δz, i.e.

$$X'\delta x + Y'\delta y + Z'\delta z = 0 \qquad (2.11c)$$

<u>The work done by the force of constraint in any virtual displacement is zero.</u>

Degrees of Freedom

If we have a system consisting of n particles, its configuration may be described by giving the 3n coordinates of the particles. However, a smaller number of variables may be sufficient to determine the configuration if the system is holonomic. Consider, for example a pendulum or a top, where the particle (or c.m. of the top) is at a constant distance a from a fixed point. The condition of constraint

$$r = (x^2 + y^2 + z^2)^{1/2} = a = f(x, y, z)$$

defines a spherical surface on which the particle must move. It is convenient to take $q_1 = r = a$, $q_2 = \theta$, $q_3 = \phi$ as a new set of coordinates. Then any pair of increments $\delta q_2 = \delta\theta$, $\delta q_3 = \delta\phi$ will constitute a virtual displacement. In general, the minimum number of generalized coordinates which may be varied arbitrarily and independently without violating the constraints (such as rigidity) is known as the number of <u>degrees of freedom</u>. In the example above, of a pendulum or top not rotating about its axis, there are two degrees of freedom.

Lagrange's Equations - Holonomic Systems

We shall now demonstrate that <u>Lagrange's Equations in the form</u> (2.7) <u>are valid for any holonomic system</u>, and that <u>the forces X, Y, Z can be taken to be just the external applied forces.</u>

To start with, Newton's equations for the kth particle are

$$m_k \ddot{x}_k = X_k + X_k' \qquad (2.14)$$

where the forces of constraint satisfy the condition

$$\sum_k X_k' \, \delta x_k = 0 \qquad (2.11c)$$

Combining (2.14) and (2.11c)

$$\sum_k (m_k \ddot{x}_k - X_k) \, \delta x_k = 0 \qquad (2.15)$$

Carrying through the same sort of calculation which led to (2.7), we have, since $\delta x_k = \sum_r (\partial x_k / \partial q_r) \, \delta q_r$

$$\sum_r \left\{ \frac{d}{dt} \frac{\partial T}{\partial \dot{q}_r} - \frac{\partial T}{\partial q_r} - Q_r \right\} \delta q_r = 0 \qquad (2.16)$$

where

$$Q_r = \sum_k X_k \frac{\partial x_k}{\partial q_r} \qquad (2.17)$$

Equation (2.16) is another statement of the fact that the forces of constraint do no work in a virtual displacement. But this equation holds for arbitrary values of the δ's, so the quantity in brackets is a coefficient which must be zero. Hence, finally, for a holonomic system

$$\frac{d}{dt} \frac{\partial T}{\partial \dot{q}_r} - \frac{\partial T}{\partial q_r} = Q_r \qquad (2.18)$$

The quantity Q_r is a <u>given</u>, generalized, external force defined by (2.17). As before T is the kinetic energy, i.e.

$$T = \frac{1}{2} \sum_k m_k \dot{x}_k^2 \qquad (2.19)$$

The equations of motion corresponding to the various q's do not involve the forces of constraint, which is what was to be shown.

Particle in Central Field

For the case of a particle moving in a central field with potential U(r), it is natural to use spherical coordinates. The line element ds is given by $(ds)^2 = (dr)^2 + r^2(d\theta)^2 + r^2 \sin^2 \theta (d\phi)^2$, so that the kinetic energy is $T = \frac{1}{2} m \dot{s}^2$
$= \frac{1}{2} m (\dot{r}^2 + r^2 \dot{\theta}^2 + r^2 \sin^2 \theta \dot{\phi}^2)$
and the Lagrange equations are

$$m\ddot{r} - mr (\dot{\theta}^2 + \sin^2 \theta \dot{\phi}^2) = - dU/dr \qquad (2.19)$$

$$\frac{d}{dt} (mr^2 \dot{\theta}) - mr^2 \sin \theta \cos \theta \dot{\phi}^2 = 0 \qquad (2.20)$$

$$\frac{d}{dt} (mr^2 \sin^2 \theta \dot{\phi}) = 0 \qquad (2.21)$$

The transformation has been carried out at the start by writing down the expression for the line element, and recognizing that the force is central. Equations (2.20 and (2.21) state essentially that the time rate of change of the angular momentum J equals the applied torque. Since the torques are here zero, the corresponding components of angular momentum are constant. Equation (2.21) states that the z-component $J_z = mr^2 \sin^2\theta \dot{\phi}$ is constant in time, or a

constant of the motion. If this expression is substituted in
Equation (2.20), and the equation multiplied by $2mr^2\dot{\theta}$, we
obtain

$$\frac{d}{dt}\,(mr^2\dot{\theta})^2 = 2J_z^2\,\frac{\cos\theta}{\sin^3\theta}\,\dot{\theta} = -\,\frac{d}{dt}\,\frac{J_z^2}{\sin^2\theta}$$

From this, the quantity

$$(mr^2\dot{\theta})^2 + \frac{J_z^2}{\sin^2\theta} = (mr)^2[(r\dot{\theta})^2 + (r\sin\dot{\phi})^2] = (mv_\perp)^2$$

(2.22)

is another constant of the motion, where $v_\perp$ is the component
of velocity perpendicular to the radius vector from the origin.
The total angular momentum $J = mrv_\perp$ and is thus constant from
Equation (2.20) and the above argument.

Since there is no applied torque, the angular momentum
$\vec{J} = \vec{r} \times m\vec{v}$ is fixed in direction, and the motion (characterized by $\vec{r}$) must be in the plane perpendicular to the direction
$\vec{J}$.

Another constant of the motion is found by substituting
(2.22) in (2.19), multiplying by $2m\dot{r}$, and integrating with
respect to time. We find

$$2m\dot{r}\,\frac{d}{dt}\,(m\dot{r}) - \frac{2J^2}{r^3}\,\dot{r} = -2m\,\frac{\partial U}{\partial r}\,\dot{r}$$

or

$$\frac{d}{dt}\,[(m\dot{r})^2 + \frac{J^2}{r^2} + 2mU] = 0$$

The quantity $\frac{1}{2}m\dot{r}^2 + (J^2/2mr^2) + U = T + U$ is thus constant,
and is called the total energy h. It is composed of radial
kinetic energy, rotational kinetic energy, and potential
energy.

The variation of r with the time is given by the equation

$$\dot{r} = \pm \ (2/m)^{1/2} \ [h - U - (J^2/2mr^2)]^{1/2} \ (2.23)$$

For the radial velocity to be real, we must have $h \geqslant U + (J^2/2mr^2)$. The equation

$$h - U - (J^2/2mr^2) = 0 \qquad\qquad (2.24)$$

may have no roots, one root, or more than one root. If it has none, the motion is always in one direction. If there is one root, as for a repulsive potential, the motion will reverse in direction at this value of r. The particle could come in from infinity and then return after the collision. If there are two roots, the particle will move back and forth, or librate, between them. The half-period of the libration can be obtained by integrating Equation (2.23) between these two limits.

Example   Consider a 3-dimensional symmetrical harmonic oscillator. Here $U = m\omega_o^2 r^2/2$, where $m\omega_o^2$ is the restoring constant. The limits of the motion are given, if $s = r^2$, from (2.24) by

$$(2h/m\omega_o^2)s - s^2 - (J/m\omega_o)^2 = 0$$

$$s = (h/m\omega_o^2) \pm [(h/m\omega_o^2)^2 - (J/m\omega_o)^2]^{1/2}$$

$$= s_o \pm s_1, \ \text{say}$$

Integrating (2.23), we have

$$t = (m/2)^{1/2} \int [h - (m\omega_o^2 r^2/2) - (J^2/2mr^2)]^{-1/2} dr$$

$$= (m/2) \int (2mhs - m^2\omega_o^2 s^2 - J^2)^{-1/2} \ ds$$

$$= (1/2\omega_o) \int [- (s - s_o)^2 + s_1^2]^{-1/2} \ ds$$

or $\qquad t = (1/2\omega_o) \sin^{-1} (s - s_o)/s_1$

Inverting, $s - s_o = s_1 \sin 2\omega_o t = r^2 - s_o$ describes the
motion if it starts halfway between the extremes of r^2,
namely at $s_o = h/m\omega_o^2$. The argument $2\omega_o t$ arises because we
have used the variable r^2, whereas for the one dimensional
harmonic oscillator we use the coordinate x and obtain
$x = x_o \sin\omega_o t$ instead.

Lagrange's Equations - Conservative Fields

When a potential function exists, then Lagrange's Equations
are somewhat simpler in form. Equation (2.17), with $X_k = -\partial U/\partial x_k$, becomes

$$Q_r = -\Sigma \frac{\partial U}{\partial x_k} \frac{\partial x_k}{\partial q_r} = - \frac{\partial U}{\partial q_r}$$

Substitution in (2.18) gives

$$\frac{d}{dt} \frac{\partial T}{\partial \dot{q}_r} - \frac{\partial T}{\partial q_r} = - \frac{\partial U}{\partial q_r} \qquad (2.25)$$

If now $L = T - U$, the function L is called the _Lagrangian_,
which is a function of the variables q_r, $\dot{q}_r$, and t (r = 1 ...
n). In terms of it, (2.25) becomes

$$\frac{d}{dt} \frac{\partial L}{\partial \dot{q}_r} = \frac{\partial L}{\partial q_r} \qquad (2.26)$$

Defining a _generalized momentum_ p_r as

$$p_r = \partial L/\partial \dot{q}_r , \qquad (2.27)$$

the equations of motion (2.26) may be written as

$$\dot{p}_r = \partial L / \partial q_r \qquad (2.28)$$

For the example of the central field just discussed, the generalized momenta are

$$p_r = \partial T / \partial \dot{r} = m\dot{r} \qquad (2.29)$$

$$p_\theta = \partial T / \partial \dot{\theta} = mr^2 \dot{\theta} \qquad (2.30)$$

$$J_z \equiv p_\phi = \partial T / \partial \dot{\phi} = mr^2 \sin^2\theta \, \dot{\phi} \qquad (2.31)$$

The radial quantity p_r is just the product of the mass with the r-component of velocity, and has the same form as the linear momentum in Cartesian coordinates.

However, the angular quantities p_θ and p_ϕ are each the product of mass with a distance times velocity and are angular momenta, or moments of momentum. Their dimensions are different from p_r because the differentiation has been taken with respect to angular velocity rather than linear velocity.

Equations (2.27) and (2.28) are particularly easy to remember. The momentum is defined as the derivative of the Lagrangian with respect to a $\dot{q}_r$ and its time rate of change equals the derivative of L with respect to the corresponding q_r.

Conservation of Energy

Search for constants of the motion leads us to inquire how the Lagrangian L varies with time.

$$\frac{dL}{dt} = \sum_r (\frac{\partial L}{\partial q_r} \dot{q}_r + \frac{\partial L}{\partial \dot{q}_r} \ddot{q}_r) + \frac{\partial L}{\partial t}$$

$$\frac{dL}{dt} = \sum_r (\dot{p}_r \dot{q}_r + p_r \ddot{q}_r) + \frac{\partial L}{\partial t}$$

$$= \frac{d}{dt} \sum_r p_r \dot{q}_r + \frac{\partial L}{\partial t} \qquad (2.32)$$

If the Lagrangian does not depend explicitly on the time, i.e. if $\partial L/\partial t = 0$, then the quantity

$$H = \sum_r p_r \dot{q}_r - L \qquad (2.33)$$

is a constant of the motion, by (2.32).

This quantity H, thus far a function of q_r, $\dot{q}_r$, and $t(r = 1 \ldots n)$, is known as the <u>Hamiltonian</u>. If Equations (2.27) are solved for $\dot{q}_r$ in terms of p_r and q_r, then we can alternatively write the Hamiltonian as $H(q_r, p_r, t)$, $(r = 1 \ldots n)$.

When the Hamiltonian is constant in time, its value may be denoted by h, and the equation

$$\sum_r p_r \dot{q}_r - L = h \qquad (2.34)$$

which is called the <u>Jacobi integral</u> or energy integral, is an expression of the conservation of energy. It is a more general form than our previous equation $T + U = h$, which is valid only when the kinetic energy is of particular type (homogeneous quadratic function of the velocities). The details will be given below.

We have already given (Chapter I) several examples involving time-dependent forces and have seen that the function $T + U$ was not constant in time, i.e. energy was not conserved. The requirement that $\partial L/\partial t = 0$ is one which means that the forces depend only on the space (and perhaps velocity) coordinates, and it ensures conservation of energy.

As stated, Equation (2.34) is the general conservation equation and holds whenever there is a Lagrangian and $\partial L/\partial t = 0$. Its usefulness becomes apparent already when the motion is referred to a set of <u>rotating axes</u>. Putting

$$x = r \cos (\theta + \omega t), \quad y = r \sin (\theta + \omega t),$$

then

$$\dot{x} = \dot{r} \cos (\theta + \omega t) - r (\dot{\theta} + \omega) \sin(\theta + \omega t)$$

$$\dot{y} = \dot{r} \sin (\theta + \omega t) + r (\dot{\theta} + \omega) \cos(\theta + \omega t)$$

The kinetic energy is

$$T = \frac{m}{2} (\dot{x}^2 + \dot{y}^2) = \frac{m}{2} [\dot{r}^2 + r^2 (\dot{\theta} + \omega)^2]$$

$$= \frac{m}{2} [\dot{r}^2 + r^2 \dot{\theta}^2 + 2r^2\omega \dot{\theta} + r^2\omega^2] \qquad (2.35)$$

It involves not only terms quadratic in $\dot{r}$ and $\dot{\theta}$, but also a term linear in $\dot{\theta}$ and a constant term.

For 2 degrees of freedom, the general kinetic energy will be taken to be

$$T = (1/2) (a_{11} \dot{q}_1^2 + 2a_{12} \dot{q}_1 \dot{q}_2 + a_{22} \dot{q}_2^2)$$

$$+ a_1 \dot{q}_1 + a_2 \dot{q}_2 + a = T_2 + T_1 + T_0$$

where T_n is a homogeneous function of degree n in the velocities. The momenta are

$$p_1 = \partial T/\partial \dot{q}_1 = a_{11} \dot{q}_1 + a_{12} \dot{q}_2 + a_1$$

and

$$p_2 = \partial T/\partial \dot{q}_2 = a_{12} \dot{q}_1 + a_{22} \dot{q}_2 + a_2$$

Then

$$p_1 \dot{q}_1 + p_2 \dot{q}_2 - T + U = 2T_2 + T_1 - T + U$$

$$= T_2 - T_o + U = h \qquad (2.36)$$

[From the manner of derivation, we see that (2.36) holds for n degrees of freedom.]

If, now, <u>the kinetic energy is a homogeneous quadratic function of the velocities</u> $\dot{q}$, then $T_o = 0$, $T = T_2$, and the sum of the kinetic energy and the potential energy is constant (provided that $\partial U/\partial t = 0$), i.e. $T + U = h$ \qquad (2.37)

Hamilton's Canonical Equations

In dealing with an ordinary differential equation of the nth order, it is often convenient to replace it by a system of n first order equations. This can be done in the present instance by using p_r and q_r as independent variables instead of q_r and $\dot{q}_r$. If Δ denotes a small increment as before, and subscripts are omitted,

$$\Delta H = \Delta(\Sigma\, p\dot{q} - L) = \Sigma(p\, \Delta q + \dot{q}\, \Delta p - \frac{\partial L}{\partial q}\, \Delta q - \frac{\partial L}{\partial \dot{q}}\, \Delta\dot{q} - \frac{\partial L}{\partial t}\, \Delta t$$

$$= \Sigma(\dot{q}\, \Delta p - \dot{p}\, \Delta q) - \frac{\partial L}{\partial t}\, \Delta t$$

by virtue of (2.27) and (2.28). Hence

$$\dot{q} = \frac{\partial H}{\partial p}, \quad \dot{p} = -\frac{\partial H}{\partial q}, \quad \text{and} \quad \frac{\partial L}{\partial t} = -\frac{\partial H}{\partial t} \qquad (2.38)$$

These equations are the <u>Hamilton canonical equations</u>. They are of first order in time, as compared with the Lagrange equations of motion (2.28), which are of second order. For one degree of freedom, let $m\ddot{x} = -\partial U/\partial x$ be the Newton or Lagrange equation. This can be replaced by the system

$\dot{x} = p/m$, $\dot{p} = -\partial U/\partial x$ or by $\dot{x} = \partial H/\partial p$, $\dot{p} = -\partial H/\partial x$, where $H = (p^2/2m) + U$. Whether one uses the Lagrange or the Hamiltonian formulation is mostly a matter of taste, depending on whether one prefers to consider one second-order differential equation or an equivalent system of two first-order differential equations. The latter may sometimes be handled with a more elegant mathematical notation.

Existence of a Lagrangian

For a particle moving in an electromagnetic field, it proves to be possible to recast the equations of motion in Lagrangian form, where the generalized forces are derivable from a potential function depending on the velocities as well as on the coordinates. What conditions of integrability must generalized forces satisfy in order that a Lagrangian exist? For motion in 2 dimensions, when can we have a potential function U $(x, \dot{x}, y, \dot{y}, t)$ such that

$$X = (dU_{\dot{x}}/dt) - U_x \text{ and } Y = (dU_{\dot{y}}/dt) - U_y ?$$

(The subscripts in this section denote partial differentiation.)

This becomes

$$X = \dot{g} - U_x \text{ and } Y = \dot{h} - U_y \qquad (2.39)$$

where we have set $g = U_{\dot{x}}$ and $h = U_{\dot{y}}$. Both g and h are functions of x, y, $\dot{x}$, $\dot{y}$, and t and will be regarded as the x- and y- components of a vector $\vec{P}$.

Now any function of the form $f(q, \dot{q}, t)$ has as time derivative

$$\dot{f} = f_q \dot{q} + f_{\dot{q}} \ddot{q} + f_t = \dot{f}\ (q, \dot{q}, \ddot{q}, t)$$

Accordingly, X and Y can depend at most on the accelerations $\ddot{x}$ and $\ddot{y}$ (as well as on velocities and position). We shall now find cross partials of X and Y and their mutual relationships.

The following auxiliary equations may be verified (by the use of the variations Δf and $\Delta\dot{f}$)

$$\frac{\partial}{\partial q}\,\dot{f} = \frac{d}{dt}\,f_q; \quad \frac{\partial \dot{f}}{\partial \dot{q}} = \frac{\partial f}{\partial q} + \frac{d}{dt}\,\frac{\partial f}{\partial \dot{q}}; \quad \frac{\partial \dot{f}}{\partial \ddot{q}} = \frac{\partial f}{\partial \dot{q}} \quad (2.40)$$

Now differentiate (2.39) note that U does not depend on these accelerations, and use (2.40) to obtain

$$X_{\ddot{y}} = (\dot{g})_{\ddot{y}} = U_{x\dot{y}} = h_{\dot{x}} = g_{\dot{y}} \quad (2.41)$$

Therefore,

$$X_{\ddot{y}} = Y_{\ddot{x}} \quad (2,\ I)$$

Again

$$X_{\dot{y}} = \dot{g}_{\dot{y}} - h_x = g_y - h_x + dg_{\dot{y}}/dt$$

and

$$Y_{\dot{x}} = \dot{h}_{\dot{x}} - g_y = h_x - g_y + dh_{\dot{x}}/dt$$

then, by (2.41)

$$X_{\dot{y}} + Y_{\dot{x}} = d(g_{\dot{y}} + h_{\dot{x}})/dt = d(X_{\ddot{y}} + Y_{\ddot{x}})dt \quad (2,II)$$

and

$$X_{\dot{y}} - Y_{\dot{x}} = 2(g_y - h_x) = -2\ \text{curl}_z\,\vec{P} \quad (2.42)$$

Finally,

$$X_y - Y_x = \dot{g}_y - \dot{h}_x = d(g_y - h_x)/dt$$

$$= (1/2)\ d\ (X_{\dot{y}} - Y_{\dot{x}})/dt \quad (2,III)$$

by (2.40) and (2.42).

Equations (2, I), (2, II), and (2, III) are the conditions that a generalized function exists. When X and Y depend only on x and y, (2, I) and (2, II) are automatically satisfied and (2, III) is the ordinary condition that the curl vanish.

Uniqueness of the Lagrangian

Once a Lagrangian L_1 has been found, there arises the question whether or not some other function L_2 would yield the same Lagrangian derivative. If this does happen, then $L_1 = L_1 (x \; \dot{x} \; t)$, $L_2 = L_2 (x \; \dot{x} \; t)$, and $L_2 - L_1 \equiv L$ satisfies

$$\dot{p}_k = \partial L / \partial x_k \qquad (2.43)$$

identically in

$$x, \; \dot{x}, \; \ddot{x}, \text{ and } t$$

where

$$p_k = \partial L / \partial \dot{x}_k$$

Expanding (2.43), we have

$$\sum_\ell [(\partial p_k / \partial x_\ell) \; \dot{x}_\ell + (\partial p_k / \partial \dot{x}_\ell) \; \ddot{x}_\ell + (\partial p_k / \partial t)] \equiv \partial L / \partial x_k$$
$$(2.44)$$

Since this is satisfied identically, the coefficient of $\ddot{x}_\ell$ must vanish, i.e. $\partial p_k / \partial \dot{x}_\ell = 0$, p_k is independent of velocities, and so L is a linear function of the velocities, i.e.

$$L = L_o (x, t) + \Sigma p_\ell \dot{x}_\ell$$

Substituting in (2.44) and equating coefficients,

$$\partial p_k / \partial x_\ell = \partial p_\ell / \partial x_k, \; \partial p_k / \partial t = \partial L_o / \partial x_k$$

which are the conditions that

$$L \, dt = L_o \, dt + \Sigma p_\ell dx_\ell$$

is an exact differential or that $L = d\chi/dt$, where χ is an arbitrary function of x and t.

The Lagrangian derivatives determine the Lagrangian to within an arbitrary additive function, which is the total derivative of a function of x and t alone.

MOMENTS OF DISTRIBUTIONS

When the dynamical system consists either of a discrete set of particles or is a continuous distribution of mass, it is frequently convenient to consider various moments, and it may be possible to characterize important aspects of the motion in terms of these. Particular success with this method has been achieved by S. Chandrasekhar and N. R. Lebovitz in their researches on ellipsoidal figures of equilibrium (of a uniformly rotating mass distribution). We shall begin the discussion with a treatment of some elementary cases and then follow it with an indication of how one can determine the above figures of equilibrium.

Motion in Three Dimensions

Let us suppose that there are n particles, of which the i-th has mass m_i and is acted on by a force $\vec{F}_i$, where

$$\vec{F}_i = \vec{F}_{ie} + \sum_{\substack{i \neq j}}^{n} \vec{F}_{ij} \qquad (2.45)$$

with $\vec{F}_{ie}$ the external force on particle i, and $\vec{F}_{ij}$ the force exerted by particle j on particle i.

The equations of motion are

$$m_i \ddot{\vec{r}}_i = \vec{F}_i \qquad (i = 1 \ldots n)$$

$$(2.46)$$

By Newton's Third Law, the internal forces cancel out in pairs, so that

$$\sum_{i=1}^{m} m_i \ddot{\vec{r}}_i = \Sigma \vec{F}_i = \Sigma \vec{F}_{ie} = \vec{F}_e$$

or alternatively, when the masses m_i are constant,

$$\frac{d^2}{dt^2} \Sigma\, m_i \vec{r}_i = \vec{F}_e$$

If the external force $\vec{F}_e$ is zero, then

$$\Sigma\, m_i \vec{v}_i = const$$

The quantity on the left-hand side is the <u>total momentum</u> of the system, equal to the vector sum of the individual momenta $m_i \vec{v}_i$, and is constant.

The quantities

$$I_1 = \Sigma\, m_i x_i, \quad I_2 = \Sigma\, m_i y_i, \quad I_3 = \Sigma\, m_i z_i$$

are called the <u>moments of first order</u>. Weighted averages of the coordinates are

$$x_c = \frac{\Sigma\, m_i x_i}{\Sigma\, m_i} \qquad y_c = \frac{\Sigma\, m_i y_i}{\Sigma\, m_i} \qquad z_c = \frac{\Sigma\, m_i z_i}{\Sigma\, m_i}$$

and these locate the <u>center of mass</u> at $\vec{r}_c$: (x_c, y_c, z_c).

With $M = \Sigma\, m_i$, the equation of motion of this center can be written as

$$M \ddot{\vec{r}}_c = \vec{F}_e \qquad\qquad (2.47)$$

If the external force F_e is zero, then the center of mass moves with constant velocity $\dot{\vec{r}}_c$.

The coordinates x_i', y_i', and z_i' relative to the center of mass are

$$x_i' = x_i - x_c, \ y_i' = y_i - y_c, \ z_i' = z_i - z_c$$

The first order moments relative to the center of mass will be

$$I_1' = \Sigma \ m_i \ (x_i - x_c) = \Sigma \ m_i x_i'$$

$$I_2' = \Sigma \ m_i \ (y_i - y_c) = \Sigma \ m_i y_i'$$

$$I_3' = \Sigma \ m_i \ (z_i - z_c) = \Sigma \ m_i z_i'$$

and will all be <u>zero</u>, by definition of x_c, y_c, and z_c.

Motion of a Rigid Rotator

<u>Moments</u> of the <u>second order</u> are encountered when we deal with a rigid body which rotates with constant angular velocity ω about some axis fixed in the body. Let us consider the body as made up of a discrete set of n particles which are fixed relative to each other. For the ith particle, if we take the moment of (2.46), using (2.45)

$$\vec{r}_i \ x \ m_i \ddot{\vec{r}}_i = \vec{r}_i \ x \ \vec{F}_i = \vec{r}_i \ x \ (\vec{F}_{ie} + \sum_{i \neq j}^{n} \vec{F}_{ij})$$

Summing over the whole configuration, since the internal torques sum to zero,

$$\frac{d}{dt} \sum_i \vec{r}_i \ x \ m_i \ \dot{\vec{r}}_i = \Sigma \ \vec{r}_i \ x \ \vec{F}_{ie} \qquad (2.48)$$

The time rate of change of the total angular momentum is thus equal to the total torque on the system.

Substituting $\vec{r}_i = \vec{r}_c + \vec{r}_i{}'$ and expanding the left-hand side, we have, noting that $\Sigma\, m_i \vec{r}_i{}' = 0$,

$$\frac{d}{dt} \underset{i}{\Sigma} \,[m_i\, (\vec{r}_c + \vec{r}_i{}') \times (\dot{\vec{r}}_c + \dot{\vec{r}}_i{}')]$$

$$= \frac{d}{dt}\, \{\Sigma m_i\, (\vec{r}_c \times \dot{\vec{r}}_c) + \Sigma m_i\, (\vec{r}_i{}' \times \dot{\vec{r}}_i{}')\}$$

$$= M\, (\vec{r}_c \times \ddot{\vec{r}}_c) + \frac{d}{dt}\, \Sigma m_i\, (\vec{r}_i{}' \times \dot{\vec{r}}_i{}')$$

$$= \vec{r}_c \times \vec{F}_e + \frac{d}{dt}\, \Sigma m_i\, (\vec{r}_i{}' \times \dot{\vec{r}}_i{}')$$

from (2.47). Consequently, from (2.48),

$$\frac{d}{dt}\, \Sigma m_i\, (\vec{r}_i{}' \times \dot{\vec{r}}_i{}') = \underset{i}{\Sigma}\, (\vec{r}_i{}' \times \vec{F}_{ie}) \qquad (2.49)$$

Referred to the center of mass, the time rate of change of the angular momentum equals the total applied torque.

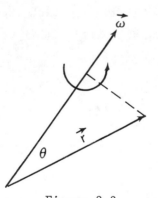

Figure 2.2

Rotation about an axis

For a rigid body, let us take the origin 0 at the center of mass, which will be regarded as fixed. If the body rotates about an axis through this origin (see Figure 2.2) the velocity of any point will be

$$\dot{\vec{r}} = \vec{\omega} \times \vec{r} \qquad (2.50)$$

with magnitude $|\dot{\vec{r}}| = \omega r \sin\theta$.

Using this expression in (2.49), dropping primes, we

have for the total angular momentum $\vec{J}$

$$\vec{J} = \Sigma m_i \ [\vec{r}_i \times (\vec{\omega} \times \vec{r}_i)] = \Sigma m_i \ [\vec{\omega} \, r_i^2 - \vec{r}_i \ (\vec{\omega} \cdot \vec{r}_i)]$$

with components

$$J_x = \omega_x \ \Sigma m_i \ (y_i^2 + z_i^2) - \omega_y \ \Sigma m_i x_i y_i - \omega_z \ \Sigma m_i x_i z_i$$

$$J_y = -\omega_x \ \Sigma m_i x_i y_i + \omega_y \ \Sigma m_i \ (x_i^2 + z_i^2) - \omega_z \ \Sigma m_i y_i z_i$$

$$J_z = -\omega_x \ \Sigma m_i x_i z_i - \omega_y \ \Sigma m_i y_i z_i + \omega_z \ \Sigma m_i \ (x_i^2 + y_i^2)$$

In the case where the total torque is zero, the components of angular momentum are constant. They are linear combinations of the second order moments Σmx^2, Σmy^2, Σmz^2, Σmxy, Σmxz, and Σmyz, which can be evaluated for both discrete systems of particles and continuous mass distributions, provided that the particles are at fixed positions or the distributions do not vary with time.

The moments are gross properties of the body. Once they have been found, they may be regarded as parameters in the description of the motion, just as the mass of a single particle can be taken as a parameter when that particle moves under the influence of a given force field.

If we set $R^2 = x^2 + y^2$, then $\Sigma mR^2 = \Sigma m(x^2 + y^2)$ is called the moment of inertia about the z-axis. Similarly, $\Sigma m(y^2+z^2)$ and $\Sigma m \ (x^2 + z^2)$ are the moments of inertia about the x-and y-axis, respectively. The quantities Σmxy, Σmxz, and Σmyz are called products of inertia. For a systematic notation, set $x_1 = x$, $x_2 = y$, $x_3 = z$ and define the second moments by

$$I_{ij} = \int \rho \ x_i x_j \ d\tau \quad (i, \ j = 1,2,3) \qquad (2.52)$$

The aggregate of these nine quantities is called the inertia tensor, and $I_{ji} = I_{ij}$.

If the body has the z-axis as an axis about which the mass distribution is symmetric (not a function of the angle ϕ about the axis), then contributions from $-x$ cancel out those from $+x$, similarly for y, and the direction x is on a par with the direction y. The products of inertia then vanish, and the moment of inertia about the x-axis equals that about the y-axis, since $I_{11} = I_{22}$.

ELLIPSOIDAL FIGURES OF FLUID EQUILIBRIUM

For various astrophysical applications, it is of interest to study the gravitational equilibrium of homogeneous masses which rotate with a uniform angular velocity ω. For small values of ω, it can be shown that an oblate spheroid (ellipsoid of revolution) will constitute a figure of equilibrium. The earth, which is slightly flattened at the poles, is a good example. In the following treatment, we shall use second order moments to obtain a relation between angular velocity and the eccentricity of the spheroid.

Consider an ideal fluid described by a density $\rho(x, t)$, and isotropic pressure $p(x, t)$, and that an element thereof is acted on only by a pressure gradient and the force of gravitation due to the other elements of the fluid. The equation of motion for this element, referred to an inertial frame of reference, is

$$\rho \frac{du_i}{dt} = - \frac{\partial p}{\partial x_i} - \rho \frac{\partial U}{\partial x_i} \qquad (2.53)$$

where u_i is the velocity component and U is the potential

$$U(x) = - G \int \frac{\rho(x)'}{|x - x'|} \, dx'dy'dz' \qquad (2.54)$$

[The net force in the x-direction on a volume element $\Delta x \Delta y \Delta z$ is equal to $(\partial p/\partial x)\Delta x \Delta y \Delta z$, and hence the first term on the right-hand side of (2.53)].

Now multiply (2.53) by x_j, integrate over the volume, and note that $p = 0$ at the surface. The result is

$$\int \rho \frac{du_i}{dt} x_j \, d\tau = \int \rho [\frac{d}{dt} (u_i x_j) - u_i u_j] \, d\tau$$

$$= \frac{d}{dt} \int (\rho u_i x_j \, d\tau - 2 T_{ij}) \, d\tau$$

$$= \delta_{ij} \Pi + W_{ij} \qquad (2.55)$$

where $T_{ij} = 1/2 \int \rho \, u_i u_j \, d\tau$, $W_{ij} = - \int \rho x_i (\partial U/\partial x_j) \, d\tau$ and $\Pi = \int p \, d\tau$.

(A detailed examination yields $W_{ij} = W_{ji}$).

Equilibrium of Rotating Ellipsoid

If we have a configuration rotating uniformly with angular velocity ω, then it is convenient to refer the equations of motion to a frame of reference rotating with this angular velocity. If in this frame $u_i = 0$, then (2.53) is to be modified by adding the centrifugal force $(\rho/2)(\partial/\partial x_i)|\omega \times r|^2$ to the right-hand side, since the centrifugal potential is represented by $(1/2)|\omega \times r|^2$.

Then, if we have a steady state and $u_i = 0$, and the x_3-axis is chosen to be along the direction of ω, (2.55) is modified to become

$$\delta_{ij} \Pi + W_{ij} + \omega^2 (I_{ij} - \delta_{i3} I_{3j}) = 0 \qquad (2.56)$$

where
$$\delta_{ij} = 1 \text{ if } i = j$$
$$= 0 \text{ if } i \neq j$$

Writing out (2.53) explicitly for the various components, and using $W_{ij} = W_{ji}$ and $I_{ij} = I_{ji}$, we have

$$W_{11} + \omega^2 I_{11} = W_{22} + \omega^2 I_{22} = W_{33} \qquad (2.57)$$

$$W_{12} + \omega^2 I_{12} = 0, \; W_{13} = W_{23} = 0, \; I_{13} = I_{23} = 0$$
$$(2.58)$$

From (2.57), we have

$$\omega^2 = (W_{33} - W_{11})/I_{11} \qquad (2.59)$$

If a rotating mass is in equilibrium under its own gravitational forces, and its shape is given, then the angular velocity may be calculated from (2.59). Starting with a sphere, for which $\omega = 0$, one may examine first spheroids and then tri-axial ellipsoids.

A spheroid may be characterized by the eccentricity of the ellipse which is rotated to generate the figure. Note that $W_{12} = 0 = I_{12}$, $W_{11} = W_{22}$, and $I_{11} = I_{22}$.

The calculation gives

$$\omega^2 = \frac{2 \, (1-e^2)^{1/2}}{e^3} \, (3-2e^2) \, \sin^{-1} e - \frac{6}{e^2} \, (1-e^2)$$
$$(2.60)$$

where ω is measured in units of $(\pi G \rho)^{1/2}$, with G the gravitational constant.

Equation (2.60) gives ω^2 as a function of e and describes the sequence of MacLaurin spheroids. This curve is plotted in Figure 2.3, and has a maximum at e = 0.92995. If the mass is deformed slightly, it will tend to return to its original shape, or be stable, if e < 0.95289. (This may be substantiated by a study of the equations of motion for small displacements from equilibrium, and the interested reader is referred to Chandrasekhar, pp 28-37, 80-88).

Jacobi found that one could have ellipsoidal figures, with three unequal axes, which are in equilibrium, the mass density being uniform. Defining A_i, A_{12}, and B_{12} as

$$A_i = a_1 a_2 a_3 \int_0^\infty \frac{du}{\Delta(a_i^2 + u)}$$

$$A_{12} = a_1 a_2 a_3 \int_0^\infty \frac{du}{\Delta(a_1^2 + u)(a_2^2 + u)}$$

$$B_{12} = a_1 a_2 a_3 \int_0^\infty \frac{u \, d u}{\Delta(a_1^2 + u)(a_2^2 + u)}$$

where the semi-axes are a_i and
$$\Delta^2 = (a_1^2 + u)(a_2^2 + u)(a_3^2 + u).$$

Equation (2.57) becomes

$$\omega^2 a_1^2 - 2A_1 a_1^2 = \omega^2 a_2^2 - 2A_2 a_2^2 = - 2A_3 a_3^2$$

Adding $2 a_1^2 a_2^2 A_{12}$ to each side,

$$a_1^2 (\omega^2 - 2B_{12}) = a_2^2 (\omega^2 - 2B_{12}) = 2 (A_{12} a_1^2 a_2^2 - A_3 a_3^2)$$

Solutions with $a_1 \neq a_2$ are possible if and only if

$$a_1^2 a_2^2 A_{12} = e_3^2 A_3 \qquad\qquad (2.62)$$

and

$$\omega^2 = 2B_{12} \qquad\qquad (2.63)$$

Equation (2.62) determines a <u>unique relation</u> between a_2/a_1 and a_3/a_1 in order that equilibrium be possible, and this

characterizes the Jacobian ellipsoids. The value of the
associated angular velocity is given by (2.63).

As $a_2 \to a_1$, the Jacobian ellipsoids will approach a Mac-
Laurin spheriod. For this,

$$\omega^2 = 2B_{11}$$

which occurs when e = .81267. The branching off of the
Jacobi sequence from the MacLaurin sequence is shown in
Figure 2.3. The Jacobi figures are stable up to a_2/a_1 =
0.432232, a_3/a_1 = 0.345069, ω^2 = 0.284030, after which pear-
shaped figures of equilibrium are possible. These in turn
may evolve into two separated masses. (For details, the
reader should consult Chandrasekhar and Jeans.) The evolution is shown schematically in Figures 2.4 - 2.6.

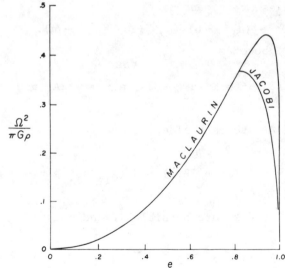

Figure 2.3 (From Chandrasekhar, S., 1964: "Ellipsoidal
Figures of Equilibrium," with permission from Yale University Press). The square of the angular velocity (in
the unit $\pi G\rho$) along the MacLaurin and the Jacobian
sequences. The abscissa, in both cases, is the eccentricity of the (1,3)-section.

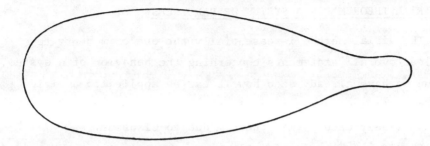

Figure 2.4   (From Jeans, J. H., 1929: "Astronomy and
Cosmogomy," Cambridge University Press, with permission). Pear-shaped figure.

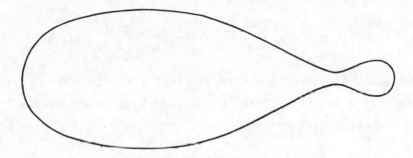

Figure 2.5   (From Jeans, J. H., 1929: "Astronomy and
Cosmogomy," Cambridge University Press, with permission). Pear-shaped figure with deep furrow.

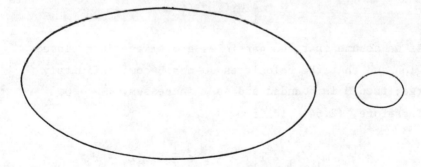

Figure 2.6   (From Jeans, J. H., 1929: "Astronomy and
Cosmogomy," Cambridge University Press, with permission). Figure of equilibrium with masses separated.

VIRIAL THEOREM FOR A SYSTEM OF PARTICLES

The virial method is essentially the use of moments to
glean overall statements concerning the behavior of a system,
and we have already seen how it can be applied to a rigid
body. Another application is to general motions, and results
in a widely used virial theorem due to Clausius.

Consider the scalar function

$$M = \sum_i \frac{1}{2} m_i \vec{r}_i \cdot \vec{r}_i$$

Its second time derivative is

$$\ddot{M} = \sum_i m_i \dot{\vec{r}}_i \cdot \dot{\vec{r}}_i + \sum_i m_i \vec{r}_i \cdot \ddot{\vec{r}}_i \qquad (2.64)$$

Integrate this over the time interval $t = 0$ to $t = \tau$,
where τ is arbitrarily long, and divide by τ to obtain the
time average $\langle \ddot{M} \rangle$, namely

$$\langle \ddot{M} \rangle = \frac{1}{\tau} \int_o^\tau \ddot{M} \, dt = \frac{1}{\tau} (\dot{M}_\tau - \dot{M}_o)$$

where

$$\dot{M} = \sum_i m_i \vec{r}_i \cdot \dot{\vec{r}}_i$$

If we assume that the particles are always in a closed
region and that the velocities do not become infinitely
large, then $\dot{M}$ is bounded and as τ increases, $\langle \ddot{M} \rangle \to 0$.

Therefore, (2.64) yields

$$\left\langle \frac{1}{2} \sum_i m_i \, (\dot{\vec{r}}_i \cdot \dot{\vec{r}}_i) \right\rangle = - \left\langle \frac{1}{2} \sum_i m_i \vec{r}_i \cdot \ddot{\vec{r}}_i \right\rangle$$

$$= - \left\langle \frac{1}{2} \sum_i \vec{r}_i \cdot \vec{F}_i \right\rangle \qquad (2.65)$$

where $\vec{F}_i$ is the force acting on the ith particle. The
left-hand side of (2.65) is the time average of the kinetic
energy of the system and is equal to the right-hand side,
called the virial of the system.

As a simple example, suppose that we have a single particle
in a central Coulomb field, so that $U(r) = -|k|/r$ represents
the potential energy. Then $\vec{r} \cdot \vec{F} = U$ and the virial theorem
states that the average kinetic energy equals half the negative of the average potential energy. This is also true for
a system of particles interacting according to Coulomb's
Law.

MOMENTS AND PLASMAS

Some useful conservation theorems about plasmas, which are
collections of charged particles, can be obtained by taking
appropriate moments, as we shall show. For a great number
of particles, one specifies the density of the particles in
$(\vec{r}, \vec{v})$ space as $f(\vec{r}, \vec{v}, t)$. By considering the flow across the
boundaries of a small volume $dxdydzdv_x dv_y dv_z$, we arrive at
the equation of continuity

$$\frac{\partial f}{\partial t} + \frac{\partial}{\partial x_i} (f \dot{x}_i) + \frac{\partial}{\partial v_i} (f \dot{v}_i) = 0$$

(where we sum over an index which occurs twice)
or, differentiating,

$$\frac{\partial f}{\partial t} + f(\frac{\partial \dot{x}_i}{\partial x_i} + \frac{\partial \dot{v}_i}{\partial v_i}) + \frac{\partial f}{\partial x_i} \dot{x}_i + \frac{\partial f}{\partial v_i} \dot{v}_i = 0$$

Now $\dot{v}_i = F_i/m$, where F_i is the force acting on the particle
and is velocity-independent for non-electromagnetic forces.

For electromagnetic forces, $F_i = q (\vec{v} \times \vec{B})_i$ and $\partial F_i / \partial v_i = 0$.
Also, v_i is independent of x_i, so that both terms in the
parenthesis vanish, and, putting everything together,

$$\frac{\partial f}{\partial t} + \frac{\partial f}{\partial x_i} \dot{x}_i + \frac{F_i}{m} \frac{\partial f}{\partial v_i} = 0 \qquad (2.66)$$

This is the <u>collisionless Boltzmann equation</u>. If colli-
sions occur, their effect is to add a term $(\partial f / \partial t)_{coll}$ to the
right-hand side. However, the total mass, momentum, and
kinetic energy are conserved in binary collisions, so that
the corresponding velocity moments of $(\partial f / \partial t)_{coll}$ vanish.

If we integrate the above equation over velocity space, and
assume that f falls off very rapidly as $v \to \infty$, so that surface integrals over a sphere $v \to \infty$ vanish, then we obtain

$$\frac{\partial n}{\partial t} + \frac{\partial}{\partial x_i} (n u_i) = 0 \qquad (2.67)$$

where $n = \int f \, d^3v$ and $u_i = (1/n) \int v_i \, d^3v$. The quantity
u_i is the average velocity in the direction x_i. Equation
(2.67) is the <u>equation of continuity</u> in configuration space
(x_1, x_2, x_3).

If we multiply (2.66) by mv_k and integrate, the result is

$$\frac{\partial}{\partial t} (nmu_k) + \frac{\partial}{\partial x_i} (nm < v_i v_k >) = n < F_k > \qquad (2.68)$$

where the angular brackets indicate an average over velocity
space, i.e.

$$n <F_k> = \int F_k \, f \, d^3v.$$

Equation (2.68) is an expression for <u>change of momentum</u>.
The first term is the rate of increase of momentum per unit

volume, while the second is a loss of momentum due to particles escaping across the surface enclosing the volume. The
right-hand side represents the force density.

If we multiply (2.66) by $mv^2/2$ and integrate, we obtain an
equation for the change of energy, namely

$$\frac{\partial}{\partial t} (n < \frac{1}{2} mv^2 >) + \frac{\partial}{\partial x_i} (n < \frac{1}{2} mv^2 v_i >) = nqE_i v_i$$

$$(2.69)$$

where q is the charge and E_i is the ith component of the
electric field. The first term is the rate of change of
energy density, the second is the heat transfer across the
boundary, and the right-hand side is the power fed in by the
electric field.

(If there is more than one species of particle, as will be
so for a real plasma, then appropriate summations must be
made to generalize the above equations.)

Equations (2.67) - (2.69) may be regarded as basic for
fluid dynamics. If one introduces the assumption that the
rate of flow ρu is proportional to the gradient of the mass
density, then (2.67) becomes the diffusion equation

$$\frac{\partial n}{\partial t} - \frac{\partial}{\partial x_i} (D \frac{\partial n}{\partial x_i}) \qquad (2.70)$$

where D is a coefficient which may depend on position.

The heat conduction equation (2.69) may be reduced to a
form similar to (2.70), i.e.

$$\rho \, c_v \frac{\partial T}{\partial t} = K \, \nabla^2 T \qquad (2.71)$$

where c_v is the specific heat at constant volume, T is the
absolute temperature, and K is the thermal conductivity. To

do this (see Huang, 1966), one assumes no mass motion ($<v>=0$), and that the particles have a velocity distribution not far from a Maxwellian one, i.e.

$$f^{(o)} = n \ (m/2 \ \pi kT)^{1/2} \ e^{-mv^2/2kT}$$

The <u>momentum equation</u> (2.68) is perhaps the most useful of the above equations for hydrodynamics, which will perhaps be evident after we have recast it into other forms. First, note that, with $u_i \equiv <v_i>$

$$<v_i v_j> = <(v_i - u_i)(v_j - u_j)> + <v_i>u_j + u_i<v_j> - u_i u_j$$

$$= <(v_i - u_i)(v_j - u_j)> + u_i u_j \qquad (2.72)$$

Setting $P_{ij} \equiv \rho<(v_i - u_i)(v_j - u_j)>$ and substituting (2.72) into (2.68), we have

$$\rho \ (\frac{\partial u_i}{\partial t} + u_j \frac{\partial u_i}{\partial x_j}) = n < F_i > - \frac{\partial P_{ij}}{\partial x_j} \qquad (2.73)$$

or

$$\frac{\partial u_i}{\partial t} + u_j \frac{\partial u_i}{\partial x_j} = \frac{1}{m} < F_i > - \frac{1}{\rho} \frac{\partial P_{ij}}{\partial x_j} \qquad (2.74)$$

The quantity $\frac{DX}{Dt} \equiv (\frac{\partial}{\partial t} + u \cdot \nabla) X$ is known as the material derivative, or total derivative, of X, because it represents the total rate of change of X attached to a point moving with average velocity u. In the case where F_i does not depend on velocity, which will now be assumed, then $<F_i> = F_i$. The quantity (P_{ij}) is called the <u>pressure tensor</u>, for its ij component represents relative momentum $\rho(v_i - u_i)$ in the i-direction being transported with relative velocity in the j-direction (or conversely) and thus setting up a stress (shear or

pressure) on the surface normal to j (or i). To be more
specific, let us consider a rectangular
sides dx_1, dx_2, dx_3, acted on by the surrounding fluid, and
regard P_{ij} as the force in the 1-direction on unit area normal to the j-direction. The net force on the parallelopiped
will be the sum of the individual net forces on surfaces normal to three directions, namely

$$(-\frac{\partial P_{i1}}{\partial x_1} - \frac{\partial P_{i2}}{\partial x_2} - \frac{\partial P_{i3}}{\partial x_3})dx_1 dx_2 dx_3 \qquad (2.75)$$

But this, divided by the volume, is precisely the last term
of (2.73). Hence, the material time rate of change of momentum of a small volume equals the total external force plus
the force exerted by the surrounding medium on this volume,
the latter being represented by (2.75).

If one now assumes that the velocity distribution is approximately Maxwellian, then (see Huang, 1966) the momentum equation can be put into the form

$$(\frac{\partial}{\partial t} + u \cdot \nabla) u = \frac{F}{m} - \frac{1}{\rho} \nabla (P - \frac{\mu}{3} \nabla \cdot u) + \frac{\mu}{\rho} \nabla^2 u \quad (2.76)$$

where μ is the coefficient of viscosity. This expression,
which is called the Navier-Stokes Equation, has several interesting applications, which we proceed to discuss.

First, suppose that there is steady irrotational flow in a
conservative external force field, so that $F = -\nabla\Phi$, $\partial u/\partial t = 0$,
and $\nabla \times u = 0$. If the viscosity is zero, i.e. $\mu = 0$, then
(2.76) becomes

$$\nabla(\frac{1}{2} u^2 + \frac{P}{\rho} + \frac{\Phi}{m}) = -\frac{kT}{m} \frac{\nabla\rho}{\rho} \qquad (2.77)$$

where the equation of state $P/\rho = kT/m$ has been used.

  (a) For uniform density, (2.77) integrates to

$$\frac{1}{2} u^2 + \frac{P}{\rho} + \frac{\Phi}{m} = \text{const} \qquad (2.78)$$

which is commonly expressed as: The sum of velocity head
plus pressure head plus potential head is constant.

(b) For uniform temperature, (2.77) integrates to

$$\rho = \rho_o \exp \left[-\frac{1}{kT} \left(\frac{1}{2} mu^2 + \Phi \right) \right]$$

with ρ_o an arbitrary constant. If there is zero mass motion,
then

$$\rho = \rho_o \exp \left(-\Phi/kT \right) \qquad (2.79)$$

Other applications of the Navier-Stokes Equation include
(a) finding the effective mass of a sphere moving with velocity u_o in an infinite, nonviscous, incompressible fluid of
constant density and (b) obtaining Stokes' Law $F = 6\pi\mu a u_o$
for the force on a sphere of radius a moving with constant
velocity u_o in a medium with viscosity coefficient μ. For
the details, which involve a knowledge of spherical harmonics,
the interested reader may consult Huang (1966).

In summary, the calculation of velocity moments has enabled
us to derive for plasmas and for neutral fluids the basic
equations of diffusion, heat conduction and momentum transfer.
This is just another demonstration of the power of the moment
method.

Exercises

1. Given that a force field is axially symmetric. Write
down the appropriate Lagrange Equations. Specialize to the
case where the particle moves in a magnetic field which is
constant in time.

2. Since Newton's Equations are valid for an inertial

system of coordinates, then Lagrange's Equations must be valid
for such a system. How, then, do we find the motion in a
plane of a particle in a rotating system of coordinates, using
Lagrange's Equations? Show, in particular, how these equations give automatically the same equations of motion that
are deduced by elementary vector methods.

3. A particle is dropped from rest at a height of 400
meters above the earth. Neglecting atmospheric resistance,
where will it strike the ground?

4. Assume that the earth is a uniform sphere rotating about
the polar axis. Let a particle be dropped into a straight
smooth tube, through the center of the earth, at the equator.
Knowing that the gravitational effect on the particle is the
same as if the mass interior to it is concentrated at the
center, find the subsequent motion. Does the particle return
to its starting point? If so, calculate the elapsed time for
this.

5. What central law of force will result in a spiral orbit
$r = a \exp b\theta$ for a particle moving in such a force field?

III

Pendulum and Top

As an application of Lagrange's Equations, we shall consider the problems (a) where a heavy particle is constrained to move in a spherical bowl, and (b) where a symmetrical spinning top moves with one point fixed. The first type of problem is a special case of the second, and hence the same sort of mathematics must govern. There can be positions of equilibrium, about which librations in an effective potential will take place. A typical libration is not now sinusoidal, but has a wave form describable in terms of Jacobian elliptic functions, which are essentially distorted sine and cosine functions of the time. More precisely, they are sine and cosine functions of a variable ϕ which depends on the time but is not proportional to it. We give a short account of these functions, since they are important here and in a later chapter.

ELLIPTIC FUNCTIONS

In place of a function $y = \sin t$, let us study the function $y = \sin \alpha$, where the angle α varies with time according to the differential equation

$$\left(\frac{d\alpha}{dt}\right)^2 = 1 - k^2 \sin^2 \alpha \qquad (3.1)$$

and k is a constant less than unity. (The form (3.1) is
chosen because it is the one actually found for the plane
pendulum, see below.) Since k < 1 and $\sin^2 \alpha < 1$, the right-
hand side of (3.1) will always be positive. Therefore, the
equation can be integrated in the form

$$t = \int_0^\alpha \frac{d\alpha}{\sqrt{1 - k^2 \sin^2 \alpha}} \equiv f(\alpha) \qquad (3.2)$$

The inverse function $\alpha = f^{-1}(t)$ is called the amplitude
and written $\alpha = am\ t$.

(We have chosen t = 0 to be at α = 0).

From (3.2), there will be a certain time t = K for α to
increase from 0 to $\pi/2$. If we now consider the functions
y = sin α and z = cos α as functions of t, we may write

$$y = \sin am\ t \equiv sn\ t \qquad (3.4)$$

$$z = \cos am\ t \equiv cn\ t \qquad (3.5)$$

where sn and cn are convenient labels for these <u>Jacobian</u>
<u>elliptic functions</u>. The periods of y and z are 2π with
respect to α, and 4K with respect to t.

Forms alternative to (3.1) are

$$\left(\frac{dy}{dt}\right)^2 = (1 - y^2)(1 - k^2 y^2) \qquad (3.6)$$

$$\left(\frac{dz}{dt}\right)^2 = 1 - k^2 + (2k^2 - 1) z^2 - k^2 z^4 \qquad (3.7)$$

$$\frac{d^2y}{dt^2} = - (1 + k^2) y + 2k^2 y^3 \qquad (3.8)$$

$$\frac{d^2z}{dt^2} = (2k^2 - 1) z - 2k^2 z^3 \qquad (3.9)$$

To obtain the time derivatives of (3.4) and (3.5) is simple.

We have

$$\frac{dy}{dt} = \cos \alpha \frac{d\alpha}{dt} \equiv \text{cnt dn t}$$

$$\frac{dz}{dt} = - \sin \alpha \frac{d\alpha}{dt} \equiv \text{snt dn t}$$

where we use the abbreviation dn t = $d\alpha/dt$ (3.10)
and where, from (3.1)

$$\text{dn}^2 \, t = 1 - k^2 \, \text{sn}^2 \, t \qquad (3.11)$$

Two extreme cases are of interest. When k = 0, then
y = cos t; when k = 1, y = coth t. This may be readily verified by substituting back into (3.6). When k^2 = 1/2, then
(3.9) becomes

$$\frac{d^2 z}{dt^2} + z^3 = 0 \qquad (3.12)$$

(This equation has been discussed by Bartlett (1968),
Bibliography, Ch. X.)

For some applications, the kinetic energy term (3.6) is a
cubic in y instead of fourth degree. This offers no obstacle,
because we can introduce substitutions which transform to
(3.6). Let the dependent variable be u instead of y, and
suppose the cubic has 3 real roots $u_1 < u_2 < u_3$. The equation for the change of u shall be

$$\left(\frac{du}{dt}\right)^2 = \beta \, (u - u_1) \, (u - u_2) \, (u - u_3) \qquad (3.13)$$

Letting $u' = u - u_1$, we have

$$\left(\frac{du'}{dt}\right)^2 = \beta \, u' \left(1 - \frac{u'}{u_2 - u_1}\right) \left(1 - \frac{u'}{u_3 - u_1}\right) (u_2 - u_1)(u_3 - u_1)$$

$$= \beta \, u' \, (1 - y^2) \, (1 - k^2 \, y^2) \, (u_2 - u_1) \, (u_3 - u_1)$$
$$(3.14)$$

where we have set

$$y^2 = \frac{u'}{u_2 - u_1} = \frac{u - u_1}{u_2 - u_1} \qquad (3.15)$$

and

$$k^2 = \frac{u_2 - u_1}{u_3 - u_1} \qquad (3.16)$$

Proceeding further, with $du' = 2(u_2 - u_1)y\, dy$, we obtain

$$\frac{4}{\beta} \left(\frac{dy}{dt}\right)^2 \frac{1}{u_3 - u_1} = (1 - y^2)(1 - k^2 y^2) \qquad (3.17)$$

If we now introduce a new time variable

$$x = \frac{t}{2} [\beta (u_3 - u_1)]^{1/2} \qquad (3.18)$$

then Equation (3.17) reduces to

$$\left(\frac{dy}{dx}\right)^2 = (1 - y^2)(1 - k^2 y^2) \qquad (3.19)$$

Since this has as solution $y = sn\,(x + \delta)$, the original equation (3.13) will have as its solution

$$u = u_1 + (u_2 - u_1)\, sn^2\,(x + \delta) \qquad (3.20)$$

where x is given by (3.18) and δ is some constant.

THE PLANE PENDULUM

The plane pendulum may be regarded as a particle constrained to move without friction in a vertical circle of radius a (Figure 3.1). If the origin is taken at the center of the circle, and θ is the angle from the downward vertical, the kinetic energy per unit mass is

$$T = 1/2\ a^2\ \dot{\theta}^2$$

The potential energy per unit mass, referred to the lowest

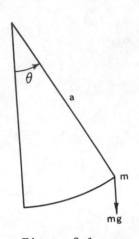

Figure 3.1

The Plane Pendulum

point is

$$U = ga(1-\cos \theta) = 2ga \sin^2 (\theta/2)$$

Since this does not contain the time explicitly, energy is conserved, i.e.

$$\frac{1}{2} a^2 \dot{\theta}^2 + 2ga \sin^2 \frac{\theta}{2} = h \qquad (3.21)$$

Solving

$$\left(\frac{\dot{\theta}}{2}\right)^2 = \omega^2 \left(\frac{h}{2ga} - \sin^2 \frac{\theta}{2}\right) \qquad (3.22)$$

where

$$\omega^2 = g/a \qquad (3.23)$$

In order for any motion to be possible, h must be positive, as we see from (3.21). If $h > 2ga$, then the quantity $\dot{\theta}^2$ is always greater than zero, and the particle keeps going around in a circle. If $h = 2ga$, the velocity is zero at the top of the circle. This is a position of unstable equilibrium, which may be approached as $t \to \infty$. If $h < 2ga$, there is libration.

I. Libration

If $h < 2ga$, there will be a maximum value of θ, call it θ_o, such that

$$\sin^2 \frac{\theta_o}{2} = \frac{h}{2ga} \qquad (3.24)$$

We shall show that the motion is a periodic libration between θ_o and $-\theta_o$. If we set

$$\sin \frac{\theta_o}{2} = k \qquad (3.25)$$

and

$$\sin \frac{\theta}{2} = k \sin \alpha, \qquad (3.26)$$

then (3.22) becomes, with the use of (3.24),

$$\frac{\dot{\theta}}{2} = \pm \omega k \cos \alpha \qquad (3.27)$$

But, differentiating (3.26),

$$\frac{\dot{\theta}}{2} \cos \frac{\theta}{2} = k \cos \alpha \, \dot{\alpha}$$

and hence, using (3.27),

$$\dot{\alpha} = \pm \omega \cos \frac{\theta}{2}$$

Squaring, we obtain, with the aid of (3.26)

$$\dot{\alpha}^2 = \omega^2 (1 - k^2 \sin^2 \alpha) \qquad (3.28)$$

which is of the form of Equation (3.1).

The motion is thus given by

$$\sin \frac{\theta}{2} = k \sin \alpha = k \text{ sn } \omega t \qquad (3.29)$$

This is periodic in α with period 2π and in ωt with period $4K$, where, from (3.2),

$$K = \int_0^{\frac{\pi}{2}} \frac{d\alpha}{\sqrt{1 - k^2 \sin^2 \alpha}} \qquad (3.30)$$

The functions sn and K are tabulated vs. argument and k, and so are known functions.

If the maximum deflection angle θ_o is small, then $k = \sin (\theta_o/2)$ is small, and (3.30) may be expanded in powers of k. Approximately, $k \cong \theta_o/2$, and

$$K \cong \int_0^{\frac{\pi}{2}} (1 + \frac{1}{2} k^2 \sin^2 \alpha) \, d\alpha$$

$$= \frac{\pi}{2} (1 + \frac{k^2}{4})$$

The period of the pendulum is thus, from (3.23)

$$\tau = 2\pi \ (a/g)^{1/2}(1 + \frac{\theta_o^2}{16}) \qquad (3.31)$$

This period varies slightly with the amplitude θ_o, but the variation can be neglected if θ_o is really small and excessive accuracy is not required.

II. Critical Motion, h = 2ga

Suppose that the particle can just reach the top of the circle, and that it starts at the bottom, with $\dot{\theta} > 0$, at time $t = 0$. From (3.22), with $h = 2ga$, we have

$$\frac{\dot{\theta}}{2} = \omega \ \cos \frac{\theta}{2} \qquad (3.32)$$

Integrating,

$$\omega t = \ln \ (1 + \sin \frac{\theta}{2}) - \ln \cos \frac{\theta}{2} \qquad (3.33)$$

When $\theta \to \pi$, the latter term on the right-hand side becomes infinite, so that it requires an infinite time for the particle to reach the top of the circle. Now, if $\theta' = \pi - \theta$, then, near $\theta = \pi$, $\cos \frac{\theta}{2} \cong \frac{\theta'}{2}$ and (3.33) becomes $\omega t = \ln (4/\theta')$ which gives

$$\theta' = 4 \ e^{-\omega t} \qquad (3.34)$$

If one wishes a formula for all time, then (3.33) can be exponentiated to yield

$$\sin \frac{\theta}{2} = \tanh \omega t \qquad (3.35)$$

III. Circular Motion

When the energy is sufficiently great, i.e. $h > 2ga$, then the particle will execute periodic motion in a circle, as we shall now show. Equation (3.22) may be written

$$\left(\frac{\dot{\theta}}{2}\right)^2 = \frac{\omega^2}{k^2} \left(1 - k^2 \sin^2 \frac{\theta}{2}\right) \qquad (3.36)$$

where

$$k^2 = 2ga/h < 1$$

This is now of the form of (3.1) and has for solution

$$\sin \frac{\theta}{2} = sn\ (\omega t/k)$$

From (3.36), the period in $2\omega t/k$ is $4K$, so that the period in t is $\tau = 2\ k\ K/\omega$. The points θ, $\theta + 2\pi$, $\theta + 4\pi$, are all equivalent, with the same velocity, determined by (3.36).

THE SPHERICAL PENDULUM

Let us now consider a particle which moves under gravity on a smooth sphere of radius a. The angle of rotation around the vertical axis shall be ϕ, and θ is measured as before from the downward polar axis.

The kinetic energy is now

$$T = \frac{1}{2}\ a^2\ (\dot{\theta}^2 + \sin^2\theta\ \dot{\phi}^2) \qquad (3.37)$$

The constant angular momentum J is

$$J = a^2 \sin^2\theta\ \dot{\phi} \qquad (3.38)$$

The equation for the conservation of energy is, using (3.38) to eliminate $\dot{\phi}$

$$\frac{1}{2}\ a^2\ \dot{\theta}^2 + 2ga\ \sin^2 \frac{\theta}{2} + \frac{J^2}{2a^2\sin^2\theta} = h \qquad (3.39)$$

We shall now show that (3.39) can be reduced to an equation of the form of (3.13), which can then be solved in terms of known elliptic functions. To facilitate the comparison, let us set $u = -\cos \theta$, $\dot{u} = \sin \theta\ \dot{\theta}$, and $2 \sin^2(\theta/2) = 1 + u$.

Multiplying (3.39) by $(2/a^2) \sin^2\theta$,

$$\dot{u}^2 = \frac{2}{a^2} (1 - u^2) [h - ga (1 + u)] - \frac{J^2}{a^4}$$

$$(3.40)$$

or

$$\dot{u}^2 = \frac{2g}{a}[(1 - u^2) (H - u) - \frac{J^2}{2ga^3}] \qquad (3.41)$$

where we have set

$$H = \frac{h}{ga} - 1 \qquad (3.42)$$

Equation (3.41) is similar to (3.13) and comparison shows that $\beta = 2g/a$. We can regard the motion as a libration of u with kinetic energy $(a/2g) \dot{u}^2$, effective potential energy

$$U^* = (H - u) (u^2 - 1) \qquad (3.43)$$

and total energy $- J^2/2ga^3$. Figure 3.2 shows a plot of U^* vs u, for $H < 1$. U^* has zeros at -1, H, and $+1$. The slope dU^*/du is zero when

$$-3u^2 + 2uH + 1 = 0,$$

which has roots

$$u_o = \frac{1}{3} \{H \pm (H^2 + 3)^{1/2}\} \qquad (3.44)$$

The second derivative here is

$$\frac{d^2U^*}{du^2} = 2 (H - 3u_o)$$

$$= \mp 2 (H^2 + 3)^{1/2} \qquad (3.45)$$

Consequently, there is a minimum of U^* at the lesser value of u_o, and a maximum at the greater value, as shown.

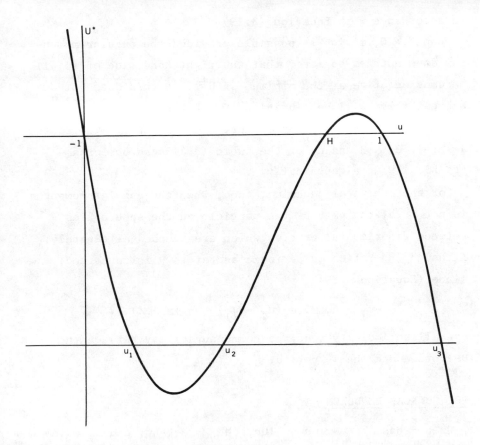

Figure 3.2

U^* vs u

When $J = 0$, there can be libration in this potential well
from $u_1 = -1$ to $u_2 = H$ (<1). This is described by (3.20),
with $u = -\cos\theta$, $\delta = 0$, $k^2 = h/2ga$, $H + 1 = h/ga$, and $u_3 = 1$,
as

$$1 - \cos\theta = \frac{h}{ga} \, \text{sn}^2 \, \sqrt{\frac{g}{a}} \, t$$

or

$$\sin\frac{\theta}{2} = k \, \text{sn} \, \sqrt{\frac{g}{a}} \, t$$

in accordance with Equation (3.29)

When $J > 0$, motion is possible provided the angular momentum does not get so large that the right-hand side of (3.41)
becomes negative at the minimum of U^*. If this does not
happen, there will be libration of u between the 2 values u_1
and u_2, shown in the figure, where $U^* = -J^2/2ga^3$. The roots
$u_1 \geqslant -1$, $u_2 \ll H$ and u_3 of the cubic (3.41) can be found
graphically or algebraically.

For the spherical pendulum, then, when the angular momentum
is not zero, the path of the particle on the sphere lies
between 2 horizontal circles which are touched alternately.
Although the motion is periodic as far as θ is concerned,
since (Equation 3.20)

$$-u_1 - \cos \theta = (u_2 - u_1) \, sn^2 \, (x + \delta),$$

overall periodicity requires commensurability between the
θ- period and the ϕ- period.

THE CONICAL PENDULUM

When J has its maximum value, the 2 roots u_1 and u_2 coin-
cide and $\dot{u} = 0$. The value of θ is a constant, $\theta = \theta_o$, and
this is the <u>conical pendulum</u>.

The equation of motion in the θ- direction may be obtained
by differentiating (3.39), and is

$$a \, \ddot{\theta} = -g \sin \theta + \frac{J^2}{a^3} \frac{\cos \theta}{\sin^3\theta}$$

The first term on the right-hand side is the θ-component of
the gravity force, and the second term is the θ-component of
the centrifugal force, $a \sin \theta \, \dot{\phi}^2 = J^2/a^3 \sin^3\theta$. These will
balance, and θ will be constant, when

$$\frac{J^2}{a^2 \sin^4\theta_o} \cos \theta_o = ga$$

or $\qquad \dot{\phi}^2 = g/a \cos \theta_o$

The period of the motion is then $T = 2\pi (a \cos\theta_o /g)^{1/2}$ and depends only on the amount $a \cos\theta_o$ by which the horizontal circle is below the center of the sphere

Small Oscillations

If J is just below its maximum value, small oscillations about θ_o can occur. Their magnitude and frequency are determined by U^*, which has the form

$$U^* = U_o^* + (H - 3u_o)(u - u_o)^2 + \ldots \qquad (3.46)$$

The equation of motion in u is, from (3.41)

$$\ddot{u} = - \frac{g}{a} \frac{\partial U^*}{\partial u} = - \frac{2g}{a}(H - 3u_o)(u - u_o)$$

The angular frequency ω of the oscillation is, with (3.46),

$$\omega = (\tfrac{g}{a})^{1/2} (2H - 6u_o)^{1/2} = (\tfrac{2g}{a})^{1/2} (H^2 + 3)^{1/4}$$
$$(3.47)$$

The amplitude of oscillation A is found from (3.41) and (3.46) by setting

$$U^* - U_o^* = \left(\frac{J_o^2 - J^2}{2ga^3} \right) = (H - 3u_o)(u_2 - u_o)^2$$

$$= (H^2 + 3)^{1/2} (u_2 - u_o)^2 \qquad (3.48)$$

from which

$$A \equiv u_2 - u_o = \left(\frac{J_o^2 - J^2}{2ga^3} \right)^{1/2} (H^2 + 3)^{-1/4}$$
$$(3.49)$$

The motion is thus $u - u_o = A \sin \omega t$ with A given by (3.49) and ω by (3.47).

THE SPINNING SYMMETRICAL TOP

If now we consider a rigid body with axial symmetry to be
fixed at one point and to be spinning about its axis through
this point, then the motion will be a generalization of that
of the spherical pendulum, which is the case for zero spin.
We now take the top axis to be at an angle χ with the upward
vertical. There will be an extra kinetic energy term due to
the spin. If ψ denote the angle of rotation about the axis,
and $\dot{\psi}$ the spin, the total angular velocity ω of a point about
this axis will equal

$$\omega = \dot{\psi} + \dot{\phi} \cos \chi \qquad (3.50)$$

(If there were no spin, $\omega = \dot{\phi} \cos \chi$, just the projection of
$\dot{\phi}$ of the top. To verify this, suppose there is a vertical
plane rotating around the z-axis, and that it momentarily

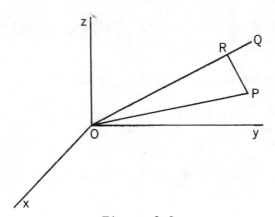

Figure 3.3

Vertical Plane ORP Rotating about the z-axis

coincides with the y-z plane, and represented by QOP, where
OQ is the top axis (Figure 3.3). Let OR be the projection
of OP on this axis, and suppose the plane be rotated by an

amount $d\phi$, The difference in path travelled by P and by R
will be, if $\alpha = \angle$ POQ and $r = OP$,

$$r \sin (\chi + \alpha) \, d\phi - r \cos \alpha \sin \chi \, d\phi = r \sin \alpha \cos \chi \, d\phi$$

$$= RP \cos \chi \, d\phi$$

Thus the rate at which the point P goes around the axis OQ
is just $\cos \chi \, \dot{\phi}$, which verifies our assertion.)

If one particle goes around in a circle of radius r, with
angular velocity ω, its kinetic energy is $mv^2/2$, equal to
$mr^2\omega^2/2$. If there are several particles, the kinetic energy
is

$$T = \sum_i m_i r_i^2 \, (\omega^2/2)$$

If there is a continuous distribution of mass, then the
kinetic energy of rotation about an axis is

$$T = I\omega^2/2$$

where

$$I = \int r^2 \, \rho \, d\tau$$

with ρ the mass density and r the distance from the axis.

In the case of a top which is symmetrical about an axis,
we can denote by A the moment of inertia about the fixed
point and by C the moment of inertia about the axis.

The total kinetic energy of such a top is

$$T = \frac{1}{2} A \, (\dot{\chi}^2 + \dot{\phi}^2 \sin^2\chi) + \frac{1}{2} C\omega^2 \qquad (3.51)$$

where ω is as given in (3.50). This is a generalization of
Equation (3.37), for which $A = a^2$. Let us specify that A and
C are to be for a unit mass.

The potential energy per unit mass, referred to the bottom
of the sphere of radius a (where a is the distance from peg

to center of gravity of the top) is

$$U = ga + ga \cos \chi \qquad (3.52)$$

Conservation of Angular Momenta

Since $T - U$ does not involve either ϕ or ψ explicitly, there are 2 constant angular momentum components, namely those around the z-axis and the top axis, i.e.

$$J_\phi = \partial T/\partial \dot{\phi} = A \sin^2\chi \, \dot{\phi} + C\omega \cos \chi \qquad (3.53)$$

and

$$J_\psi = \partial T/\partial \dot{\psi} = C\omega \qquad (3.54)$$

Solving for $\dot{\phi}$,

$$\dot{\phi} = \frac{J_\phi - J_\psi \cos \chi}{A \sin^2\chi} \qquad (3.55)$$

Conservation of Energy

Since the time does not enter T or U explicitly, and since T is a homogeneous quadratic in $\dot{\chi}$ and $\dot{\phi}$, energy is conserved, and $T + U = h$, or

$$\frac{A}{2} (\dot{\chi}^2 + \sin^2\chi \, \dot{\phi}^2) + \frac{C}{2} \omega^2 + ga + ga \cos \chi = h$$
$$(3.56)$$

Putting $u = \cos \chi = - \cos \theta$, $\dot{u} = - \sin \chi \, \dot{\chi}$ (as before), this becomes

$$\frac{A}{2} (\dot{\chi}^2 + \sin^2\chi \, \dot{\phi}^2) = ga \, (H - u) \qquad (3.57)$$

where we have set

$$H = \frac{1}{ga} (h - ga - \frac{C}{2} \omega^2) \qquad (3.58)$$

(This reduces to the previous definition when C = 0.

Equation of θ Motion

From (3.57), since the left-hand side is positive or zero, we must have H ≥ u for any actual motion. Using Equations (3.53), (3.54), and (3.55) in (3.57), the equation of motion for u is

$$\frac{A}{2} \dot{u}^2 = ga(1 - u^2)(H-u) - \frac{1}{2A} (J_\phi - J_\psi u)^2 \tag{3.59}$$

$$= \sin^2 \chi [ga (H - u) - \frac{A}{2} \sin^2 \chi \dot{\phi}^2] \tag{3.60}$$

Equation (3.59) may be regarded as a generalization of (3.41), the equation for the spherical pendulum. The right-hand side is a cubic in u, and with different coefficients. We obtain (3.41) from (3.59) by setting C = 0 and A = a^2.

The motion described by (3.59) can therefore be expected to be a libration, a limitation motion, or a motion for which $\dot{\theta}$ does not change sign. The limitation motion will be toward the point χ = 0, the upward vertical. Setting $\dot{\chi}$ = 0 there, we find from (3.56) and (3.58) that H = 1. We shall restrict ourselves here to librations, which will occur between two values, u_1 and u_2, of u, where $-1 \le u_1 < u_2 < H < 1$.

Libration Limit

The roots of the cubic in (3.59) may be found by intersecting the two curves (see Figure 3.4)

$$y_1 = \frac{2gaA}{J_\psi^2} (H - u) (u^2 - 1) = \frac{2gaA}{J_\psi^2} U^* \tag{3.61}$$

$$\text{and } y_2 = - (u - \frac{J_\phi}{J_\psi})^2 = - (u - u_c)^2, \text{ where } u_c = J_\phi/J_\psi. \tag{3.62}$$

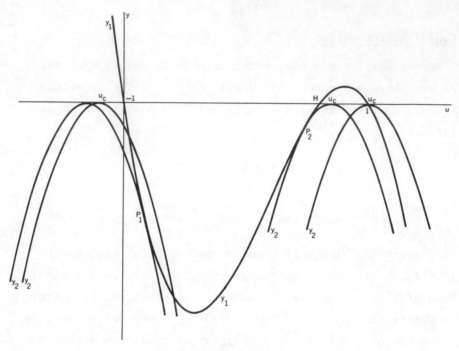

Figure 3.4

y_1 and y_2 vs. u

The curve represented by (3.61) has the same shape as that for the effective potential for the spherical pendulum. Its minimum is at

$$u_o = (1/3) \{H - (H^2 + 3)^{1/2}\} \qquad (3.44)$$

and its zeros at -1, H, and +1. It is to be intersected now not by a horizontal line but by a downward parabola which is tangent to the line y = 0 at the point $u = u_c = J_\phi/J_\psi$. In general, there will be two intersections, u_1 and u_2, the libration limits, where

$$-1 \leq u_1 < u_2 < H < 1.$$

The complete category of librational motions will be covered if we let u_c range from sufficiently negative values to sufficiently positive values with H held fixed. When the curves (3.61) and (3.62) first touch, $u_c < -1$, u_1 equals u_2 and there is <u>steady precession</u> at P_1. When u_c now increases, two intersections appear, $u_1 < u_2$. When $u_c = -1$, then $u_1 = -1$ and $-1 < u_2 < H$. When $u_c = H$, $-1 < u_1 < H$. As u_c increases still further, the two intersections remain for a while, $-1 < u_1 < u_2 < H$, but finally coalesce, and at this point there is again steady precession at $P_2(u_1 = u_2 < H)$ For values of u_c greater than this, no libration is possible.

<u>Steady Precession</u>

For steady precession, we must have both angular velocity and acceleration in the χ-direction equal to zero. Differentiating (3.59) with respect to the time, we have, setting $\dot{u} = 0$,

$$A\ddot{u} = -ga\,[(1 - u_1^2) + 2u_1\,(H - u_1] + (J_\psi/A)\,(J_\phi - J_\psi\,u_1)$$

$$= -\sin^2\chi_1\,[ga - J_\psi\,\dot{\phi} + Au_1\,\dot{\phi}^2]$$

$$= -\sin^2\chi_1\,F(\dot{\phi}_1) \qquad\qquad (3.63)$$

where we have used (3.55), set $\dot{\phi} = \dot{\phi}_1$, its initial value, $\chi = \chi_1$, and

$$F(\dot{\phi}) \equiv Au\,\dot{\phi}^2 - J_\psi\,\dot{\phi} + ga$$

The function $F(\dot{\phi})$ is quadratic in $\dot{\phi}$, with minimum at $J_\psi/2Au_1$ where $F = ga - (J_\psi^2/4Au_1)$. When $u_1 > 0$ and the spin is large, this is negative and $F(\dot{\phi}) = 0$ has two real roots

$$\dot{\Phi}_a = \frac{J_\psi}{2Au_1} \left[1 - \left(1 - \frac{4gaAu_1}{J_\psi^2} \right)^{1/2} \right] \quad (3.64)$$

$$\dot{\Phi}_b = \frac{J_\psi}{2Au_1} \left[1 + \left(1 - \frac{4gaAu_1}{J_\psi^2} \right)^{1/2} \right] \quad (3.65)$$

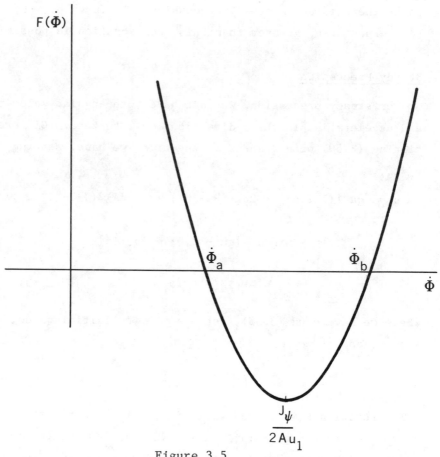

Figure 3.5

Roots of $F(\dot{\Phi}) = 0$

These are the angular velocities for steady precession, since by (3.63) the acceleration in the χ direction is zero. The latter value is for a fast precession, approximately

$$\dot{\phi}_b \simeq J_\psi / A u_1$$

The slow precession has angular velocity

$$\dot{\phi}_a \simeq g a / J_\psi$$

and is what is generally observed.

Since the radicand must be positive, we must have

$$J_\psi^2 > 4 g a A u_1$$

which imposes a slight restriction on J_ψ if $\chi_1 < \pi/2$, but which is always satisfied for $\chi_1 > \pi/2$ (pendulum).

Now, from (3.63), the top will rise whenever $F(\dot{\phi}) < 0$. But we know that it must rise from its lowest point $u = u_1$ to keep the kinetic energy positive. Accordingly, $F(\dot{\phi}_1) < 0$ at $u = u_1$, so that $\dot{\phi}_1$ must lie in the range between $\dot{\phi}_a$ and $\dot{\phi}_b$. (By the same argument, since the top must fall from $u = u_2$, $F(\dot{\phi})$ must be positive at this limit of the libration.)

Survey of Possible Motions

Let us now investigate a sequence of motions which all start from the lower libration limit $u = u_1$, $\dot{u} = 0$, but which have different initial angular velocities $\dot{\phi} = \dot{\phi}_1 > 0$. There are two general relations which involve $\dot{\phi}_1$, which from (3.55) and (3.60) are

$$2 g a \, (H - u_1) = A \sin^2 \chi_1 \, \dot{\phi}_1^{\,2} \qquad (3.66)$$

$$u_c - u_1 = \frac{A \sin^2 \chi_1}{J_\psi} \dot{\phi}_1 \qquad (3.67)$$

The point $u = u_c = J_\phi/J_\psi$ is, from (3.55), that point at which the angular velocity is zero, i.e. $\dot{\phi} = 0$. This will occur in the actual motion if u_c lies between u_1 and u_2, where $-1 \le u_2 \le H$. Equation (3.55) may be recast as

$$A \sin^2 \chi \, \dot{\phi} = J_\psi (u_c - u) \qquad (3.68)$$

Consequently, whenever u becomes greater than u_c, the angular velocity $\dot{\phi}$ becomes negative, which results in a loop on the sphere of unit radius.

When $u_c < -1$, $u_c - u$ is negative throughout the motion, and so is the angular velocity, by (3.68). If $u_c = -1 = u_1$, we have a limitation motion, not a libration. But if $-1 < u_c < H$, then the angular velocity $\dot{\phi}$ is positive for $u_1 < u < u_c$, zero for $u = u_c$, and negative for $u_c < u < u_2 < H$. If $u_c = H < 1$, the angular velocity is positive for $u_1 < u < u_2 = u_c$, but becomes zero for $u = H$, and simultaneously $\dot{u} = 0$. The initial angular velocity at u_1 is found by combining (3.66) and (3.67) into the form

$$2ga (H - u_c) = 2ga (H - u_1 + u_1 - u_c)$$

$$= A \sin^2 \chi_1 \, \dot{\phi}_1 \, (\dot{\phi}_1 - \frac{2ga}{J_\psi}) \qquad (3.69)$$

whence

$$\dot{\phi}_1 = 2ga/J_\psi, \text{ for } u_c = H \qquad (3.70)$$

The motion is between two horizontal circles, with cusps at the upper one $u_2 = H$. If $H < u_c$, the angular velocity is always positive during the motion $u_1 < u < u_2 < H < 1$, by (3.68). If

$u_c = 1$, the angular velocity is, from (3.55),

$$\dot{\phi} = J_\psi/A(1 + u)$$

with initial value

$$\dot{\phi}_1 = J_\psi/A(1 + u_1) \qquad\qquad (3.71)$$

The initial angular velocity may be increased above its value in (3.71) until it reaches $\dot{\phi}_b$, the value for fast steady precession. For any value within this range, the motion will be a libration (nutation) in u, with the angular velocity always positive.

(At the start, we kept H fixed and varied u_c, which allows u_1 to vary. This procedure of sliding a parabola from left to right is perhaps preferable when one desires to see how the intersections with U^* change. Alternatively, one can keep u_1 constant in (3.66) and (3.67) and let $\dot{\phi}_1$ vary, thereby determining u_c and H. The viewpoint to be adopted is largely a matter of taste.)

Small Oscillations

If the top is slightly disturbed from the steady conical motion, it will oscillate about it, just as the spherical pendulum does about its steady motion. To find the equation for oscillation, we merely have to vary equations (3.53) and (3.63), i.e. to substitute

$$\chi = \chi_1 + \delta\chi, \quad \dot{\phi} = \dot{\phi}_1 + \delta\dot{\phi}, \text{ and to use } F(\dot{\phi}_1) = 0.$$

We have

$$J_\phi = A\sin^2\chi\ \dot{\phi} + J_\psi \cos \chi \qquad\qquad (3.53)$$

and

$$0 = A \sin^2\chi\ \delta\dot{\phi} + (2A\dot{\phi} \sin \chi \cos \chi - J_\psi \sin \chi)\ \delta\chi$$

from which

$$\delta\dot{\phi} = \frac{J_\psi - 2A\dot{\phi}\cos\chi}{A\sin\chi}\,\delta\chi \qquad (3.72)$$

Furthermore,

$$A\ddot{\chi} = \sin\chi\,[A\cos\chi\,\dot{\phi}^2 - J_\psi\,\dot{\phi} + ga] \qquad (3.63)$$

and

$$A\delta\ddot{\chi} = \sin\chi\,[-A\sin\chi\,\dot{\phi}^2\,\delta\chi + (2A\dot{\phi}\cos\chi - J_\psi)\,\delta\dot{\phi}] \qquad (3.73)$$

Substituting (3.72) in (3.73)

$$\delta\ddot{\chi} + k^2\,\delta\chi = 0$$

where

$$k^2 = \sin^2\chi_1\,\dot{\phi}_1^2 + \frac{1}{A^2}\,(J_\psi - 2A\dot{\phi}_1\cos\chi_1)^2$$

Since $k^2 > 0$, we do have oscillations, and they are just the librations near the bottom of the effective potential. When the spin is very large, the angular frequency k is approximately $J_\psi/A = C\omega/A$.

Exercises

1. A particle of mass m is constrained to move along the parabola $z = x^2$ and the force of gravity acts in the $-z$-direction. If the particle is initially at rest at (x_o, z_o), find the expression for the subsequent motion as a function of time. What is the period?

2. If a particle of mass m moves on the paraboloid of revolution $z = x^2 + y^2$, starting from some point (x_o, y_o, z_o) with arbitrary angular velocity but zero velocity normal to the z-direction, and gravity still acts in the $-z$-direction, what will be the subsequent motion as a function of time? Will it be periodic?

3. A particle of unit mass moves in a plane, being attracted by two fixed Newtonian centers of force in the plane, one
of mass m_1 at $x = 1$, $y = 0$ and the other of mass m_2 at $x = -1$,
$y = 0$. Using the elliptic coordinates defined by
$x = \text{ch } u \cos v$, $y = \text{sh } u \sin v$, set up the Lagrange Equations,
and obtain u and v as Jacobi elliptic functions of a parameter,
and thus a representation of an orbit. Show qualitatively
what types of orbits may be expected, and how their motions
are limited, if they are. How may one obtain the periods?

4. For the symmetrical top, derive a general expression,
as a generalization of (3.20), for the motion in the
χ-direction. With $H = 0.6$ and $u_c = 0.5$, plot both the
χ-motion and the ϕ-motion, and show that loops appear. With
$H = 0.6$ and $u_c = 0.6$, find the ϕ-interval between successive
cusps. With $H = 0.6$, find $u_c = u_{c1}$ for which there is steady
precession (P_2 in Figure 3.4). If now u_c is chosen to be
halfway between u_{c1} and $H = 0.6$, find the limits u_1 and u_2 of
the motion and the angular velocities at these two limits.
Plot the motion as above, and observe that the angular
velocity does not change sign.

The Two-Body Problem

When two bodies interact with each other according to an inverse-square force law, and there is no external applied force, the resulting motions are fairly simple. It will turn out, from the discussion below, that each body moves either in an ellipse, a parabola, or an hyperbola, with the center of mass at one focus. To show this, it is first necessary to discuss the motion of the center of mass, and then the relative motion with respect to it.

ROLE OF CENTER OF MASS

In general, let there be given a system of particles, the kth of which is acted on by an internal force $\vec{F}_k^{\,i}$ and an external force $\vec{F}_k^{\,e}$. The internal force is the resultant of the forces due to the other particles. Because action and reaction are equal and opposite in Newtonian mechanics, the sum of the internal forces is zero, i.e. $\Sigma\ \vec{F}_k^{\,i} = 0$. Let the total mass be $M = \Sigma\ m_k$. The vector distance $\vec{R}$ from an arbitrary fixed origin to the center of mass is defined by the relation

$$\vec{R} = \Sigma\ m_k\ \vec{r}_k / M.$$

The acceleration of this point is given by

$$\ddot{\vec{MR}} = \Sigma\, m_k\, \ddot{\vec{r}}_k = \Sigma\, \vec{F}_k^{\,i} + \Sigma\, \vec{F}_k^{\,e} = \vec{F}^{e}$$

where $\vec{F}^{e}$ is the total external force on the system, namely

$$\vec{F}^{e} = \Sigma\, \vec{F}_k^{\,e}.$$

The motion of the center of mass of a system of particles, with total mass M, is the same as that of a particle of mass M acted upon by the total external force exerted on the system. If this force is zero, then the center of mass moves with constant velocity.

If there are only two particles interacting with each other, and the external force is zero, then the relative motion obeys a simple rule. Let mass m_2 be situated at P, mass m_1 be at Q, and denote the vector $\vec{PQ}$ by $\vec{r} = \vec{r}_1 - \vec{r}_2$ (See Figure 4.1). The relative acceleration is, by Newton's Equations,

$$\ddot{\vec{r}} = \ddot{\vec{r}}_1 - \ddot{\vec{r}}_2 = \frac{\vec{F}_1^{\,i}}{m_1} - \frac{\vec{F}_2^{\,i}}{m_2}$$

$$= \vec{F}_1^{\,i}\left(\frac{1}{m_1} + \frac{1}{m_2}\right)$$

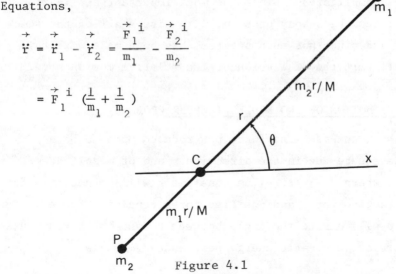

Figure 4.1
Geometry for Two Masses with Center of Mass at C

since

$$\vec{F}_2{}^i = -\vec{F}_1{}^i \text{ (action-reaction law)}$$

Now, if we introduce a "reduced" mass μ defined by

$$\frac{1}{\mu} = \frac{1}{m_1} + \frac{1}{m_2},$$

the equation of motion is

$$\mu\ddot{\vec{r}} = \vec{F} \ (\equiv \vec{F}_1{}^i)$$

The relative motion of m_1 with respect to m_2 is that of a fictitious particle of "reduced" mass μ acted on by the force of m_2 on m_1.

If the origin be chosen at the center of mass C, then R = 0 gives $m_1r_1 = m_2r_2$, which then yields m_2r/M for the distance CQ and m_1r/M for PC.

(When more than two particles interact, the relative motion is more complicated. Since the most important application would be to the 3-body problem, with masses all of the same order, and since not much progress has been made with this, we shall omit the discussion of the relative motion here.)

RELATIVE MOTION OF TWO BODIES (KEPLER PROBLEM)

We shall consider the force per reduced mass to be an inverse square one in the direction $\vec{r}$ and of magnitude $\kappa/r^2 = -\partial U/\partial r$, where $U = \kappa/r$. (The constant κ will be negative for an attractive force and positive for a repulsive force.) The angle θ will denote the angle between PQ and the fixed x-axis.

The relative kinetic energy per reduced mass is

$$T = \frac{1}{2} (\dot{r}^2 + r^2\dot{\theta}^2).$$

Conservation of energy $(T + U = h)$ is expressed by the equation

$$\frac{1}{2} (\dot{r}^2 + r^2\dot{\theta}^2) + \frac{\kappa}{r} = h \qquad (4.1)$$

Conservation of angular momentum is given by

$$\dot{\theta} = J/r^2 \qquad (4.2)$$

where J is a constant angular momentum per reduced mass. Substituting (4.2) in (4.1), we have after rearrangement

$$\dot{r}^2 = 2h - \frac{2\kappa}{r} - \frac{J^2}{r^2} \qquad (4.3)$$

Dividing through by $\qquad \dot{\theta}^2 = J^2/r^4$

gives $\qquad\qquad \left(\frac{dr}{d\theta}\right)^2 = \frac{r^4}{J^2} \left(2h - \frac{2\kappa}{r} - \frac{J^2}{r^2}\right) \qquad (4.4)$

This is the equation which determines how r varies with the angle θ. We shall choose $\theta = 0$, the initial line, such that r is an extremum, i.e. $dr/d\theta = 0$ at this value.

Integration of (4.4) is accomplished most readily by substituting $u = 1/r$, or $du = -dr/r^2$. Then

$$\left(\frac{du}{d\theta}\right)^2 = \frac{2h}{J^2} \left(1 - \frac{\kappa}{h} u - \frac{J^2}{2h} u^2\right)$$

$$= - \left(u + \frac{\kappa}{J^2}\right)^2 + \frac{2hJ^2 + \kappa^2}{J^4} \qquad (4.5)$$

We readily see that this has the trigonometric solution

$$u + \frac{\kappa}{J^2} = A \cos\theta \qquad (4.6)$$

where

$$A^2 = (2hJ^2 + \kappa^2)/J^4 \qquad (4.7)$$

Equation (4.6) is the polar equation of a conic section, with the origin at one focus. (Development of necessary formulae involving conic sections is carried out in Appendix B, to which we shall refer frequently).

The General Motion

When the force is attractive, (4.6) may be compared with (B2) in the form

$$u = (1/\ell) \ (1 + e \cos \theta) \qquad \text{(B2)}$$

The result is

$$\ell = - J^2/\kappa \text{ and } e = - A J^2/\kappa \qquad \text{(4.8)}$$

The nature of the motion will depend on the value of e.

On the other hand, when the force is repulsive, (4.6) can be compared with (B16) in the form

$$u = (1/\ell)(e \cos \theta - 1) \qquad \text{(B16)}$$

to yield

$$\ell = J^2/\kappa \text{ and } e = A J^2/\kappa \qquad \text{(4.9)}$$

The motion is along an hyperbola A'L' (Figure 4.4) with repulsion from opposite focus F.

Both ℓ and e depend on the absolute value of κ.

The eccentricity is given by (see 4.7 and 4.8)

$$e^2 = \frac{A^2 \ J^4}{\kappa^2} = \frac{2 \ h \ J^2}{\kappa^2} + 1$$

or

$$e = (1 + \frac{2hJ^2}{\kappa^2})^{1/2} \qquad \text{(4.10)}$$

If, then, <u>the total energy h is negative</u>, the eccentricity
is less than unity (e < 1) and <u>the motion is in an ellipse</u>
with the origin at one focus. <u>If the total energy h is posi-</u>
<u>tive</u>, then the <u>motion is along an hyperbola with the origin</u>
at one focus, since e > 1. Parabolic motion occurs for h = 0,
when e = 1.

The semimajor axis a may be found, using $\ell = a \left| 1 - e^2 \right|$.
Combining (4.8) or (4.9) with this and (4.10), we have

$$\ell = a \left| 1 - e^2 \right| = a \left| 2h \right| J^2 / \kappa^2 = J^2 / |\kappa|$$

from which

$$a = |\kappa| / |2h| \qquad\qquad (4.11)$$

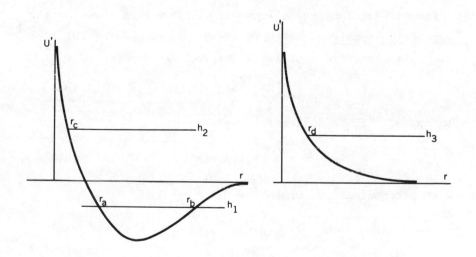

Attractive Field   Repulsive Field

Figure 4.2

Effective Potential Energy vs. Radius

The limits of the motion can be found by setting r = 0 in Equation (4.3). If we define an effective potential energy U' by

$$U' = \frac{\kappa}{r} + \frac{J^2}{2r^2} \qquad (4.12)$$

then $\dot{r} = 0$ when U' = h. Plots of U' vs. r are given in Figure 4.2 for both an attractive and a repulsive field. The attractive potential has a minimum for $r = -J^2/\kappa$. For $h = h_2 > 0$, the kinetic energy is positive when r is greater than a minimum value r_c at which U' = h_2. For $h = h_1 < 0$, the motion is confined between a minimum $r = r_a$ and a maximum $r = r_b$, at both of which U' = h_1. Finally, the effective repulsive potential is always positive, and the motion is only possible for $r > r_d$, where U' = $h_3 > 0$ at $r = r_d$.

Elliptic Motion

The motion along the ellipse may be found in terms of the eccentric anomaly ε, which is itself a function of the time t. This eccentric anomaly is defined by

$$\cos \varepsilon = x/a, \ \sin \varepsilon = y/b \qquad (B7)$$

In Appendix B, we derive the expression

$$r = a (1 - e \cos \varepsilon) \qquad (B9)$$

with extreme values, where $\dot{r} = 0$,

$$r_1 = a (1 - e) \text{ and } r_2 = a (1 + e) \qquad (B10)$$

Starting with known values of h and of J, we may obtain a and e from the relations

$$a = |\kappa| \ / \ (-2h) \qquad (4.11)$$

$$1 - e^2 = (-2h) \ J^2/\kappa^2 \qquad\qquad (4.10)$$

Equation (4.3) may be rewritten as

$$r^2 \ \dot{r}^2 = 2h \ (r - r_1) \ (r - r_2)$$

$$= -2h \ a^2 \ e^2 \ (1 - \cos \varepsilon) \ (1 + \cos \varepsilon)$$

$$= -2h \ (ae)^2 \ \sin^2 \varepsilon$$

where the expressions (B9) and (B10) have been substituted. But, differentiating (B9) re t,

$$\dot{r} = ae \ \sin \varepsilon \ \dot{\varepsilon}$$

and combination of these two equations and (4.11) gives

$$r\dot{\varepsilon} = (-2h)^{1/2} = (|\kappa|/a)^{1/2}$$

or, using (B9) and defining $n = (|\kappa|/a^3)^{1/2}$

$$(1 - e \cos \varepsilon) \ \dot{\varepsilon} = n$$

which may be integrated to yield

$$\varepsilon - e \sin \varepsilon = nt \qquad\qquad (4.13)$$

Figure 4.3 is a plot of the eccentric anomaly versus nt, proportional to the time. The time t_m to the minor axis is less than a quarter period, as we see by setting $\varepsilon = \pi/2$, which results in

$$nt_m = \frac{\pi}{2} - e$$

or

$$t_m = (\frac{\pi}{2} - e) \ \frac{\tau}{2\pi}$$

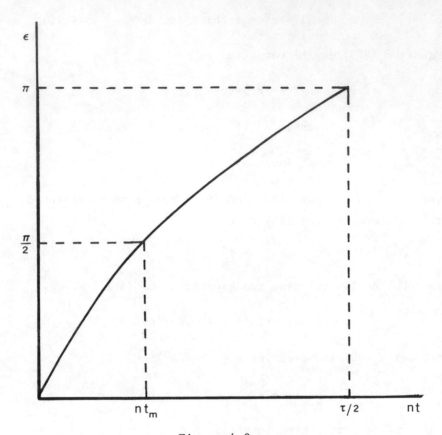

Figure 4.3

Eccentric Anomaly vs. Time

The period τ, the time from $\varepsilon = 0$ to $\varepsilon = 2\pi$, equals $2\pi/n$, so that n is called the <u>mean motion</u> (angular frequency). From its definition, the mean motion is related to the major axis by

$$n = |\kappa|^{1/2} / a^{3/2} \qquad (4.14)$$

and the period is

$$\tau = 2\pi\, a^{3/2}\, |\kappa|^{-1/2} \qquad (4.15)$$

The square of the period is proportional to the cube of the major axis. This relationship is known as Kepler's Law.

Hyperbolic Motion

For an attractive field, the particle is constantly pulled toward the focus at $r = 0$ and the path is concave toward this focus; for a repulsive field, the force is away from the focus and the path is convex toward this focus. Consequently, for $h > 0$, the path of an attracted particle will be along the branch AL of the hyperbola nearest to F and described by (B2), namely $\dfrac{1}{r} = \dfrac{1}{\ell} (1 + e \cos \theta)$ (B2)

where $\ell = J^2/\kappa$. (See Figure 4.4.)

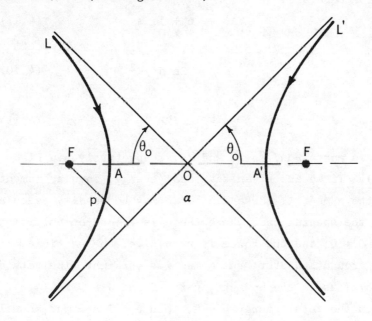

Figure 4.4

Hyperbolic Motion Near F

LA Attractive Field and L'A'   Repulsive Field

On the other hand, the path of a repelled particle will be along the branch A'L' farthest from F and described by (B16), namely

$$\frac{1}{r} = \frac{1}{\ell} \ (e \cos \theta - 1)$$

where

$$\ell = J^2/\kappa$$

Distance of Closest Approach

The distance of closest approach for attraction will be

$$FA = (e - 1) \ a \qquad\qquad (4.16)$$

and for repulsion

$$FA' = (e + 1) \ a \qquad\qquad (4.17)$$

where

$$e^2 - 1 = 2h \ J^2/\kappa^2 \qquad\qquad (4.10)$$

and

$$a = |\kappa|/2h \qquad\qquad (4.11)$$

These correspond to r_c and r_d, respectively, in Figure (4.2). If we are given the total energy, angular momentum and the number κ, these distances may be readily calculated.

In the special case where there is a head-on collision, i.e. $J = 0$, and the force is repulsive, the particle will turn around and start retracing its straight line path at $r = h/\kappa$, as we see by setting $\dot{r} = 0$ and $J = 0$ in (4.3).

When the total energy $h = 0$, then $e = 1$ and motion will be in a parabola if the force is attractive. (If the force is repulsive, the particle cannot come in, see Figure 4.2.) The distance of closest approach is found by setting $\dot{r} = 0$ and $h = 0$ in (4.3), and is $r = - J^2/2\kappa$. From the geometry, this should be one half the latus rectum, and we see from (4.8) that it is.

Angle of Deflection

If a particle come in from infinity, go along an hyperbola and leave toward infinity, the total angle of deflection α will be $\alpha = \pi - 2\theta_o$ (Figure 4.4).

From (B18) and the definition of b, with (4.10) and (B 12),

$$\tan \theta_o = (b/a) = (e^2 - 1)^{1/2} = (2h)^{1/2} \; J/|\kappa| \tag{4.18}$$

But if the initial velocity is v_o at $r = \infty$, the kinetic energy there is $h = \frac{1}{2} v_o^2$, so that

$$\tan \theta_o = v_o \; J/|\kappa|. \tag{4.19}$$

If the distance from focus to asymptote, called the __impact parameter__, is denoted by p, then the angular momentum always equals $v_o p$, its initial value. Substituting into (4.19), we obtain

$$\tan \theta_o = v_o^2 p/|\kappa| \tag{4.20}$$

Rutherford Scattering

If a particle come in from infinity with impact parameter p, then it will have been deflected through an angle $\alpha = \pi - 2\theta_o$ by the time it has gone along the hyperbola. Experimentally, one can send in a beam of particles and measure the fraction which is scattered in space into the angular region from α to $\alpha + d\alpha$ and ϕ to $\phi + d\phi$. If $\sigma(\alpha,\phi)$ is the outgoing fraction per unit solid angle, then, if this does not depend upon ϕ, the total fraction coming off in the angular region from α to $\alpha + d\alpha$ will be

$$2\pi \sin \alpha \; d\alpha \; \sigma(\alpha)$$

This must equal in magnitude the number incident between p and p + dp, which is $2\pi \; p \; dp$.

So we have

$$2\pi \sin \alpha d\alpha \ \sigma(\alpha) = -2\pi p \ dp$$

where the minus sign enters because α decreases as p increases. Consequently, using (4.20) and its consequence $\sec^2\theta_o \ d\theta_o = v_o^2 \ dp/|\kappa|$,

$$\sigma(\alpha) = \frac{1}{2 \sin 2\theta_o} \ p \ \frac{dp}{d\theta_o} = \frac{\kappa^2}{2v_o^4} \ \frac{\tan\theta_o \ \sec^2\theta_o}{\sin 2\theta_o}$$

$$= \frac{\kappa^2}{4v_o^4} \ \frac{1}{\cos^4\theta_o} = \frac{\kappa^2}{4v_o^4} \ \frac{1}{\sin^4 (\alpha/2)}$$

which is the famous Rutherford Scattering Formula. It gives the fraction of the incident particles which is scattered through a deflection angle α, as a function of the angle α and the initial velocity v_o. Since this agrees very well with experiment when α-particles are scattered by a target, it is strong presumptive evidence that the force between an α-particle and a scattering center is of the inverse square type (Coulomb interaction) assumed in the derivation of the formula. From the manner of its derivation, we see that the formula is the same for an attractive force ($\kappa < 0$) as it is for a repulsive force ($\kappa > 0$). Strictly speaking, one does not expect to be able to describe the motion of a charged particle, in the presence of an atomic nucleus, as along a hyperbolic orbit, because the particle can be diffracted. However, inclusion of diffraction effects makes no difference for an inverse square force, and one still obtains the Rutherford Scattering Formula.

Exercises

1. As unit of distance, astronomers take the semimajor axis of the earth's orbit as 1 astronomical unit (A.U.). Given that Mercury has an orbit with semimajor axis 0.387 A.U. and eccentricity $e = 0.206$, calculate its period. Calculate and plot the time dependence of the polar coordinates (r, θ) and (r, ε), referred to the Sun.

2. An α-particle, with energy 5 Mev. is deflected by a gold nucleus $(Z = 79)$ through an angle of 45°. What is the value of the impact parameter? What is the distance of closest approach? What is the ratio of scattering intensity at 90° to that at 45°?

V

Electromagnetic Forces

Whether one wishes to design accelerators for high-energy
particles or just studies the motion of particles in natural
electromagnetic fields (such as that of the earth), it is
necessary to know the properties of such fields. Accordingly,
we shall give a brief account of electromagnetic theory, just
enough so that one can handle the motion of high-velocity
charged particles. This material will serve as background
for a discussion of relativistic effects, for motion in high-
energy accelerators, and for motion in the magnetic field of
the earth. We shall treat in order (1) Coulomb's Law and its
consequences (2) Ampere's Laws of magnetism and (3) Faraday's
Law of induction. This will culminate in Maxwell's Equations
and the force on a charged particle. Vector methods will be
used throughout, for these should be familiar to the average
beginning graduate student of today.

Coulomb's Law

The force $\vec{F}$ on a charge q, situated at a point P(xyz), due
to a charge q' at the point P'(x'y'z'), obeys an inverse-
square law, called Coulomb's Law and similar to Newton's Law
of Gravitational Interaction between 2 masses. This force

$\vec{F}$ may be expressed, if both particles are at rest, by

$$\vec{F} = \frac{qq'}{R^3}\, \vec{R} \qquad (5.1)$$

where R is the distance from P' to P, i.e.

$$R^2 = \{(x-x')^2 + (y-y')^2 + (z-z')^2\}^{1/2} \qquad (5.2)$$

The electric intensity $\vec{E}$ is defined to be $\vec{F}/q$, or the force per unit charge, and is a vector function of both space and time coördinates, i.e.

$$\vec{E} = \vec{E}\,(x,y,z,t) = \vec{F}/q \qquad (5.3)$$

The magnitude of $\vec{E}$, from (5.1) is

$$|E| = q'/R^2$$

Hence, if we surround the point P' by a sphere of radius r and integrate $|E|$ over its surface, we have

$$\int E_n\, dS = 4\pi q' \qquad (5.4)$$

because the total solid angle is 4π.

The ratio of this surface integral to the enclosed volume $\Delta\tau$ is defined as the $\underline{\text{divergence}}$ of $\vec{E}$, and is

$$\operatorname{div} \vec{E} = \lim_{\Delta\tau \to 0} \frac{\int E_n\, dS}{\Delta\tau} = 4\pi\rho \qquad (5.5)$$

where ρ is the charge density in general.

Equation (5.4) may be shown to hold when the surface is not that of a sphere, and is the essence of $\underline{\text{Gauss' Theorem}}$: The total outward flux, i.e., the integral of the normal component of the electric intensity over the surface, equals 4π times the total charge inside.

(For simplicity, we assume that we deal with matter which is non-polarized, where the dielectric constant is that of empty space.) Since

$$\vec{E} = \frac{q'}{R^3} \vec{R} = - q' \vec{\nabla} \left(\frac{1}{R}\right)$$

the quantity

$$\Phi = q'/R \qquad\qquad (5.6)$$

may be taken as the electric potential, and

$$\vec{E} = - \vec{\nabla}\Phi = - \overrightarrow{grad} \ \Phi \qquad\qquad (5.7)$$

If there are many charges, the resulting fields superpose and

$$\Phi = \sum_i \frac{q_i'}{R_i} \qquad\qquad (5.8)$$

or

$$\Phi(xyz) = \int \frac{\rho'(x'y'z')}{R} \, d\tau' \qquad\qquad (5.9)$$

may be used to calculate the total potential, when the charges are fixed and constant in time.

Combining (5.5) and (5.7),

$$\nabla^2 \ \Phi \equiv div \ grad \ \Phi = - 4\pi\rho \qquad\qquad (5.10)$$

This is a differential equation, known as <u>Poisson's Equation</u>, which connects the potential Φ and the charge density ρ.

Now the work done on the charge q in going around a closed path is

$$\oint \vec{F} \cdot \vec{ds} = q \oint \vec{E} \cdot \vec{ds} = - q \oint \vec{\nabla}\Phi \cdot \vec{ds}$$

$$= - q \oint \frac{d\Phi}{ds} = 0$$

The integral vanishes because one gets back to the same
value of potential that was started with, after having traversed the closed loop.

It is now useful to define the <u>curl</u> of a vector $\vec{A}$. We take
a small closed loop with a normal $\vec{n}$ to its plane. The ratio
of $\oint \vec{A} \cdot \vec{ds}$ around the loop to the area of the loop, in the
limit of zero area, will be called the component of $\overrightarrow{curl\ A}$ in
the direction $\vec{n}$. Thus

$$\text{curl } \vec{A} = \lim_{\Delta S \to o} \frac{\oint \vec{A} \cdot \vec{ds}}{\Delta S} \qquad (5.11)$$

Since $\oint \vec{E} \cdot \vec{ds} = 0$, it follows that

$$\text{curl } \vec{E} = 0 \qquad (5.12)$$

In fact, curl grad f $\equiv$ 0 for any scalar function f.

$$(5.13)$$

Dipoles

Since matter is ordinarily neutral and atomic, it is useful
to consider a small collection of charges consisting of an
equal number of positives and negatives. To calculate the
potential at any point P, we have only to add up the individual contributions due to the separate charges. Let 0 be an
arbitrary origin within the collection, let a charge q_i' be
located at Q, (see Figure 5.1) and denote the distances by
$\ell_i' = $ OQ, $r = $ OP, and $R_i = $ QP. The angle POQ shall be called
θ_i'.

The potential at P due to the charge q_i' is q_i'/R_i. But if
$\ell_i' \ll r$, then $R_i \cong r - \ell_i' \cos \theta_i'$. Consequently the potential
is

$$\frac{q_i'}{r - \ell_i' \cos\theta} \cong \frac{q_i'}{r} \left(1 + \frac{\ell_i'}{r} \cos \theta\right). \qquad (5.14)$$

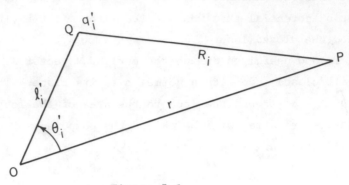

Figure 5.1

Charge q'_i in a Small Collection

Summing over all the charges, the total potential at P is

$$\Phi = \frac{\Sigma q'_i}{r} + \frac{1}{r^2} \; \Sigma \; q'_i \; \ell'_i \; \cos \theta'_i \; \ldots \qquad (5.15)$$

The first term on the right-hand side is the potential due to a collection with total charge $Q' = \Sigma \; q'_i$. If we take this to be zero, the most important term in the potential equals $1/r^2$ times the quantity $\Sigma \; q'_i \; \ell'_i \; \cos \theta_i$. This may be regarded as the sum of components along the direction OP of individual vectors $\vec{p_i}' = q'_i \; \vec{\ell'_i}$. The sum will thus be the component of the resultant vector $\vec{p}' = \Sigma \; \vec{p'_i} = \Sigma \; q'_i \; \vec{\ell'_i}$.

The vector $\vec{p}'$ is defined to be the <u>dipole moment</u> of the collection. An individual contribution to the potential, $q'_i \; \ell'_i \; \cos \theta_i /r^2$, can be regarded as the potential due to a charge q'_i at Q and another charge $-q'_i$ at 0, hence the name dipole.

Another way of writing (5.15) is, for the neutral case,

$$\Phi = \frac{\vec{p}' \cdot \vec{r}}{r^3} = - \vec{p}' \cdot \vec{\nabla} \left(\frac{1}{r}\right) \qquad (5.16)$$

(Note that $\vec{p}'$ does not change if the origin be displaced by a small vector amount $\vec{a}$, for $\vec{a} \, Q' = 0$ if the collection is neutral).

Force and Torque on a Dipole

Now suppose that a dipole is in an electric field, making an angle of θ with it (Figure 5.2). If the charge $-q$ is at x,y,z, and $+q$ at $x + \Delta x$, $y + \Delta y$, $z + \Delta z$. then the total force is

$$\vec{F} = q \, [\vec{E} \, (x + \Delta x, \, y + \Delta y, \, z + \Delta z) - \vec{E}(x,y,z)]$$

$$= q \, (\vec{\ell} \cdot \vec{\nabla}) \; \vec{E} = (\vec{p} \cdot \vec{\nabla}) \; \vec{E}$$

$$= \nabla \, (\vec{p} \cdot \vec{E}) - \vec{p} \times (\vec{\nabla} \times \vec{E}) \qquad\qquad (5.17)$$

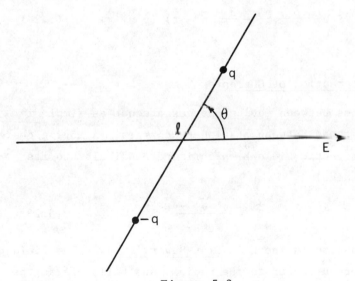

Figure 5.2

Electric Dipole in Electric Field

where $\vec{\ell}$ has components Δx, Δy, Δz and $\vec{p} = e\vec{\ell}$. When the field $\vec{E}$ is irrotational, $\vec{F} = - \vec{\nabla U} = \nabla \, (\vec{p} \cdot \vec{E})$ (5.18)
where

$$U = - (\vec{p} \cdot \vec{E}) \qquad (5.19)$$

This gives the <u>potential energy</u> of a <u>dipole</u> in an <u>electric field</u>.

The <u>torque</u> on the dipole is

$$\vec{T} = \vec{p} \times \vec{E} \qquad (5.20)$$

If the source of the field is a dipole, then the field is, from (5.16),

$$\vec{E} = - \vec{\nabla \Phi} = \vec{\nabla}[\vec{p}' \cdot \vec{\nabla} \frac{1}{r}] \qquad (5.21)$$

The potential energy then works out to be

$$U = \frac{\vec{p} \cdot \vec{p}'}{r^3} - \frac{3}{r^5} \, (\vec{p} \cdot \vec{r}) \, (\vec{p}' \cdot \vec{r}) \qquad (5.22)$$

Mutual Interaction of Currents

The forces between moving charges are not as simple in nature as those between stationary charges. A moving charge produces magnetic field $\vec{B}$ which is perpendicular to its velocity, namely

$$\vec{B} = \frac{q' \, \vec{v}' \times \vec{R}}{c \, R^3} \qquad (5.23)$$

This equation is known as <u>Biot-Savart's Law</u>. The field $\vec{B}$ is also perpendicular to the radius, and falls off as the square of the distance. Actually one may regard this field as a purely intermediate quantity, and say that the force on another moving charge is the only physical concept, because

for low velocities it is proportional to the acceleration,
which is observable. This force is found to be given by

$$\vec{F} = \frac{q}{c} \, (\vec{v} \times \vec{B})$$ (5.24)

It is perpendicular to both the velocity $\vec{v}$ and the field $\vec{B}$.
Thus, the interaction between source and other particle is
much more complicated than when the particles are stationary.
In that case, both the electric field and the force are <u>longi-
tudinal</u>, i.e., along the line joining the particles. However,
when the source particle moves, it gives rise to the <u>induction</u>
field $\vec{B}$ which is transverse to the radius. The force $\vec{F}$ is
transverse to $\vec{B}$ and is in the ($\vec{v}'$, $\vec{R}$) plane (see Figure 5.3).
Call this the xy plane, with $\vec{R}$ along the y-axis, and v' at an
angle ϕ' with the x-axis.

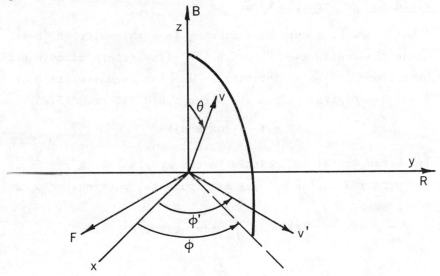

Figure 5.3

Source charge with velocity $\vec{v}'$ exerts force $\vec{F} \sim \vec{v} \times \vec{B}$ on
charge with velocity $\vec{v}$.

The induction $\vec{B}$ will be in the z-direction and proportional to v' cos ϕ'. If the velocity vector $\vec{v}$ is at an angle θ with the z-axis, the force $\vec{F}$ will be proportional to v'v sinθ cosϕ', and will be in the xy plane, in the direction making an angle ϕ − (π/2) with the x-axis, as shown in the figure.

If $\vec{v}$' is along the + x-axis, the induction will be a maximum. If, in addition, $\vec{v}$ is along the + x-axis, the force will be in the direction − $\vec{R}$ and therefore attractive; if $\vec{v}$ is along the − x-axis, the force will be repulsive.

If one current flows in a circle in a plane ⊥ $\vec{R}$, and another does likewise at a distance R from the first, the 2 loops will attract if the currents flow in the same direction and repel if in opposite directions.

Field due to Straight Wire

Let there be a current I flowing in a thin wire which is along the z-axis (see Figure 5.4). (The return circuit will be presumed to be at infinity.) We wish to calculate B at a point P distant ρ from the wire. This is, from (5.23)

$$\vec{B} = I \int (\vec{dz} \times \vec{R})/R^3$$

It is tangential to a circle in the xy plane with center on the wire and radius ρ. Its magnitude is, putting z = ρ tanθ, R = ρ secθ

$$|B| = 2I \int_0^\infty (dz) \cos \theta/R^2$$

$$= \frac{2I}{\rho} \int_0^{\frac{\pi}{2}} \cos \theta \, d\theta = \frac{2I}{\rho}$$

Integrating around the circle,

$$\int \vec{B} \cdot \vec{ds} = \int_0^{\frac{\pi}{2}} (2I/\rho) \, \rho d\phi = 4\pi I$$

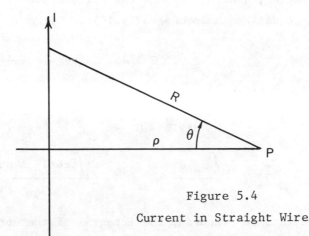

Figure 5.4

Current in Straight Wire

Then

$$\text{curl } \vec{B} = \lim_{\Delta S \to 0} \frac{\oint \vec{B} \cdot \vec{ds}}{\Delta S} = 4\pi \frac{\vec{I}}{\Delta S} = 4\pi \vec{j} \quad (5.25)$$

where $\vec{j}$ is the current density.

For the following, it is useful to express $\vec{B}$ in another way. From (5.23),

$$\vec{B} = \frac{q'}{c} \vec{\nabla} \frac{1}{R} \times \vec{v}' = \frac{q'}{c} \vec{\nabla} \times \frac{\vec{v}'}{R} = \overrightarrow{\text{curl}} \frac{q' \vec{v}'}{cR}$$

$$= \overrightarrow{\text{curl}} \vec{A}, \quad (5.26)$$

where the vector $\vec{A}$ is defined in general by

$$\vec{A} = \sum_i \frac{q_i' \vec{v}_i'}{cR_i} \to \int \frac{\vec{j}'}{R} d\tau' \quad (5.27)$$

and is called the vector potential.

It bears the same relation to the currents $q_i' \vec{v}_i'/c$ that the scalar potential Φ, defined in (5.8), bears to the charges q_i'.

In particular, since (5.10) is the differential form of (5.9), the equation analogous to (5.10) is, from (5.27),

$$\nabla^2 \vec{A} = - 4\pi \vec{j} \qquad (5.28)$$

From (5.26), it immediately follows that

$$\text{div } \vec{B} = 0 \qquad (5.29)$$

For

$$\text{div } \vec{B} = \lim_{\Delta\tau \to 0} \frac{\int B_n \, dS}{\Delta\tau} = \lim_{\Delta\tau \to 0} \frac{\int \text{curl}_n \, A \, dS}{\Delta\tau} = 0$$

because, by Stokes' Theorem, the integral of the normal component of the curl over a <u>closed</u> surface is zero.

For later reference, we include here a relation on the vector potential, namely div $\vec{A}$ = - (1/c) $\partial\Phi/\partial t$. This may be demonstrated by using the identity

$$\text{div } u \vec{C} \equiv u \text{ div } \vec{C} + (\vec{C} \cdot \nabla u)$$

where u is any scalar and $\vec{C}$ is some vector.

From (5.27), with repeated application of this identity,

$$\text{div } \vec{A} = \int \text{div } \frac{\vec{j}'}{R} \, d\tau' = \int (\vec{j}' \cdot \vec{\nabla} \frac{1}{R}) \, d\tau'$$

$$= - \int (\vec{j}' \cdot \vec{\nabla}' \frac{1}{R}) \, d\tau'$$

$$= - \int \text{div}' \frac{\vec{j}'}{R} \, d\tau' + \int \frac{\text{div}' \, \vec{j}'}{R} \, d\tau'$$

$$= - \int \frac{j_n'}{R} \, dS' - \frac{1}{c} \int \frac{\partial\rho'}{\partial t} \frac{1}{R} \, d\tau'$$

$$= - \frac{1}{c} \frac{\partial\Phi}{\partial t} \qquad (5.30)$$

We have used (a) the fact that j_n' = 0 over the surface

containing the current, (b) the equation of charge conservation

$$\text{div } \vec{j} = -\frac{1}{c}\frac{\partial \rho}{\partial t} \qquad (5.31)$$

and (c) the definition of potential in Equation (5.9).

Small Current Loops

The special case where a current flows in a circle of
radius small compared with the distance to a neighboring
current loop is of considerable interest, because the vector
potential is then approximated closely by one term.

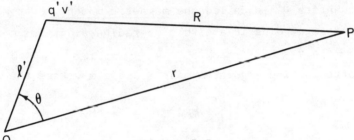

Figure 5.5

Current q'v' in a Small Collection

The expression (5.15) is replaced by

$$\vec{A} = \frac{\Sigma \, q'_i \, \vec{v}'_i}{c \; r} + \frac{1}{cr^2} \Sigma \, q'_i \, \vec{v}'_i \, \ell'_i \, \cos \theta'_i \qquad (5.32)$$

In the case of the current loop, the net current in any
direction vanishes, and so does the first term on the right-
hand side. The second term may be transformed as follows.
Note that

$$\frac{d}{dt} \, \vec{\ell}' \, (\vec{\ell}' \cdot \vec{r}) = \vec{v}' \, (\vec{\ell}' \cdot \vec{r}) + \vec{\ell}' \, (\vec{v}' \cdot \vec{r})$$

and

$$\vec{r} \times (\vec{\ell}' \times \vec{v}') = -\vec{v}' \, (\vec{\ell}' \cdot \vec{r}) + \vec{\ell}' \, (\vec{v}' \cdot \vec{r})$$

Subtracting the second equation from the first and summing
over the charges

(5.32')

$$\Sigma\ q'\vec{v}'\ (\vec{\ell}'\cdot\vec{r}) = \frac{1}{2}\ \frac{d}{dt}\ \Sigma\ q'\vec{\ell}'\ (\vec{\ell}'\cdot\vec{r}) - \frac{1}{2}\ \vec{r}\ \times\ \Sigma\ \vec{\ell}'\ \times\ q'\vec{v}'$$

For a periodic motion, the first term on the right-hand
side averages to zero. Consequently, the vector potential in
(5.32) becomes

$$\vec{A} = -\frac{1}{r^3}\ (\vec{r}\ \times\ \vec{\mu}') = \vec{\mu}'\ \times\ \vec{\nabla}'\frac{1}{r} \qquad (5.33)$$

where

$$\vec{\mu}' = \frac{1}{2c}\ \Sigma\ (\vec{\ell}'\ \times\ q'\vec{v}') \qquad (5.34)$$

This quantity $\vec{\mu}'$ is called the <u>magnetic moment</u>. For a
single charge q going in a circle of radius a, it is in magnitude $\frac{1}{2c}$ a q v.

The corresponding angular momentum $\vec{J} = a\ m\ \vec{v}$, and the ratio
is

$$\left|\frac{\mu'}{J}\right| = \frac{q}{2mc}$$

The field is now

$$\vec{B} = \vec{\nabla}\ \times\ (\vec{\mu}'\ \times\ \vec{\nabla}'\ \frac{1}{r})$$

$$= (\vec{\mu}'\cdot\vec{\nabla})\ \vec{\nabla}\ \frac{1}{r} - \vec{\mu}'\nabla^2\frac{1}{r}$$

$$= (\vec{\mu}'\cdot\vec{\nabla})\ \vec{\nabla}\ \frac{1}{r} \qquad \text{at } r \neq 0 \qquad (5.35)$$

since $\nabla^2\frac{1}{r} = 0$ for $r \neq 0$. But in this case we can write

$$\vec{B} = \vec{\nabla}\ (\vec{\mu}'\cdot\vec{\nabla}\ \frac{1}{r}) \qquad (5.36)$$

and hence define a magnetic scalar potential

$$\phi_m = -\ \vec{\mu}'\ \cdot\ \vec{\nabla}\ \frac{1}{r} \qquad (5.37)$$

So that $\vec{B} = -\ \vec{\nabla}\phi_m$ and $\overrightarrow{\text{curl}}\ \vec{B} = 0 \qquad (5.38)$

Since Equations (5.36) and (5.38) are analogous to (5.21), we see that the magnetic moment $\vec{\mu}'$ acts as a source of $\vec{B}$ in the same way that the electric moment $\vec{p}'$ acts as a source of $\vec{E}$.

Next, we need to obtain an expression for the force, analogous to (5.18), on a small current loop. With the same notation as before

$$\vec{F} = \frac{1}{c} \Sigma \, q_i \vec{v}_i \times \vec{B}_i$$

$$\cong \frac{1}{c} \Sigma \, q_i \vec{v}_i \times \vec{B}_o + \frac{1}{c} \Sigma [q_i \vec{v}_i \times (\vec{\ell}_i \cdot \vec{\nabla}) \vec{B}_o] \tag{5.39}$$

If the total current is zero, as here, the first term on the right-hand side vanishes.

For what follows, let us note that (5.32') can be generalized, in the case of a periodic motion, to

$$\Sigma \, q\vec{v} \, (\vec{\ell} \cdot \vec{C}) = - \vec{C} \times \vec{\mu}c \tag{5.40}$$

where C is an arbitrary vector function.

The x-component of (5.39) is, using (5.40) and $\overrightarrow{\text{curl}} \, \vec{B} = 0$,

$$\frac{1}{c} \Sigma \, q_i [\dot{y}_i \, (\vec{\ell}_i \cdot \vec{\nabla} B_z) - \dot{z}(\vec{\ell}_i \cdot \vec{\nabla} B_y)]$$

$$= (\vec{\mu} \times \vec{\nabla} B_z)_y - (\vec{\mu} \times \vec{\nabla} B_y)_z$$

$$= - \mu_x \, \text{div} \, \vec{B} + (\vec{\mu} \cdot \vec{\nabla}) \, B_x \tag{5.41}$$

But the divergence of $\vec{B}$ is zero. Hence the force on the current loop is, from (5.39) and (5.41),

$$\vec{F} = (\vec{\mu} \cdot \vec{\nabla}) \, \vec{B} = \vec{\nabla} \, (\vec{\mu} \cdot \vec{B}) \tag{5.42}$$

This is of the same form as (5.18), with the magnetic moment

$\vec{\mu}$ interacting with $\vec{B}$ in the same way as the electric moment $\vec{p}$ interacts with $\vec{E}$, to give the force $\vec{F}$. We can, therefore, say that the magnetic moment behaves as if it were composed of 2 magnetic poles of opposite signs and a small distance apart, and that Coulomb's Law would be valid between such magnetic monopoles. Particles of this nature have not been observed, so this way of thinking may be just a mathematical convenience.

The torque on a current loop is, by (5.39),

$$\vec{T} = \Sigma \; \vec{\ell}_i \; x \; (\frac{q_i \vec{v}_i}{c} \; x \; B)$$

$$= \Sigma \; \frac{q_i \vec{v}_i}{c} \; (\vec{\ell}_i \cdot \vec{B}) - \vec{B} \Sigma \vec{\ell}_i \cdot \frac{q_i \vec{v}_i}{c}$$

$$= \vec{\mu} \; x \; \vec{B} \qquad\qquad (5.43)$$

since the second right-hand term averages to zero. This equation is the analogue of (5.20).

To sum up, when one loop carrying a steady current acts on another similar loop, the force exerted is

$$\vec{F} = \nabla \; (\vec{\mu} \cdot \vec{B}) \qquad\qquad (5.42)$$

where

$$\vec{B} = \nabla \; (\vec{\mu}' \cdot \nabla \frac{1}{r}) \qquad\qquad (5.36)$$

and the magnetic moment of a current loop is defined as a vector normal to the plane of the loop and with magnitude

$$(1/2c) \; \Sigma \; (\vec{\ell} \; x \; q\vec{v}). \qquad\qquad (5.44)$$

Now suppose that there are N electrons, each with charge q, in an element with length ds and cross-section S at any time, and that they move with velocity $\vec{v}$. The charge density will be $\rho = Nq/Sds$. The total current, or charge crossing a

given cross-section per unit time, arises from charges in a
length v, and is

$$I = \rho v \, S/c = Nqv/cds$$

where the factor c puts the current in electromagnetic units.

The above summation in (5.44) can be rewritten, so that for
a circular current loop

$$\vec{\mu} = \frac{1}{2} I \oint \vec{\ell} \times \vec{ds} = \vec{\mu}_1 I \, (\pi a^2) \qquad (5.45)$$

where a is the radius of the circle and $\vec{\mu}_1$ is a unit vector
normal to its plane. The <u>magnitude</u> of the <u>magnetic moment</u>
thus <u>equals</u> the <u>current multiplied</u> by the <u>area</u> of the <u>loop</u>.

Non-Steady Fields

Up until now, we have been discussing the action of stationary charges on stationary charges and of steady currents on
steady currents. This has led to the introduction of the
electric intensity $\vec{E}$ and the magnetic induction $\vec{B}$, satisfying
the equations

$$\vec{E} = - \vec{\nabla}\phi \qquad (5.7)$$

$$\phi = \sum_i q'_i/R_i \rightarrow \int \frac{\rho' \, d\tau'}{R} \qquad (5.9)$$

$$\vec{B} = \overrightarrow{\text{curl}} \; \vec{A} \qquad (5.26)$$

$$\vec{A} = \sum_i q'_i \vec{v}'_i/cR_i \rightarrow \int \frac{\vec{j}' \, d\tau'}{R} \qquad (5.27)$$

with differential expressions

$$\text{div } \vec{E} = 4\pi\rho, \quad \overrightarrow{\text{curl}} \; \vec{E} = 0 \qquad (5.5),(5.12)$$

$$\text{div } \vec{B} = 0, \quad \overrightarrow{\text{curl}} \; \vec{B} = 4\pi\vec{j} \qquad (5.29),(5.25)$$

The force on a moving charge is

$$\vec{F} = q\ \{\vec{E} + \frac{\vec{v} \times \vec{B}}{c}\} \qquad (5.3), (5.24)$$

Futhermore, charge and current densities are related to the scalar and vector potentials by the relations

$$\nabla^2\ \Phi = -\ 4\pi\rho;\ \nabla^2\vec{A} = -\ 4\pi\vec{j} \qquad (5.10), (5.28)$$

We now wish to generalize these expressions to include the case where the quantities can vary with time. The simplest experimental relation which enables this to be done is Faraday's Law. This states that an electromotive force is induced in a current loop if the flux of induction through the loop changes. Quantitatively,

$$\oint \vec{E} \cdot d\vec{s} = -\ \frac{d}{cdt} \int B_n\ dS \qquad (5.46)$$

where B_n is the component of $\vec{B}$ normal to the plane of the loop.

The differential form of (5.46) is obtained by making the loop small, i.e.

$$\text{curl}_n\ \vec{E}\ =\ \lim_{\Delta S \to 0} \frac{d}{dt}\ \frac{\int B_n\ dS}{c\Delta S}\ =\ -\ \frac{1}{c}\ \frac{\partial B_n}{\partial t}$$

or

$$\overrightarrow{\text{curl}}\ \vec{E}\ =\ -\ \frac{1}{c}\ \frac{\partial \vec{B}}{\partial t} \qquad (5.47)$$

But from (5.26), $\vec{B} = \overrightarrow{\text{curl}}\ \vec{A}$, so that

$$\overrightarrow{\text{curl}}\ (\vec{E} + \frac{1}{c}\ \frac{\partial \vec{A}}{\partial t})\ =\ 0$$

Hence, the quantity in parentheses must be the gradient of a scalar, which equals the potential in the case when $\vec{A}$ does not depend on time.

Accordingly, we write

$$\vec{E} = -\frac{1}{c}\frac{\partial A}{\partial t} - \vec{\nabla}\Phi \qquad (5.48)$$

When $\vec{A}$ and Φ do not involve time, this reduces to Equation (5.7). Faraday's Law thus shows that, when the magnetic field varies with time, it can give rise to an acceleration of a particle through the contribution to the electric field.

Suppose an electron moves in a circle of radius ρ_o with center on the axis, and plane perpendicular to the axis, of a cylindrically symmetrical magnetic field B_z (which depends on the time). In cylindrical coordinates ρ, z, ϕ, the z-component of $\vec{B} = \mathrm{curl}\ \vec{A}$ is

$$B_z = \frac{1}{\rho}\left[\frac{\partial(\rho A_\phi)}{\partial \rho} - \frac{\partial A_\rho}{\partial \phi}\right]$$

But A_ρ cannot depend on ϕ. The field B_z can be regarded as arising from a vector potential with a component in the ϕ direction only, given by

$$\rho A_\phi = \int_o^\rho \rho\ B_z\ d\rho \qquad (5.49)$$

This equation has direct application to motion in the betatron (see (6.20) below).

In the non-steady state, Equation (5.25) needs to be modified. The condition for charge conservation is, using (5.5),

$$c\ \mathrm{div}\ \vec{j} = -\ \partial\rho/\partial t = -\frac{1}{4\pi}\frac{\partial}{\partial t}\ \mathrm{div}\ \vec{E}$$

or

$$\mathrm{div}\ (\vec{j} + \frac{1}{4\pi c}\frac{\partial \vec{E}}{\partial t}) = 0 \qquad (5.50)$$

Instead of (5.25), we replace the current by the expression

in the parentheses, and <u>have</u>

$$\overrightarrow{\text{curl}} \ \vec{B} = 4\pi \ \vec{j} + \frac{1}{c} \frac{\partial \vec{E}}{\partial t} \tag{5.51}$$

The additional term $\dfrac{1}{4\pi c} \dfrac{\partial \vec{E}}{\partial t}$ is called the <u>displacement</u>
<u>current</u> density. Now, substituting (5.26), (5.30), and (5.48)
into (5.51), we find

$$\overrightarrow{\text{curl}} \ \overrightarrow{\text{curl}} \ \vec{A} \equiv \overrightarrow{\text{grad}} \ \text{div} \ \vec{A} - \nabla^2 \vec{A}$$

$$= 4\pi \vec{j} + \frac{1}{c} \frac{\partial}{\partial t} \{ - \frac{1}{c} \frac{\partial \vec{A}}{\partial t} - \vec{\nabla}\Phi \}$$

from which

$$\nabla^2 \vec{A} - \frac{1}{c^2} \frac{\partial^2 \vec{A}}{\partial t^2} = - 4\pi \vec{j} \tag{5.52}$$

Also, substituting (5.30) and (5.48) into (5.5), we have

$$\text{div} \ \vec{E} = 4\pi\rho = - \frac{1}{c} \frac{\partial}{\partial t} \ \text{div} \ \vec{A} - \nabla^2 \Phi$$

$$= \frac{1}{c^2} \frac{\partial^2 \Phi}{\partial t^2} - \nabla^2 \Phi$$

or

$$\nabla^2 \Phi - \frac{1}{c^2} \frac{\partial^2 \Phi}{\partial t^2} = - 4\pi\rho \tag{5.53}$$

Equations (5.52) and (5.53) are satisfied by the potentials

$$\Phi \ (x,y,z,t) = \int \frac{\rho(x'y'z',t - \frac{R}{c})}{R} \ d\tau' \tag{5.54}$$

$$\vec{A} \ (x,y,z,t) = \int \frac{\vec{j}(x'y'z',t - \frac{R}{c})}{R} \ d\tau \tag{5.55}$$

as may in fact be verified by differentiation. These potentials express the idea that the effects of the source charges
and currents situated at (x'y'z') are felt at (xyz) at a
time which is later by the amount R/c, i.e., the time required

for a signal to travel the distance

$$R = \{(x-x')^2 + (y-y')^2 + (z-z')^2\}^{1/2}$$

with the velocity of light c. For this reason, the quantities Φ and $\vec{A}$ in (5.54) and (5.55) are termed <u>retarded</u> <u>potentials</u>.

Summing up, the equations for non-polarizable matter are the <u>Maxwell Equations</u>:

$$\text{div } \vec{E} = 4\pi\rho, \quad \overrightarrow{\text{curl }} \vec{E} = -\frac{1}{c}\frac{\partial \vec{B}}{\partial t}$$

$$(5.5),(5.47)$$

$$\text{div } \vec{B} = 0, \quad \overrightarrow{\text{curl }} \vec{B} = 4\pi\vec{j} + \frac{1}{c}\frac{\partial \vec{E}}{\partial t}$$

$$(5.29),(5.51)$$

$$\vec{E} = -\frac{1}{c}\frac{\partial \vec{A}}{\partial t} - \vec{\nabla}\Phi, \quad \vec{B} = \overrightarrow{\text{curl }} \vec{A}$$

$$(5.48),(5.26)$$

$$\vec{F} = q\{\vec{E} + \frac{\vec{v} \times \vec{B}}{c}\} \qquad (5.3),(5.24)$$

where Φ and $\vec{A}$ are the retarded potentials of (5.54) and (5.55).

If, then, we know the source charge and current densities as a function of time, we can calculate the potentials from (5.54) and (5.55), the fields from (5.48) and (5.26), and the force on a moving charge q from (5.3) and (5.24). This force will then govern the motion of this charge in accordance with Newton's Equations. The further discussion in this book shows how the various methods of mechanics can be applied to find this motion.

VI

Motion in an Electromagnetic Field

Since there are many occasions where one must know how a
charged particle moves in an electromagnetic field, we shall
discuss this case in considerable detail. It will be shown
how one can construct a Lagrangian function, and its explicit
form will be given. From this, it is then an easy matter to
set up the equations of motion. In the simpler examples,
these equations may be integrated readily when there are constants of the motion such as the energy and the angular momentum. Stability criteria may then be constructed without
difficulty, even if the speeds are relativistic. Even for
more complicated situations, especially periodic non-autonomous systems, one can determine whether or not there are
stable regions, as will be shown in Chapter X.

  One of the first examples to be treated extensively was that
of a particle in the field of a magnetic dipole. Störmer
found periodic orbits for such particles, and also obtained
criteria which would allow particles to reach in from infinity and arrive at a given radius, say the surface of the earth.
(The periodic orbits were later discovered experimentally by
Van Allen.) Lemaitre and Vallarta, and coworkers, studied the
trajectories in great detail and were able to determine the
range of directions, as a function of energy latitude and
zenith angle, which particles arriving at the earth would

have. Since the theory is extensive, we devote an entire
chapter (Chapter XI) to it. (This chapter is deferred because
we first need to take up such concepts as characteristic
exponents, stability and asymptotic orbits.)

In the present chapter, after setting up the Lagrangian for
an electromagnetic field, we shall confine ourselves to discussing the more elementary aspects of the operation of a
betatron and of a linear accelerator. (For design details,
which are outside the scope of this book, we refer the reader
to a review by Slater, to the book by Lichtenberg, and to
Stanford accelerator reports.) The alternating gradient synchrotron and the associated strong focussing will be taken up
in Chapter X.

Lagrange Potential Function for a Lorentz Force

We shall now show how to construct a function U such that
the Lorentz force $\vec{F} = e\ (\vec{E} + (1/c)\ (\vec{v} \times \vec{B}))$ is derivable from
it, i.e.

$$X = (dU_{\dot{x}}/dt) - U_x, \text{ etc.} \qquad (6.1)$$

(The reader may verify that the integrability relations I,
II, and III (Chapter II) are satisfied.)

The x and y components of force are

$$X = q\ [E_x + \frac{1}{c}\ (\dot{y}B_z - \dot{z}B_y)]$$

$$= q\ [-\frac{1}{c}\frac{\partial A_x}{\partial t} - \frac{\partial \Phi}{\partial x} + \frac{1}{c}\ (\vec{v} \times \text{curl } \vec{A})_x]$$

$$= q\ [-\frac{1}{c}\frac{dA_x}{dt} - \frac{\partial \Phi}{\partial x} + \frac{1}{c}\frac{\partial}{\partial x}\ (\vec{v}\cdot\vec{A})] \qquad (6.2)$$

$$Y = q\ [E_y + \frac{1}{c}\ (\dot{z}B_x - \dot{x}B_z)] \qquad (6.3)$$

Calculating partial derivatives and using (2.42), with

$$P_x = U_{\dot{x}}, \quad P_y = U_{\dot{y}},$$

$$-\text{curl}_z \, \vec{P} = \frac{1}{2}\left(\frac{\partial X}{\partial \dot{y}} - \frac{\partial Y}{\partial \dot{x}}\right) = \frac{q}{c} B_z$$

$$= \frac{q}{c} \, \text{curl}_z \, \vec{A}$$

Accordingly

$$\text{curl} \, (\vec{P} + \frac{q}{c} \, \vec{A}) = 0$$

and therefore $\vec{P} = -\frac{q}{c} \vec{A} + \overrightarrow{\text{grad}} \, \chi$, where χ is an arbitrary scaler function $\chi \, (x,y,z,t)$. Since the induction $\vec{B}$ is the quantity which appears in the force expression, this function χ is without effect on the motion.

Taking the x-component and differentiating re the time, and substituting (6.1) and (6.2),

$$\frac{\partial U}{\partial x} = \dot{P}_x - X = -\frac{q}{c}\frac{dA_x}{dt} + \frac{\partial \dot{\chi}}{\partial x} - X$$

$$= \frac{\partial \dot{\chi}}{\partial x} + q\frac{\partial}{\partial x}[\Phi - \frac{(\vec{v}\cdot\vec{A})}{c}]$$

Integrating,

$$U = \dot{\chi} + q[\Phi - \frac{\vec{v}\cdot\vec{A}}{c}] + u(\dot{x},\dot{y},\dot{z},t)$$

where u is an arbitrary function of the velocities and time.

Differentiating re $\dot{x}$,

$$P_x \equiv \frac{\partial U}{\partial \dot{x}} = \frac{\partial \dot{\chi}}{\partial \dot{x}} - \frac{q}{c} A_x + \frac{\partial u}{\partial \dot{x}}$$

$$= \frac{\partial \chi}{\partial x} - \frac{q}{c} A_x + \frac{\partial u}{\partial \dot{x}}$$

Comparing this with the above expression for P_x, we have $\partial u/\partial \dot{x} = 0$ etc., so that u can be at most a function of t, which can be absorbed into $\dot{\chi}$ and subsequently neglected, since χ does not influence the motion.

So finally, the function from which the Lorentz force can be derived is

$$U = q\Phi - (q/c) \ (v \cdot A) \tag{6.4}$$

and the relativistic Lagrangian for the motion of a charged particle in an electromagnetic field is, using (1.24),

$$L = T^* - U = - m_o c^2 \ (1 - \frac{v^2}{c^2})^{1/2} - q\Phi + \frac{q}{c} \ (v \cdot A) \tag{6.5}$$

The x-component of generalized momentum is

$$P_x = \frac{\partial T^*}{\partial \dot{x}} - \frac{\partial U}{\partial \dot{x}} = m\dot{x} + \frac{q}{c} A_x \tag{6.6}$$

and the corresponding equation of motion is

$$\dot{p}_x = \frac{\partial L}{\partial x} = - q \frac{\partial \Phi}{\partial x} + \frac{q}{c} \frac{\partial}{\partial x} \ (v \cdot A) \tag{6.7}$$

The Hamiltonian is

$$H = \sum_{xyz} p_x \dot{x} - L = mc^2 + q\Phi \tag{6.8}$$

If L does not contain the time explicitly, then H will be a constant of the motion.

Now

$$(mc)^2 = (m_o c)^2 + (mv)^2$$

Accordingly

$$mc^2 = c\{(m_o c)^2 + (m\dot{x})^2 + \ (m\dot{y})^2 + (m\dot{z})^2\}^{1/2}$$

If $\Phi = 0$, then from (6.6) the Hamiltonian is

$$H = c \ \{(m_o c)^2 + \sum_{xyz} (p_x - \frac{q}{c} A_x)^2\}^{1/2} \tag{6.9}$$

Equations (6.5) – (6.9) describe the motion of a particle, of charge q and rest mass m_o, in an electromagnetic field with potentials Φ and $\vec{A}$.

PARTICLE ACCELERATORS

Most of our basic knowledge about the structure of atomic nuclei has come from collision experiments, where particles of high energy impinge on and are scattered by matter. In the early days, these particles either came from radioactive emitters or were cosmic rays. However, the low beam intensities made it almost imperative to develop machines which could accelerate charged particles to high energy, producing a beam of fairly well-defined energy and of high intensity. Such devices, which are called particle accelerators, operate on the principle of furnishing a large number of small increments of energy, successively adding them to the particle energy until the high energy is attained. This is much more practical than the earlier method of allowing particles to fall through a large difference of potential, which method suffers from troubles with insulation breakdown.

There are two main types of machines, according to whether the particle trajectory is straight or curved. The first type is the linear accelerator, which accelerates charged particles along approximately straight trajectories by means of alternating electric fields. The early machine of Sloan and Lawrence (1931) consisted of a series of cylindrical tubes, along the axis of which the particles travel. These tubes are connected to a high-frequency oscillator in such a way that successive tubes (electrodes) have alternating polarity. The electric field is zero inside the tubes, and the particles are accelerated across the gaps between the tubes. If the

length of the cylinders is chosen so that a particle which
has been accelerated in the first gap spends a time in each
tube equal to the half-period of the generator, then it always
encounters the field in each gap with the same phase as in the
first, and so is accelerated across each gap. The length ℓ
of each cylinder must equal $vT/2$, where T is the period of the
oscillator and v is the velocity of a particle. Since v
increases from each cylinder to the next, the length ℓ will
have to increase also. However, if v approaches c, as for
fast electrons, then the length approaches $cT/2$, a constant.
The most energetic linear accelerator to date is that at
Stanford, where an energy of 20 Bev is attained with a total
length of 10,000 feet.

There are two examples of machines where the particles move
in roughly circular orbits, namely the betatron and the
synchrotron. The former has been used to accelerate electrons
to energies as high as 312 Mev, which is a practical upper
limit because of the cost of iron. The latter is now the
standard machine for producing fast protons, with energies
going up to the order of 50 Bev.

In the betatron, the electrons are kept on a circular orbit
by the Lorentz force of the magnetic field, and are accelerated by an electric field produced at the orbit by a changing
magnetic flux inside (Faraday's Law). The magnet is fed by
an alternating current of frequency usually between 50 and
200 cps., which results in the induced electric field alternating with the same frequency. The electrons are accelerated
only during one half-cycle, and are then used experimentally.

In the synchrotron, the electrons are again restricted to a
circular orbit, but kept there by a ring-shaped magnet, which
provides the field only in the region of the orbit. The rf

electric field for the acceleration of the particles is loca-
lized at a certain place in the vacuum chamber (a doughnut)
and is obtained by exciting stationary electromagnetic waves
in a resonant cavity through which the particles pass. The
rf angular frequency ω_e must be an integral multiple k of the
angular frequency around the orbit, i.e. ω_e = kv/R, where v
is the velocity and R the radius of the orbit. But the
balancing of centrifugal force and Lorentz force gives
$|q|$ B = mv/R, where q is the charge, B the magnetic field,
and m the velocity-dependent mass. Elimination of v/R gives
ω_e = k $|q|$ B/m where

$$m = m_o \{1 + (\frac{mv}{m_o c})^2\}^{1/2}$$

$$= m_o \{1 + (\frac{BRq}{m_o c})^2\}^{1/2}$$

The synchrotron is designed so that B varies periodically
at low frequency and ω_e is forced to change so that the above
dependence on B is satisfied. If the particles arrive at the
resonant cavity so that the electric field accelerates them,
they can finally attain the maximum energy, being bunched in
the process.

THE BETATRON

The betatron is a device for accelerating particles by
making use of Faraday's Law of induction.

Suppose we have (1) a cylindrically symmetrical time-
dependent field B_z (ρ,t) and (2) an electron moving in a
circle $\rho = \rho_o$ whose plane is perpendicular to B_z (Figure
6.1).

From Faraday's Law, the tangential electric field produced

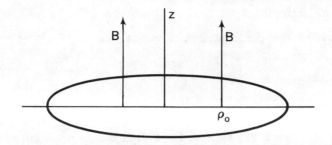

Figure 6.1

Electron in a Magnetic Field

at the circle will be

$$E_\phi = \frac{1}{2\pi\rho_o c} \frac{d}{dt} \int B_z \, dS$$

The magnetic field may be considered as arising from circular currents, and the vector potential calculated from the equation

$$\vec{A} = \int \frac{\vec{j'} \, dt'}{R}$$

Then we find that both the component in the z-direction vanishes and also the ρ-component vanishes. For

$$R^2 = (\rho'^2 + \rho^2 - 2\rho\rho' \cos\phi + z^2)^{1/2}$$

and

$$A_\rho = \int_0^{2\pi} \frac{j'\sin\phi d\phi}{(\rho'^2 + \rho^2 - 2\rho\rho' \cos\phi + z^2)^{1/2}} = 0$$

The integral vanishes because the integrand is an odd function of ϕ.

Hence the only component of $\vec{A}$ is in the ϕ-direction, i.e. A_ϕ. The Lagrangian is them, from (6.5)

$$L = - m_o c^2 \{1 - \frac{1}{c^2} (\dot\rho^2 + \rho^2\dot\phi^2 + \dot z^2)\}^{1/2} + \frac{q}{c} \rho\dot\phi \, A_\phi$$

$$(6.10)$$

Equations of Motion

From (6.10), the generalized momenta are

$$p_\rho \equiv \partial L/\partial\dot\rho = m\dot\rho$$

$$p_z \equiv \partial L/\partial\dot z = m\dot z$$

$$p_\phi \equiv \partial L/\partial\dot\phi = m\rho^2\dot\phi + (q/c)\,\rho A_\phi \qquad (6.12)$$

The Hamiltonian, with $\Phi = 0$, equals

$$H = mc^2 = c\{m_o^2\,c^2 + p_\rho^2 + p_z^2 + (\frac{p_\phi}{\rho} - \frac{qA_\phi}{c})^2\}^{1/2}$$

$$\qquad (6.13)$$

$$= c\{m_o^2\,c^2 + p_\rho^2 + p_z^2 + p^2\}^{1/2} \qquad (6.14)$$

where

$$p = \frac{p_\phi}{\rho} - \frac{qA_\phi}{c} = m\rho\dot\phi \qquad (6.15)$$

The canonical equations are then

$$\dot p_\phi = 0 \text{ or } p_\phi = \text{const} \qquad (6.16)$$

$$\dot p_\rho = -\frac{\partial H}{\partial\rho} = -\frac{p}{m}\frac{\partial p}{\partial\rho} \qquad (6.17)$$

$$\dot p_z = -\frac{\partial H}{\partial z} = -\frac{p}{m}\frac{\partial p}{\partial z} \qquad (6.18)$$

Motion in the plane z = 0

Let us first neglect any motion in the z-direction. The electrons will be injected into a plane of constant z, say z = 0, to which the lines of the magnetic field are perpendicular. There will then be no forces tending to remove the electrons from this plane. The force on an electron will be

$$\vec F = q\,\{-\frac{1}{c}\frac{\partial\vec A}{\partial t} + \frac{1}{c}\,(\vec v \times \vec B)\}$$

For z = 0, the magnetic field is in the z-direction and is

$$B_z = (\text{curl A})_z = \frac{1}{\rho} \frac{\partial}{\partial \rho} (\rho A_\phi) = \frac{\partial A_\phi}{\partial \rho} + \frac{A_\phi}{\rho} \quad (6.19)$$

(as may be readily verified from the invariant definition of the curl, with $\Delta S = \rho d\rho d\phi$).

Otherwise expressed,

$$\rho B_z = \frac{\partial}{\partial \rho} (\rho A_\phi) \quad \text{or} \quad \rho A_\phi = \int_o^\rho B_z \, \rho \, d\rho \quad (6.20)$$

The radical Lagrange equation is

$$\frac{d}{dt} (m\dot{\rho}) = m \rho \dot{\phi}^2 + \frac{q}{c} \dot{\phi} \frac{\partial (\rho A_\phi)}{\partial \rho}$$

$$= m\rho\dot{\phi}^2 + \frac{q}{c} \rho\dot{\phi} B_z \quad (6.21)$$

The radial force consists of 2 terms, the first being the centrifugal force and the second the radial component (inward for an electron) of the term $\frac{q}{c} (\vec{v} \times \vec{B})$.

The angular equation takes the form, [see (6.12), (6.16), and (6.19)]

$$\frac{d}{dt} m\rho^2\dot{\phi} = - \frac{q}{c} \frac{d}{dt} (\rho A_\phi) = - \frac{q}{c} (\frac{\partial}{\partial \rho} (\rho A_\phi) \dot{\rho} + \rho \frac{\partial A_\phi}{\partial t})$$

$$= - \frac{q}{c} (\rho\dot{\rho} B_z + \rho \frac{\partial A_\phi}{\partial t}) \quad (6.22)$$

The right-hand side is the torque, consisting of a term from $(q/c) (\vec{v} \times \vec{B})$ and one from the accelerating term

$$\vec{E} = - \partial\vec{A}/c\partial t.$$

Instantaneous and Equilibrium Orbits

We now inquire under what circumstances the electron can describe a circular orbit.

(a) Suppose that the <u>field does not change with time</u>. Then the condition $\dot{p}_\rho = 0$, from (6.21), gives $p = m\rho\dot{\phi} = - (q/c) \rho B_z$. The electron then describes a circle which will be called the instantaneous circle. The condition on p for the electron being in this instantaneous circle may also be written

$$p_i = - \frac{q}{c} B_{zi} \rho_i \qquad (6.23)$$

where the subscript i stands for the instantaneous circle.

(b) When the field changes with time, the requirement $\dot{p}_\rho = 0$ does not necessarily imply that the orbit remains of constant radius. It is then of some interest to determine what conditions must be imposed on an instantaneous circle so that its radius be constant in time. From (6.21), $\dot{p}_\rho = 0$ with (6.15) implies that

$$\frac{p_\phi}{\rho_i^2} = - \frac{q}{c} \frac{\partial A_\phi}{\partial \rho} \Big|_i \qquad (6.24)$$

Now $\frac{\partial A_\phi}{\partial \rho}$ will in general vary with time while all the other quantities in the equation are constants. The only way to satisfy (6.24) then is to require $p_\phi = 0$ and

$$\frac{\partial A_\phi}{\partial \rho} \Big|_o = 0 \qquad (6.25)$$

where ρ_o is the equilibrium circle, that is, the instantaneous circle with constant radius.

The requirement on the field, (6.25), can be expressed more
conveniently with the aid of (6.19) as

$$\rho_o \, B_{zo} = A_{\phi o} \qquad\qquad (6.26)$$

The total flux inside the equilibrium circle is then

$$\int B_z \, dS = \oint \vec{A}_{\phi o} \cdot \vec{ds} = 2\pi\rho_o A_{\phi o}$$

$$= 2\pi \, \rho_o^2 \, B_{zo} \qquad\qquad (6.27)$$

which is twice what it would be if the magnetic field were
uniform over the plane inside the circle. In other words,
the field must be designed to fall off as ρ increases from
0 to ρ_o, if the condition for the equilibrium orbit is to
be satisfied.

An electron started in this circle with $p_\phi = 0$ will continue
in this circle so long as (6.27) is obeyed while the field is
changing.

It has been noted that the instantaneous circle remains a
circle only when the field is not changing. One could say
more explicitly, for each given p_ϕ and particular time t, that
there exists an instantaneous circle with radius ρ_i such that
if an electron with this given p_ϕ were started at ρ_i it would
continue in the circle ρ_i if the field did not vary any more
in time. Thus the instantaneous circle for an electron with
given p_ϕ is a function of time. The equilibrium circle, however, is a constant of the machine. One usually has two
equilibrium circles, one stable and the other unstable, in
the standard betatron.

Radial Oscillations

We now wish to find the condition for stable oscillations in the radial direction. Let $\dot{\phi} = \omega$ be regarded as constant, and set $d/dt = \omega \, d/d\phi$, with $\omega = p/m\rho$. Then (6.21) becomes

$$\frac{d}{d\phi} \left(m\omega \, \frac{d\rho}{d\phi} \right) - m\omega\rho = \frac{q\rho}{c} B_z \qquad (6.28)$$

Expand about the equilibrium circle $\rho = \rho_o$, with

$$x = \rho - \rho_o$$

$$\Delta p = p - p_o$$

$$B_z = B_o + \Delta B + \frac{\partial B}{\partial x} x + \dots$$

where ΔB is an azimuthal variation in B and $p_o = -(q/c)\rho_o B_o$ at equilibrium. Then

$$\frac{p}{\rho} \frac{d^2 x}{d\phi^2} - p = \frac{q\rho_o}{c} (B_o + \Delta B) + \frac{q\rho_o}{c} \frac{\partial B}{\partial x} x + \frac{qx}{c} B_o$$

or

$$\frac{p}{\rho} \frac{d^2 x}{d\phi^2} - \frac{q}{c} B_o \left(1 + \frac{\rho_o}{B_o} \frac{\partial B}{\partial x} \right) x = \Delta p + \frac{q\rho_o}{c} \Delta B$$

or, approximately, setting $p/\rho = p_o/\rho_o$,

$$\frac{d^2 x}{d\phi^2} + \left(1 + \frac{\rho_o}{B_o} \frac{\partial B}{\partial x} \right) x = \frac{\rho_o}{p_o} \Delta p - \frac{\rho_o}{B_o} \Delta B \qquad (6.29)$$

There will be stable oscillatory solutions if the coefficient of x is positive, i.e.

$$-\frac{\rho_o}{B_o} \frac{\partial B_z}{\partial \rho} < 1 \qquad (6.30)$$

Vertical (z) Oscillations

From (6.18) and (6.15), the z-equation of motion is

$$- \frac{d}{dt}(m\dot{z}) = - \frac{p}{m}\frac{q}{c}\frac{\partial A_\phi}{\partial z} = \frac{q}{c}\rho\dot{\phi}\,B_\rho \qquad (6.31)$$

since

$$B_\rho = - \partial A_\phi / \partial z$$

Making the substitution $d/dt = \omega\, d/d\phi$, we have

$$- \frac{d}{d\phi}\left(\frac{1}{\rho}\frac{dz}{d\phi}\right) = \frac{q}{c}\frac{\rho}{p}B_\rho \cong - \frac{B_\rho}{B_o} \qquad (6.32)$$

Since the curl of B is approximately zero,

$$\frac{\partial B_z}{\partial \rho} = \frac{\partial B_\rho}{\partial z}$$

Expanding B_ρ about $z = 0$, $B_\rho \cong \dfrac{\partial B_\rho}{\partial z}\,z$.

Substituting in (6.32), the result is

$$\frac{d^2 z}{d\phi^2} - \frac{\rho_o}{B_o}\frac{\partial B_z}{\partial \rho}\,z = 0$$

This has stable oscillatory solutions if

$$- \frac{\rho_o}{B_o}\frac{\partial B_z}{\partial \rho} > 0 \qquad (6.33)$$

Stability in General

If we define the field index by

$$n = - \frac{\rho_o}{B_o}\frac{\partial B_z}{\partial \rho} \qquad (6.34)$$

then the condition for stable oscillations, radial and

vertical, is

$$0 < n < 1 \qquad\qquad (6.35)$$

The equation for free radial oscillations is

$$\frac{d^2x}{d\phi^2} + (1-n)\ x = 0 \qquad\qquad (6.36)$$

while that for vertical oscillations is

$$\frac{d^2z}{d\phi^2} + nz = 0 \qquad\qquad (6.37)$$

Non-Linear Effects

In the above treatment, it has been assumed that the field
index n does not vary with ϕ. This restriction need not
always apply, for it will depend on how the pole faces are
shaped. A possible generalization of (6.37) would be

$$\frac{d^2z}{d\phi^2} + n(\phi)\{1 + a\ z^2\}\ z = 0 \qquad\qquad (6.38)$$

This preserves the symmetry with respect to z and allows
for an azimuthal variation of the field index as well. Such
a system is non-autonomous, periodic in ϕ, and will be discussed below.

THE LINEAR ACCELERATOR

The linear accelerator is a device for accelerating charged
particles to high energies, by means of a longitudinal alternating electric field. This field is one which consists of
a traveling wave, the phase velocity w of which is matched
to the velocity of the particle to be accelerated. If particles have the correct phase, they can continually pick up

energy, just as they would in a constant field. We shall not give the details of how the matching of the phase velocity is done, but shall merely indicate the procedure and carry through the analysis of the motion on the assumption that w is a definite function of position, i.e. the distance z down the wave guide.

Constant Phase Velocity w

The longitudinal electric field component will be taken to be $E_z = E \sin \omega (t - z/w)$, where ω is the angular frequency.

A particle of charge q and velocity v will accelerate according to the equation

$$\frac{dp}{dt} = qE \sin \omega(t - z/w). \qquad (6.39)$$

where $p = mv$ and m is the relativistic (transverse) mass. The velocity will be $v = dz/dt$.

Now introduce moving axes traveling with the constant phase velocity w, and write the displacement z' with respect to these axes $z' = z - wt$.

Equation (6.39) becomes

$$\frac{dp}{dt} = - qE \sin \omega z'/w \qquad (6.40)$$

If a particle is at $z' = 0$, then from (6.40) its momentum will stay constant, so if its velocity is initially w, it will remain that. If the particle is at $z' > 0$, its momentum will decrease; if it is at $z' < 0$, its momentum will increase, (both statements holding as long as the sine function in (6.40) is positive, i.e. $\omega z'/w < \pi$).

The right-hand side of (6.40) represents a restoring force $\vec{F}$, which always urges the particle to the origin. If we set

$F = - \partial U/\partial z'$, the potential energy is

$$U = - \frac{qEw}{\omega} \cos \frac{\omega z'}{w} \qquad (6.41)$$

The relative velocity is

$$\frac{dz'}{dt} = \frac{dz}{dt} - w = \frac{p}{m} - w \qquad (6.42)$$

Equations (6.40) and (6.42) are the same as the usual Hamiltonian canonical equations in p and z', except for the additive constant − w in (6.42). Hence, if we set

$$\frac{dp}{dt} = - \frac{\partial H}{\partial z'}, = -qE \sin \frac{\omega z'}{w} \qquad (6.43)$$

$$\frac{dz'}{dt} = \frac{\partial H}{\partial p} = \frac{p}{m} - w, \qquad (6.44)$$

the Hamiltonian will be (relative to $m_o c^2$)

$$H = mc^2 - pw - qE(w/\omega) \cos (\omega z'/w) - m_o c^2 \qquad (6.45)$$

where the extra term −pw comes in to take care of the above additive constant −w in (6.42). Since the force does not involve the time explicitly, only z', energy is conserved, i.e.

$$H = h = const.$$

The time can be eliminated if we devide (6.43) by (6.44), i.e.

$$\frac{dp}{dz'} = - \frac{qE \sin (\omega z'/w)}{(p/m) - w} \qquad (6.46)$$

with

$$mc = (m_o^2 c^2 + p^2)^{1/2} \qquad (6.47)$$

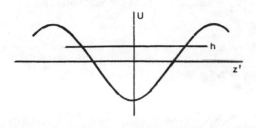

Figure 6.2

Sinusoidal Potential Field U

The motion is in the sinusoidal potential field U (Figure 6.2) and will be a libration if the amplitude is small, so that the energy constant is less than the maximum of U. If it is greater, then the motion is continually toward positive z' or toward negative z'.

At the limit of a libration, dz'/dt = 0 and p = m(w)w. Suppose that z' = a at this limit and that

$$U = U_a = - qE(w/\omega) \cos(\omega a/w).$$

Then the Hamiltonian (6.45) will be

$$m_o (c^2 - w^2) (1 - \frac{w^2}{c^2})^{-1/2} + U_a - m_o c^2 = h \qquad (6.48)$$

or

$$U_a = h + m_o c^2 - m_o c^2 (1 - \frac{w^2}{c^2})^{1/2} \qquad (6.49)$$

For a given value of a, we can calculate U_a and hence h. Thus, in the case of librations, specification of the amplitude gives h. The relation H = h represents a closed curve in the (z' p) phase plane, the slope of which is given by (6.46). A set of such closed curves is shown in Figure 6.3, for w = c/2. They resemble those for the simple plane pendulum, and would be exactly the same were it not that the particle moves with relativistic velocity. The figure also shows progressive motion to the right if the initial velocity

at $z' = 0$ is sufficiently great, and progressive motion to
the left if it is small enough.

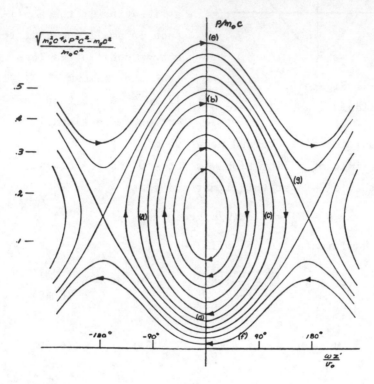

Figure 6.3

(After Slater, J. C., 1948, Rev. Mod. Phys. 20, 473),
Phase-plane trajectories for a linear accelerator bunching section, $v = c/2$.

We can obtain an approximate expression for H if we expand
the momentum about its equilibrium value, setting $p = p_o + p'$,
where $p_o = m_o w / (1 - \frac{w^2}{c^2})^{1/2}$.

Then put $f(p) = mc^2 - pw$

$$= c \, (m_o^2 c^2 + p^2)^{1/2} - pw$$

Differentiating,

$$f'(p) = cp \ (m_o^2 c^2 + p^2)^{-1/2} - w$$

$$f''(p) = c \ (m_o^2 c^2 + p^2)^{-1/2} - (cp^2)(m_o^2 c^2 + p^2)^{-3/2}$$

$$= m_o^2 c^3 \ (m_o^2 c^2 + p^2)^{-3/2}$$

$$= m_o^2/m^3 = (1/m_o) (1 - \frac{v^2}{c^2})^{3/2}$$

Consequently, the Taylor's series expansion is, with $f'(p)|_o = 0$,

$$h = H = m(w) \ c^2 + \frac{p'^2}{2m_\ell} - qE \ (w/\omega) \ \cos \ (\omega z'/w) - m_o c^2$$

$$(6.50)$$

Here $m_\ell = m_o \ (1 - \frac{w^2}{c^2})^{-3/2}$, the longitudinal mass.

Equation (6.50) contains (a) the kinetic energy of the point of equilibrium $(m - m_o) \ c^2$, (b) the kinetic energy $p'^2/2m_\ell$ relative to this, and (c) the potential energy. If we stipulate that the times involved are short, then the first term may be regarded as constant, since the velocity does not change. The Hamiltonian is then of exactly the same nature as that for the plane pendulum. The motion can be solved for completely in terms of elliptic functions. For small amplitudes, the angular frequency of libration is, by comparison with an harmonic oscillator, equal to

$$\omega_o = [qE \ \omega/wm_\ell]^{1/2}, \ by \ (6.50) \ and \ (6.43).$$

Variable Phase Velocity

If the phase velocity is constant, then a particle will receive no net acceleration, since from (6.46) the acceleration is sinusoidal in $\omega z'/w$. In order for there to be

acceleration, we have to arrange the wave guide so that the
phase velocity increases constantly, and this can be done.
The present section will be devoted to the consequences of
doing it, and will be a study of how a particle will move if
the phase velocity increases slowly as a function of distance
along the tube, $w = w(z)$.

Just as the particle could oscillate about some mean position when the phase velocity was constant, so now we expect
that the particle can do the same with phase velocity variable. In particular, the particle can be at rest with respect
to the forcing wave, i.e. $v = w$.

The phase is given by

$$\phi = \omega t - \omega \int^z \frac{dz}{w(z)} \qquad (6.51)$$

and its change with time by

$$\frac{d\phi}{dt} = \omega \left(1 - \frac{v}{w}\right) \qquad (6.52)$$

If $v = w$, then the phase does not change with time, $\phi = \phi_0$.
The equation of motion for the particle is

$$\frac{dp}{dt} = qE \sin \phi \qquad (6.53)$$

If the particle rests with respect to the wave, then
$p_0 = m(w)w$ and

$$\frac{dp_0}{dt} = qE \sin \phi_0 \qquad (6.54)$$

If the accelerator is designed so that the time rate of
change of momentum dp_0/dt is a constant, then the force term
on the right-hand side of (6.54) will be constant, and the
phase ϕ_0 will be constant. Let us assume this to be the case.
Equations (6.52) and (6.53) govern the motion, and reduce to
(6.40) and (6.42) if we set $\phi = \phi_0 - (\omega z'/w)$, $\phi_0 = 0$, $p = mv$.

The force on the particle is in general

$$qE \sin \omega(t - \int^z \frac{dz}{w(z)})$$

If we make the <u>assumptions that w does not vary during an
oscillation, and that v $\cong$ w</u>, then the force is

$$qE \sin \omega(t - \int^{z_o} \frac{dz}{w(z)} - \frac{z'}{w})$$

$$= qE \sin \omega(t_o - \frac{z'}{w})$$

$$= qE \sin (\phi_o - \frac{\omega z'}{w}) = qE \sin \phi \qquad (6.55)$$

where $z' = z - z_o$ and $\phi_o = \omega t_o$. Here z_o is the position the
wave has reached at time t_o, or the position of the non-
oscillating particle. [Note that $\phi' = \phi - \phi_o = - \omega z'/w$].
The net rate of change of momentum in the moving system is

$$\frac{dp}{dt} - \frac{dp_o}{dt} = qE \sin (\phi_o - \frac{\omega z'}{w}) - qE \sin \phi_o \qquad (6.56)$$

This differs from (6.40) by the extra force term $-qE \sin \phi_o$,
which is necessitated due to the acceleration. If the whole
right-hand side of (6.56) be taken as the negative gradient
of a potential U, then

$$U = -qE \, (w/\omega) \cos \omega(\phi_o - \frac{z'}{w}) + qE \, z' \sin \phi_o$$

$$= qE \, (w/\omega)(-\cos \phi - \phi' \sin \phi_o) \qquad (6.57)$$

This potential is plotted in Figure 6.4b, and it has an
extremum when $\sin \phi = \sin \phi_s$. A minimum occurs at $\phi = \phi_s$
and a maximum at $\phi = \pi - \phi_s$. Since, with the above approximations, the potential depends only on ϕ' and not explicitly
on time, the energy is conserved. The expression of this is

analogous to (6.45), i.e.

$$H = mc^2 - pw + U - m_o c^2 = h \qquad (6.58)$$

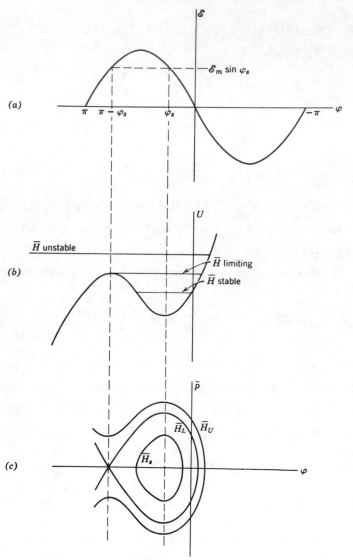

Figure 6.4

Field, potential, and phase-space diagram for a linear accelerator bunching section (From Lichtenberg, A. J., 1968, "Phase Space Dynamics of Particles," Wiley, with permission).

If the energy constant lies between $U(\phi_s)$ and $U(\pi-\phi_s)$, then librations will occur. If this constant is greater than $U(\pi-\phi_s)$, a particle will come in, slow down to zero velocity when it hits the wall $U(\phi)$ and then go back and continually retrogress. This behavior is shown in Figure 6.4(c), for motion in the phase-plane. (It should be noted that z' increases to the right.)

Synchrotron Oscillations

The librations due to the above forcing term (6.56) are known as synchrotron (or synchronous) oscillations. For small amplitudes, they are sinusoidal and we can find the angular frequency ω_0. Expanding p in powers of $v' = v - w$, we find

$$p = p_0 + \frac{dp}{dv}\Big|_v v' + \cdots$$

$$\cong p_0 + m_\ell v'$$

Then, from (6.56) with z' small,

$$\frac{d}{dt}(p - p_0) = m_\ell \frac{d^2 z'}{dt^2}$$

$$= qE\{\sin\phi_0 \cos\frac{\omega z'}{w} - \cos\phi_0 \sin\frac{\omega z'}{w} - \sin\phi_0\}$$

$$\cong -qE(\omega z'/w)\cos\phi_0$$

which is just the equation of an harmonic oscillator, with angular frequency

$$\omega_0 = \left(\frac{qE\omega}{m_\ell w}\cos\phi_0\right)^{1/2}$$

This behaves as $m_\ell^{-1/2}$, or $\omega_0 \sim (1 - \frac{v^2}{c^2})^{3/4}$ which goes to zero as $v \to c$. The period of the oscillation becomes infinite.

VII

Relativistic Dynamics

I. SPECIAL RELATIVITY

When a charged particle moves at high speed, it is found
experimentally that its mass m depends on its velocity, and
that

$$m = m_o / (1 - \frac{u^2}{c^2})^{1/2} \qquad (7.1)$$

where m_o is a constant called the <u>rest mass</u>, u is the velocity
of the particle and c is the velocity of light.

Newton's Equations then must be modified to the form

$$\frac{d}{dt} m\vec{u} = \vec{F} \qquad (7.2)$$

with $\vec{F}$ the force acting on the particle.

The above variation of mass with velocity is at first sight
quite mysterious, but an analysis with the help of the theory
of special relativity shows that it is to be expected if
the velocity of light is the same in all (inertial) systems.
We proceed, then, to outline the concepts of special
relativity and then to deduce (7.1)

Inertial Systems

According to Newton's First Law, there exist motions of a
particle in a straight line with constant velocity. When such
is the case, the frame of reference is said to be an inertial
frame, and the coordinates of the particle constitute an inertial system of coordinates. If another frame moves with constant velocity with respect to the first then the motion of
the particle will also be with constant velocity with respect
to the new frame and in a straight line, and this new system
of coordinates is thus also an inertial system.

We can be more precise in stating how the two inertial
systems are to be compared. An observer in one inertial
system, called S, can set up space coordinates by the use of
standard measuring rods. Then he can define the elapsed time
at one point, say the origin, with the aid of a motion, such
as that of a pendulum, which occurs periodically. The time
required for a light signal to travel from the origin (in
vacuum) to a point P at distance r from the origin will be
assumed equal to the time for the return trip, and equal to
r/c, where c is the constant velocity of light (in vacuum).
This enables the S observer to arrange for a set of clocks in
his system, all running at the same rate and synchronized
with each other (see below). An event is something which
happens at point P (x,y,z) and at time t. Now let there be
another inertial system S' moving with constant velocity v
(with respect to S) in the x-direction. The S' observer can
set up (with the same standard rods) his own system of space
and time coordinates (x', y', z', t') and describe events in

terms of them. The same event, then, will be described in
one way by the S observer and in another way by the S' observer.

The S observer and the S' observer agree on two things,
namely (1) the velocity of light measured by each observer in
his own system equals c, and (2) if the S observer see a particle as moving with constant velocity with respect to himself, the S' observer will see that motion as having constant
velocity with respect to himself. These points of agreement
are sufficient, as we shall show, to establish a unique
correlation between the S' coordinates (x', y', z', t') and
the S coordinates (x, y, z, t).

To state the matter quantitatively, when the motion in one
inertial system is described by the equations

$$x = x_o + u_x (t - t_o)$$

$$y = y_o + u_y (t - t_o) \qquad (7.3)$$

$$z = z_o + u_z (t - t_o)$$

where u_x, u_y, and u_z are the constant components of velocity,
then the description in another inertial system will be

$$x' = x'_o + u'_x (t' - t'_o)$$

$$y' = y'_o + u'_y (t' - t'_o) \qquad (7.4)$$

$$z' = z'_o + u'_z (t' - t'_o)$$

with u'_x, u'_y, u'_z the new constant components of velocity.

Now experimentally one finds that the velocity of light in
any inertial system is equal to c, so that if the above
straight lines   be light rays, we must have that the

two equations

$$u_x^2 + u_y^2 + u_z^2 = c^2 \qquad (7.5)$$

$$u_x'^2 + u_y'^2 + u_z'^2 = c^2 \qquad (7.6)$$

hold simultaneously.

The problem is then to determine transformations

$$x' = x'\ (x,y,z,t)$$

$$y' = y'\ (x,y,z,t) \qquad (7.7)$$

$$z' = z'\ (x,y,z,t)$$

$$t' = t'\ (x,y,z,t)$$

such that straightness of lines and the velocity of light are
unchanged in going from one system to the other. Fock has
shown quite generally that these conditions require (7.7) to
be linear. (See Appendix B.)

Derivation of the Lorentz Transformation -- Determination of Coefficients

Let us then take the transformation to be linear, and the
primed system S' to move in the + x-direction with velocity
v with respect to the original system S (see Figure 7.1).
At $t_o = 0 = t_o'$, suppose the origins to coincide.

Since the transformation is linear, write it as

$$x' = ax + bt$$
$$\qquad \qquad (7.7a)$$
$$t' = ex + ft$$

with a, b, e, f constants.

Now suppose that a light signal is sent out from the origin
at time $t = t' = 0$. At a later time t, it reaches a point

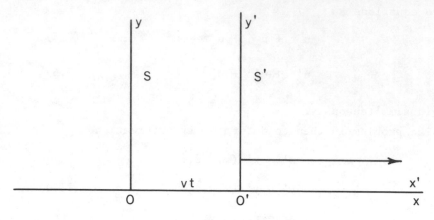

Figure 7.1

Inertial System S' moves with constant velocity re Inertial
System S

described by the S observer as (x, t) and by the S' observer
as (x', t'). Since each observer finds the velocity of light
to be c, we must have the following pairs of equations holding simultaneously,

$$w \equiv ct - x = 0 \qquad (7.8a)$$

$$w' \equiv ct' - x' = 0 \qquad (7.8b)$$

$$\bar{w} \equiv ct + x = 0 \qquad (7.9a)$$

$$\bar{w}' \equiv ct' + x' = 0 \qquad (7.9b)$$

$$ax + bt = x' = 0 \qquad (7.10a)$$

$$x - vt = 0 \qquad (7.10b)$$

Therefore, each member of a pair must be proportional to
the other, which implies that

$$b = - av \text{ and } x' = a(x - vt) \qquad (7.11)$$

$$w' = ct' - x' = Aw = A (ct - x) \qquad (7.12)$$

$$\bar{w}' = ct' + x' = B\bar{w} = B (ct + x) \qquad (7.13)$$

Equation (7.12) is associated with propagation of light
toward more positive values of x, as seen by two different
observers, and (7.13) with propagation toward more negative
values of x. Since there is assumed to be no preferred
direction, the constant A should be the same function of +v
that B is of −v, i.e.

$$A(\beta) = B(-\beta), \quad \beta = v/c \qquad (7.14)$$

Now the line w = 0 is the same as x = ct, and the line $\bar{w}$ = 0
is the same as x = − ct. Adoption of these lines as axes
amounts to rotating by $\pi/4$ in the (x, ct) plane. Solving
for x' and ct' and using (7.12) and (7.13),

$$2x' = \bar{w}' - w' = B\bar{w} - Aw$$

$$= B (ct + x) - A (ct - x)$$

$$= (B - A) ct + (B + A) x$$

$$2ct' = \bar{w}' + w' = B\bar{w} + Aw \qquad (7.15)$$

$$= (B + A) ct + (B - A) x$$

The inverse transformation is, solving,

$$x = \frac{A+B}{2AB} x' - \frac{B-A}{2AB} ct' \qquad (7.16)$$

$$ct = - \frac{B-A}{2AB} x' + \frac{A+B}{2AB} ct'$$

Comparing (7.15) with (7.11), we have (A−B) = β(A+B),
and 2x' = (A+B) (x − βct). Equation (7.16) becomes
$2x = \frac{(A+B)}{AB} (x' + \beta ct')$.
Now x should be the same function of x' + vt' that x' is
of x − vt. Observer S sees system S' moving in one direction,
while observer S' sees system S moving in the opposite direction. Consequently, AB = 1. Combining with A−B = β(A+B),

$$A = (1 + \beta)^{1/2}/(1 - \beta)^{1/2}$$

$$B = (1 - \beta)^{1/2}/(1 + \beta)^{1/2} \qquad (7.17)$$

which is consistent with (7.14). Also,

$$a = (A + B)/2 = (1 - \beta^2)^{-1/2}$$

The constant a is usually denoted by γ, and we have

$$\gamma = (1 - \frac{v^2}{c^2})^{-1/2} \qquad (7.18)$$

The final form of the transformation, called the Lorentz transformation from S to S' is (relative velocity = v)

$$x' = \gamma (x - \beta ct)$$

$$ct' = \gamma (ct - \beta x) \qquad (7.19)$$

The inverse transformation, that from S' to S, is (relative velocity = -v)

$$x = \gamma (x' + \beta ct') \qquad (7.20)$$

$$ct = \gamma (ct' + \beta x')$$

and is the Lorentz transformation when the "moving" system has velocity - v, with respect to the observer.

In ordinary Newtonian mechanics, the equations corresponding to (7.19) are

$$x' = x - vt$$

$$t' = t \qquad (7.21)$$

These can be obtained from (7.19) and (7.18) if we let v/c approach zero.

The classical theory is thus a limiting case of the special theory of relativity, when the velocity of light is regarded as infinite. Special theory tells us what the relations are

when the velocity of light is finite and the same in all
inertial frames of reference.

Recapitulation-Lorentz Transformation

1.   From the 2 assumptions that (a) straight lines remain
straight lines and (b) the velocity of light is invariant on
transformation from one inertial system to another, it follows that the transformation must be a linear one.

2.   The requirements that $x' = ct'$ when $x = ct$ and $x' = -ct'$
when $x = -ct$ specialize the transformation to

$$x' = ax + b\ ct$$

$$ct' = bx + a\ ct$$

3.   The condition that the primed system S' move with
velocity v means that $x' = 0$ when $x = vt$, provided the origins
coincide at $t' = t = 0$.   Therefore

$$x' = a\ (x - vt)$$

and $b = -\ a\ \beta$.

4.   Assuming that the coefficient a can only depend on the
magnitude of the velocity, not on direction, i.e. $a(v) = a(-v)$,
the double transformation $S \underset{v}{\to} S' \underset{-v}{\to} S'' \equiv S$ yields

$$a = (1 - \beta^2)^{-1/2} \equiv \gamma$$

5.   The general linear transformation $(x,t) \to (x',t')$ has
four arbitrary coefficients.   These are reduced to two by
conditions 2 above, to one by condition 3, and made
completely definite by condition 4.

6.   The final correspondence between (x,t) and (x',t'),
embodied in the Lorentz transformation must be a physical one
which can be checked by an observer in one system communi-
cating with an observer in the other system.   Furthermore,

the consequence such as variation of longitudinal and transverse masses with velocity, equivalence of mass and energy,
and time of decay of fast particles, are all subject to experimental check and are in agreement with the theory.

Geometry of the Lorentz Transformation. Simultaneity

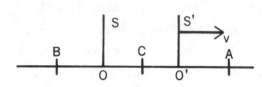

Figure 7.2

Relative Motion of Inertial
System S', with Mirrors at
A and B.

An observer at the ori-
gin 0 in S can define
clocks to be simultaneous
in the following way: Let
a clock at 0 register zero
($t = 0$) when a light signal is sent out. When the
signal arrives at a point
with coordinate x, set that
clock to read $t = x/c$.
Thereafter, if the clocks run at the same rate, they will be
simultaneous if the t readings are equal.

We now wish to compare descriptions by an S observer and
an S' observer of the same sequence of events (Figures 7.2,
7.3, 7.4). At time $t = t' = 0$, let the origins 0 and 0' coincide. Let one light signal L be sent out to the right to be
reflected from a mirror at a distance x_1' (in S') from 0'.
Let another signal L' be sent out simultaneously to the left
to be reflected from a mirror situated at $-x_1'$. Both signals
will get back to 0' at the same time $t_r' = 2t_1' = 2x_1'/c$.

How does all this look to an observer fixed in S? He will
see L be reflected from a mirror at some point A, which is
at a distance x_1 from the origin and whose clock reads
$t_1 = x_1/c$. Similarly, L' will be reflected from B, with
coordinates x_2 (< 0) and $t_2 = -\dfrac{x_2}{c}$. The reflected signals

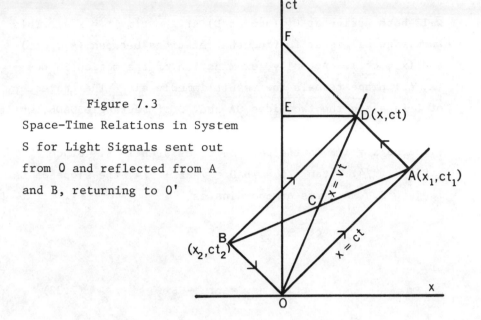

Figure 7.3
Space-Time Relations in System
S for Light Signals sent out
from 0 and reflected from A
and B, returning to 0'

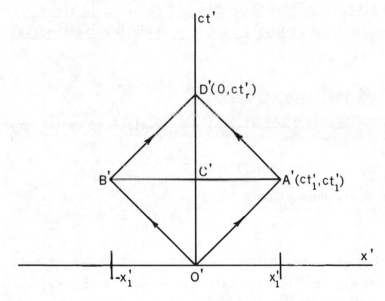

Figure 7.4
Space-Time Relations in System S' for Light Signals sent
out from 0, reflected from A and B, returning to 0'

will both arrive at O' (now at D) at the time $t = x/v$. This
knowledge allows us to find the relations between (x_1, ct_1)
and (x_2, ct_2). Actually, examination of the motion in the
(x, ct) plane reveals the result immediately. The trajectory
of L is along the two sides OA and AD of rectangle OADB, while
the trajectory of L' is along OB and BD. The diagonal OD with
slope $ct/x = 1/\beta$ is the path of O' as seen in S. Since
$\angle DOB = (\pi/4) + \tan^{-1} \beta = \angle BAD = \angle ABX + \pi/4$, the slope of BA
is just β. In terms of coordinates,

$$ct_1 - ct_2 = \beta (x_1 - x_2)$$

or

$$ct_1 - \beta x_1 = ct_2 - \beta x_2 \qquad (7.22)$$

Equation (7.22) states that, <u>if two events</u> such as A and B
<u>are simultaneous in the S' system</u>, an observer in the S sys-
tem will see them as having the same value for the function
$ct - \beta x$.

Alternative Deduction of (7.19)

The relations (7.10a), (7.10b) and (7.11) lead to the equation

$$x' = a(\beta) (x - \beta ct)$$

When $x = ct$, we must have

$$ct' = x' = a(\beta) ct (1 - \beta) \qquad (7.23)$$

But according to (7.22), $t' = 0$ and $ct - \beta x = 0$ must hold
simultaneously, so that ct' is proportional to $ct - \beta x$, which
equals $ct(1 - \beta)$ when $x = ct$. Hence, from the above equation,
the proportionality factor is $a(\beta)$, and

$$ct' = a(\beta) (ct - \beta x)$$

We can get back to the original system by a transformation with relative velocity $-v$, namely

$$x = a(-\beta)(x' + \beta \, ct')$$

$$= a(\beta) \, a(-\beta)(x - \beta ct + \beta ct - \beta^2 x)$$

from which

$$a(\beta) \, a(-\beta) \, (1 - \beta^2) = 1.$$

If now we assume $a(-\beta) = a(\beta)$, then we find

$$a(\beta) = \gamma = (1 - \beta^2)^{-1/2}.$$

This assumption is vindicated by agreement of the theoretical predictions with experiment.

Elapsed Times in S and S' (Time Dilation)

We can now compare the time t_r' to go from the origin O' to the mirror and back with the time t_r recorded by the S observer for the same sequence of events.

At A, $x_1 = ct_1$ and $x_1' = ct_1'$, so that from (7.23) with $a = \gamma$

$$t_r' = 2t_1' = (2\gamma/c) \, x_1 \, (1 - \beta) \qquad (7.24)$$

This is the time in the S' system associated with D, and is twice the time for the light signal to go from the origin O' to the mirror.

The time in the S system at D (elapsed time along OAD) may be obtained, noting that $EF = DE = x$ and $OF = 2x_1$, from

$$OE + EF = OF$$

or

$$ct_r + x = 2x_1$$

$$t_r = 2x_1/(c + v)$$

or

$$= 2x_1/c(1 + \beta) \qquad (7.25)$$

Dividing (7.24) by (7.25),

$$\frac{t'_r}{t_r} = \gamma(1 - \beta^2) = (1 - \beta^2)^{1/2} = 1/\gamma \qquad (7.26)$$

The elapsed time in the S' system from 0 to D is t' and since 0' is at rest in S', is just the time recorded by the clock at 0'. It is also the elapsed time in S' for the signal to go to the mirror and back. The elapsed time in the S system for the signal to go to the mirror and to return to meet the (moving) origin of S' is t_r, which, by (7.26) is greater than the elapsed time t'_r at the origin of the S' system, there being an amplification factor of $(1 - \beta^2)^{-1/2}$. This is known as time dilation, and is a consequence of the synchronization prescription. It should be emphasized that the time measurements in the S system are made at two distinct spatial points, and this is why the dilation depends on the synchronization.

Lorentz Contraction

Because of the fact that light signals are used to effect synchronization of clocks at different points, the comparison of lengths in the two systems appears strange at first glance. At a definite time $t = t_1$ in system S, two observers, one at $x = x_1$ and the other at $x = x_2$ make marks directly opposite themselves on system S', and call these $x' = x'_1$ and $x' = x'_2$, respectively. The distance between x'_1 and x'_2 can be measured by observers fixed in S', and at their leisure. From (7.19), with t = const,

$$x'_2 - x'_1 = \gamma \ (x_2 - x_1) \qquad (7.27)$$

According to this equation, the distance $|x_2 - x_1|$ in S is shorter by the factor $(1 - \beta^2)^{1/2}$ than the distance $|x_2' - x_1'|$ measured in S'. This contraction, the Lorentz contraction, by which a rod fixed in S' appears to be shorter in S, is an effect arising from (1) the fact that the velocity of light is the same in the two systems and (2) the positions of x_1 and x_2 are noted at the same time t.

(Reciprocally, if t' = const, one finds that

$$\Delta x = \gamma \Delta x'$$

as one must expect since the observers in S and S' are equivalent.)

If, in particular, the observers in system S make their marks at time t = 0, the times in system S' will be: For observer 1', $t_1' = -\gamma\beta x_1/c$; For observer 2', $t_2' = -\gamma\beta x_2/c = -\beta x_2'/c$. If these 2 observers, one at x_1' and the other at x_2' should make their marks, that by observer 1' at time $t_1' = -\beta x_1'/c$ and that by observer 2' at time $-\beta x_2'/c$, then these actions would be simultaneous in system S (i.e. t = 0, by 7.20) and $x_1 = \gamma (x_1' - \beta^2 x_1') = \frac{1}{\gamma} x_1'$, by (7.20), and also $x_2 = \frac{1}{\gamma} x_2'$, so that $x_2 - x_1 = (x_2' - x_1')/\gamma$ which is just (7.27) over again. Thus, if the observers in the S' system behave altruistically, and make their measurements so that there is simultaneity in the other system (S), they will find that the length in the S system will be larger by the factor γ than their own. On the other hand, if they are egoists and insist on measuring so that there is simultaneity (t' = const.) in their own system S', then the length in the S system will be smaller by the factor $\frac{1}{\gamma}$ than in their own S' system. Observers in the S system and observers in the S' system start out on an equal footing, but the very act of

making a measurement with t = const. introduces a lop-sidness
into the situation and the consequent Lorentz contraction.

Construction of the Lorentz Transformation (Figures 7.5,7.6)

We wish now to demonstrate how, using graphical methods,
we may find (ct', x') if we are given (ct, x). A simple way
to do this is to use the auxiliary variables w = ct - x,
$\bar{w}$ = ct + x, and (7.12), (7.13) and (7.17), namely

$$w' = A w$$

$$\bar{w}' = w/A \qquad\qquad (7.28)$$

with $A = (1 + \beta)^{1/2}/(1 - \beta)^{1/2}$.

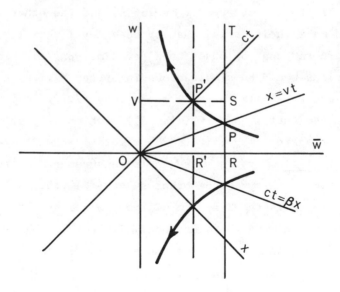

Figure 7.5

Lorentz Transformation, as a stretching in the w-direction
and a contraction in the $\bar{w}$ direction (see text).

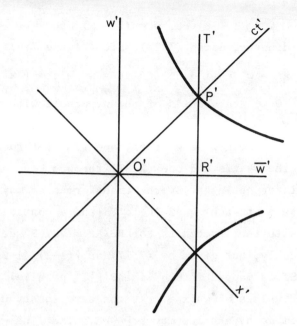

Figure 7.6

Point P (vt, ct) has been mapped to point P' (0, ct')

(See Figure 7.5)

Equations (7.28) show that the transformation is a stretching
of the w-coordinate by the factor A, together with a shrinkage of the $\bar{w}$ coordinate by the same amount, so that

$$w'\bar{w}' = w\bar{w}$$

or

$$w = \text{const.}/\bar{w}$$

or

$$c^2 t^2 - x^2 = \text{const.} \qquad (7.29)$$

This is the equation of an hyperbola with $\bar{w}$, w axes for its
asymptotes. If P is any point on the line x = vt (Figure
7.5) the track of the S' origin in the (x, ct) plane, its
($\bar{w}$, w) coordinates are $\bar{w} = ct + x$, w = ct - x   (7.8)

or $\overline{w} = (1 + \beta)\ ct,\ w = (1 - \beta)\ ct$

The new coordinates, using (7.17), are (Figure 7.6)

$$ct' - x' = w' = (1 - \beta^2)^{1/2}\ ct = \overline{w}' = ct' + x'$$

so that $x' = 0$. Intersection of the hyperbola with this line,
at an angle $\pi/4$ with the $\overline{w}'$ axis, gives the transformed point
P' (0, ct'). If we now draw lines, one (VP'S) through P'
parallel to the $\overline{w}$ axis and one (RPST) through P parallel to
the w-axis intersecting at S, the transformation may be
thought of as a stretching of RP to RS and a contraction of
VS to VP', with the stretch factor RS/RP = VS/VP' determined
here graphically, but given by (7.17) as $[(1+\beta)/(1-\beta)]^{1/2}$.
The line RPST is mapped into a line R'P'T' parallel to itself.
For negative values of w (x > ct), the same construction with
the mirror image hyperbolas may be used. The line ct = βx
then maps into the line t' = 0, points on which are simultaneous with the origin 0' in the S' system.

The Transformation as a Mapping (Figure 7.7)

It is very convenient for visualization to say that each
point P(x, ct) is mapped by the transformation (7.19) into
another point P'(x', ct'), where this has been accomplished
by stretching in the direction x = -ct and contracting in
the direction x = ct. The new point P' will lie on an
hyperbola through P with the equation

$$c^2t^2 - x^2 = \text{const.} \qquad (7.29)$$

From our previous discussion, we know that if P is at
x = β ct, the point P' lies on the vertical axis with
coordinates x' = 0, and ct' = $(c^2t^2 - v^2t^2)^{1/2}$
= ct $(1 - \beta^2)^{1/2}$ < ct. This is the _time dilation_, where the
elapsed time ct in the system S is greater than the elapsed

time in the system S', for x' fixed. Furthermore, the point
Q at (x, 0) is mapped into Q', on the line ct = −βx, with
coordinates (γx, ct' = −γβx). This is an expression of the
Lorentz contraction, where the length x' = γx in S' appears
to be only x, < γx, in the S system.

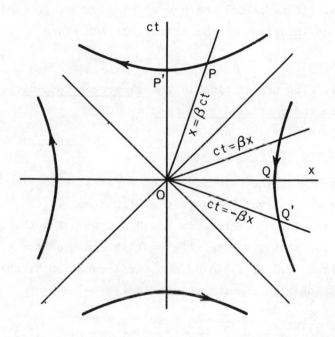

Figure 7.7

Mapping is along hyperbolas with $c^2t^2 - x^2$ = const.

Proper Time

If the frame S' is chosen to be the frame where the particle
is at rest, and x' = 0 represents the position of the particle,

then the time recorded at this point is $t' = OP'/c$. For
other frames moving with uniform velocity with respect to S',
the particle will appear to have coordinates x, β ct, where

$$c^2t^2 - x^2 = c^2(t')^2. \qquad (7.30)$$

A Lorentz transformation from one inertial frame to another
is represented by a mapping of one point into another one on
the hyperbola (7.30), and t' is the same for all such transformations. It is called the <u>proper time</u>, is designated by
τ, and is an <u>invariant</u> of the transformation since

$$c^2t^2 - x^2 = c^2(t')^2 - (x')^2$$

in general. The proper time is the <u>time recorded in the
frame where the particle is at rest</u>.

Transformation of Velocities

If we know what the velocity u'_x of a particle is with
respect to an observer fixed in the S' system, we can readily
find the velocity u observed in the S system. For this, we
need only see what the line $x' = u'_x t'$ becomes on transformation. From (7.20), dividing the first equation by the
second and substituting the value of x',

$$\frac{u_x}{c} = \frac{x}{ct} = \frac{u'_x + \beta c}{c + \beta u'_x} = \frac{(u'_x/c) + \beta}{1 + (\beta u'_x/c)} \qquad (7.31)$$

If $c \to \infty$, then $u_x = u'_x + v$, which is the ordinary Newtonian
relation between velocities in two inertial systems. Alternatively stated, the Newtonian behavior is obtained when
$v \ll c$, but when v is comparable with c, then the results are
changed appreciably because one must divide by the denominator in (7.31). In more detail, let us multiply (7.31) by
β and subtract from unity, to get

$$1 - (\beta u_x/c) = (1 - \beta^2)/[1 + (\beta u_x'/c)].$$

This tells us that, as u_x' goes from 0 to c, u_x increases steadily from v to c. In the limit, for general β, $u_x' = c$ implies $u_x = c$ and conversely, which was one of our basic postulates. Finally, when $\beta = 1$, then $u_x = c$ no matter what u_x' is.

An equation equivalent to (7.31) is obtained by squaring and subtracting both sides from unity, and is

$$1 - (\frac{u_x}{c})^2 = (1 - \beta^2) \, [1 - (u_x'/c)^2]/(1 + \beta \frac{u_x'}{c})^2$$

$$(7.32)$$

If one goes to motion in three dimensions, with S' still moving with velocity v re S, (7.20) is extended to

$$x = \gamma \, (x' + \beta ct'), \quad ct = \gamma \, (ct' + \beta x')$$

$$y = y' \qquad\qquad z = z' \qquad\qquad (7.20a)$$

The components of velocity perpendicular to the relative (x) direction are

$$u_y = \frac{u_y'}{\gamma(1+(\beta u_x'/c))} \quad u_z = \frac{u_z'}{\gamma(1+(\beta u_x'/c))} \qquad (7.31a)$$

and (7.32) takes the form

$$1 - (\frac{u}{c})^2 = (1 - \beta^2) \, (1 - (\frac{u'}{c})^2) \, / \, (1 + \beta \frac{u_x'}{c})^2$$

$$(7.32a)$$

Change of Mass with Velocity

Given that a particle momentarily at rest obeys Newton's equation

$$m_o \, (d^2x'/dt'^2) = X' \, (x' \; t') \qquad (7.33)$$

How will the equivalent equation look when the particle has the instantaneous velocity v?

To answer this question, we suppose that the particle is moving with velocity v with respect to the frame S, that the frame S' moves with constant velocity v re S, and therefore that the particle is momentarily at rest in S'.

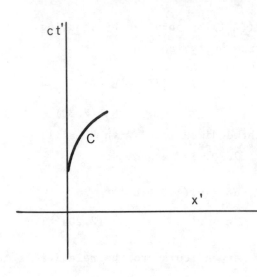

ct'

C

x'

Figure 7.8

Motion of accelerated particle, momentarily at rest in S'.

The motion of the particle near the origin will be represented approximately in the (x', ct') plane by the equation

$$x' = k \ (ct')^2$$

where k is some constant, and by a parabola C tangent to the ct' axis, as in the figure.

For a short time, it is sufficiently accurate to regard x' as equal to zero. Then (7.20) becomes

$$x = \gamma \ (x' + \beta \ ct')$$
$$(7.34)$$
$$t \cong \gamma t'$$

Differentiating re t',

$$\frac{dx}{dt'} = \gamma \ (\frac{dx'}{dt'} + \beta c) \qquad (7.35)$$

$$dt = \gamma \ dt'$$

Repeating the process, since $\beta c = $ const.

$$\frac{d^2x}{dt'^2} = \gamma \ \frac{d^2x'}{dt'^2} = \gamma \ \frac{d}{dt} \ (\gamma \ \frac{dx}{dt}) \qquad (7.36)$$

Solving and substituting in (7.33), with m_o constant,

$$\frac{d}{dt} \left(\frac{m_o}{(1-\beta^2)^{1/2}} \frac{dx}{dt} \right) = X' \ (x',t') = X(x,t) \tag{7.37}$$

(Note that β, which refers to the instantaneous velocity, is treated as a constant in (7.34), but must be taken as time-dependent in applications of (7.37).)

The left hand side arises naturally because of time dilation. We suppose that we know X', the force in system S', as a function of x' and t'. But the connections of these coordinates with x and t are given by the <u>Lorentz</u> transformation, so that we can find how the force depends on x and t.

If the Lorentz transformation is one where y' = y and z' = z, then the companion equations of motion are

$$\frac{d}{dt} \left(m_o \ \gamma \ \frac{dy}{dt} \right) = Y'/\gamma \equiv Y \tag{7.38}$$

$$\frac{d}{dt} \left(m_o \ \gamma \ \frac{dz}{dt} \right) = Z'/\gamma \equiv Z \tag{7.39}$$

where Y and Z are force functions which can be found.

We have thus shown the validity of (7.2), which is seen to be a natural consequence of the Lorentz transformation linking two inertial systems. The mode of derivation shows that it holds for all particles whether or not they possess a charge. The quantity $m_o\gamma = m_o \ (1 - \beta^2)^{-1/2}$, which reduces to the rest mass m_o when v = 0, takes the place of m_o when $v \neq 0$ and is simply called the mass and denoted by "m". Newton's Equations then express generally that <u>the time rate of change of the momentum $\overrightarrow{mv}$ equals the applied force.</u>

In the above treatment, it has been assumed that γ is real, or that v $\leq$ c. This results in real Lorentz contractions and

and time dilations. There have been speculations about the
possibility of particles (tachyons) which could move with
velocity greater than that of light, but no firm predictions
have been made so far.

Longitudinal and Transverse Mass

If a force is exerted in the direction of motion, the ratio
of the force to the resulting acceleration is called the
longitudinal mass. In the present case, the left hand side
of (7.37) is

$$m_o \left(\gamma \frac{dv}{dt} + v \frac{d\gamma}{dt} \right) = m_o \left(\gamma + \frac{\gamma^3 v^2}{c^2} \right) \frac{dv}{dt}$$

$$= m_o \gamma^3 (d^2x/dt^2)$$

The longitudinal mass therefore equals $m_o \gamma^3$. It is important for motion in a linear accelerator.

If a force is exerted normal to the direction of motion,
the ratio of this force to the resulting acceleration is
known as the transverse mass. The left-hand side of (7.38),
for initial velocity dy/dt = 0, is

$$m_o \gamma \, d^2y/dt^2.$$

The transverse mass is thus equal to $m_o \gamma$, and is the mass
measured in an apparatus such as a mass spectrometer.

Constants of the Motion

Just as in the case of slowly moving particles it is
important to know what quantities are constant during the
motion, so it is with fast particles. We have already
discussed this in Chapter I and have learned that, under
certain conditions, an energy integral exists (see 1.28).

But if a function of one set of coordinates is a constant,
this function expressed in terms of a new set of coordinates
(here linked to the old through a Lorentz transformation) will
also be constant. The simpler relations usually hold for a
system in which the particle is momentarily at rest.

Common Four-Vectors

For ease in keeping track of the transformation properties,
it is useful to introduce the concept of the 4-vector. A set
of 4 quantities which behaves as the set (x, y, z, ct) is
called a contravariant 4-vector, while a set which behaves
as $(\frac{\partial}{\partial x}, \frac{\partial}{\partial y}, \frac{\partial}{\partial z}, \frac{\partial}{c\partial t})$ is called a covariant 4-vector (7.40).
(See Appendix C.) The set dx, dy, dz, cdt $\equiv dx_o$ is a con-
travariant 4-vector. (7.41)

If the basic transformation is

$$x' = \gamma (x - \beta ct), \quad y' = y, \quad z' = z, \quad ct' = \gamma (ct-\beta x)$$

$$(7.42)$$

then

$$\frac{\partial}{\partial x} = \gamma(\frac{\partial}{\partial x'} - \beta \frac{\partial}{c\partial t'}), \quad \frac{\partial}{\partial y} = \frac{\partial}{\partial y'}, \quad \frac{\partial}{\partial z} = \frac{\partial}{\partial z'}, \quad \frac{\partial}{c\partial t} = \gamma(\frac{\partial}{c\partial t'} - \beta\frac{\partial}{\partial x'})$$

$$(7.43)$$

The transformation of the partials from the primed system S'
to the unprimed system S is the same as the transformation of
the coordinates from the unprimed system S to the primed
system S'.

The transformation from S to S' is such that $\gamma = (1 - \frac{v^2}{c^2})^{-1/2}$
is constant and that $c^2 dt^2 - dx^2$ is an invariant equal to
$c^2 d\tau^2$, with τ the proper time, or time in the rest system.
Dividing (7.41) by $d\tau$, we have that

$$u_x = \frac{dx}{d\tau}, \quad u_y = \frac{dy}{d\tau}, \quad u_z = \frac{dz}{d\tau}, \quad u_o = c \frac{dt}{d\tau} \quad (7.44)$$

is a set of quantities transforming as (7.41) and is therefore a contravariant 4-vector which is called the 4-velocity.
The invariant of the transformation is

$$u_o^2 - u_x^2 - u_y^2 - u_z^2 = c^2 \left(\frac{dt}{d\tau}\right)^2 - \left(\frac{dx}{d\tau}\right)^2 - \left(\frac{dy}{d\tau}\right)^2 - \left(\frac{dz}{d\tau}\right)^2$$

$$= c^2 \gamma^2 - \gamma^2 v^2 = c^2 \gamma^2 (1 - \beta^2) = c^2 \quad (7.45)$$

Multiplying the set (7.44) by the rest mass m_o, we obtain
the 4-momentum

$$P_x = m_o \gamma \frac{dx}{dt}, \ P_y = m_o \gamma \frac{dy}{dt}, \ P_z = m_o \gamma \frac{dz}{dt}, \ P_o = m_o \gamma c$$
$$(7.46)$$

with a corresponding invariant

$$p_o^2 - p_x^2 - p_y^2 - p_z^2 = m_o^2 c^2 \qquad (7.47)$$

which is just (7.45) times m_o^2.

Now for a particle in free space ($U = 0$) the quantity
$p^2 + m_o^2 c^2 = m^2 c^2 = p_o^2$ is a constant of the motion, from
(1.28). The kinetic energy (1.27a) is $(m - m_o) c^2$. One
defines a total energy E as the kinetic energy plus $m_o c^2$,
and then has the famous relation

$$E = mc^2 \qquad (7.48)$$

from which $p_o = mc = E/c$ and (7.47) becomes

$$(E/c)^2 - p^2 = m_o^2 c^2 \qquad (7.49)$$

The constant $m_o c^2$ is called the rest energy.
This rest energy is the energy which is internal to the
particle, or locked-in energy, and is proportional to the
rest mass m_o. For a deuteron, it will equal the sum of the
rest energies of the neutron and proton less the energy of
binding. If this latter energy is given to the deuteron, it

can disintegrate into neutron and proton, each with its own
rest energy. If a particle has zero rest energy, such as is
the case for a photon, then $E = c\,|p|$. For an elementary
particle such as an electron, meson, proton, etc., the rest
energy is a certain definite value, the theoretical basis for
which is at present lacking.

Analogous considerations apply to charge density. Suppose
this is ρ^o in the rest system. We can multiply (7.44) by ρ^o
to obtain a contravariant vector known as the 4-current,
with components

$$\rho^o\,\gamma\,\frac{dx}{dt},\ \rho^o\,\gamma\,\frac{dy}{dt},\ \rho^o\,\gamma\,\frac{dz}{dt},\ \rho^o\,\gamma\,c$$

or, if we set $\rho = \rho^o\,\gamma$,

$$\rho\,\frac{dx}{dt},\ \rho\,\frac{dy}{dt},\ \rho\,\frac{dz}{dt},\ \rho\,c \qquad (7.50)$$

The first three components are recognized as space components of an ordinary current density, while the fourth component is proportional to the ordinary charge density. The
invariant relations associated with this 4-vector are

$$c^2\,\rho^2 - \rho^2 v^2 = c^2\rho^2/\gamma^2 = c^2\,(\rho^o)^2 \qquad (7.51)$$

and

$$\int \rho\ dx\ dy\ dz = \text{constant} \qquad (7.52)$$

Since the Jacobian of (7.42) is unity and

$$dx\ dy\ dz\ dt = dx'\ dy'\ dz'\ dt'$$

and, since ρ transforms as dt,

$$\rho\ dx\ dy\ dz = \rho'\ dx'\ dy'\ dz'.$$

The total charge in a volume thus remains constant on transformation.

$$\int \rho\ d\tau = \int \rho'\ d\tau' = q$$

Transformation of Force on Charged Particles

The preceding discussion has been deliberately kept simple, so that we could show how one arrives naturally at a generalized mass which varies with the velocity. Nothing was said about how the force depends upon the position and time coordinates, nor about how its form changes with the frame of reference. Since this is interesting for a charged particle in an electromagnetic field, we proceed to discuss it.

We now wish to find how the electric field $\vec{E}'$ in the frame S' is connected with the fields $\vec{E}$ and $\vec{H}$ in the frame S.

The fields $\vec{E}$ and $\vec{H}$ may be represented in terms of scalar potential Φ and vector potential $\vec{A}$ as

$$\vec{E} = - \overrightarrow{\text{grad}} \ \Phi - \frac{1}{c} \frac{\partial \vec{A}}{\partial t} \qquad (7.53)$$

$$\vec{H} = \overrightarrow{\text{curl}} \ \vec{A}$$

Now the electromagnetic field is characterized by a scalar potential Φ and a vector potential $\vec{A}$ satisfying the equations

$$(\nabla^2 - \frac{1}{c^2} \frac{\partial^2}{\partial t^2}) \ \Phi = -4\pi\rho \qquad (7.54)$$

$$(\nabla^2 - \frac{1}{c^2} \frac{\partial^2}{\partial t^2}) \ \vec{A} = \frac{-4\pi\vec{j}}{c}$$

The differential operator in parenthesis, the dalembertian is an invariant of the transformation and will act on a 4-vector to produce another 4-vector. Consequently, we may take (Φ, A_x, A_y, A_z) to be the components of a contravariant 4-potential, which the dalembertian converts into the 4-current (7.50) up to the factor $4\pi/c$. Now these components transform as

$$A_x' = \gamma \ (A_x - \beta\Phi),$$

$$\qquad (7.55)$$

$$\Phi' = \gamma \ (\Phi - \beta A_x)$$

while

$$\frac{\partial}{\partial x'} = \gamma(\frac{\partial}{\partial x} + \frac{\beta}{c}\frac{\partial}{\partial t}), \quad \frac{1}{c}\frac{\partial}{\partial t'} = \gamma(\beta\frac{\partial}{\partial x} + \frac{1}{c}\frac{\partial}{\partial t})$$

(7.56)

Applying (7.56) to (7.55) and combining, using (7.53)

$$-E_x' = \frac{\partial\Phi'}{\partial x'} + \frac{1}{c}\frac{\partial A_x'}{\partial t'} = \frac{\partial\Phi}{\partial x} + \frac{1}{c}\frac{\partial A_x}{\partial t} = -E_x$$

$$-E_y' = \frac{\partial\Phi'}{\partial y'} + \frac{1}{c}\frac{\partial A_y'}{\partial t'} = \gamma(\frac{\partial\Phi}{\partial y} + \frac{1}{c}\frac{\partial A_y}{\partial t}) + \gamma\beta(\frac{\partial A_y}{\partial x} - \frac{\partial A_x}{\partial y})$$

$$= -\gamma E_y + (\gamma\dot{x}/c) H_z$$

$$-E_z' = \frac{\partial\Phi'}{\partial z'} + \frac{1}{c}\frac{\partial A_z'}{\partial t'} = \gamma(\frac{\partial\Phi}{\partial z} + \frac{1}{c}\frac{\partial A_z}{\partial t}) + \gamma\beta(\frac{\partial A_z}{\partial x} - \frac{\partial A_x}{\partial z})$$

$$= -\gamma E_z - (\gamma\dot{x}/c) H_y$$

These equations are equivalent to

$$E_x' = E_x$$

$$E_y' = \gamma[E_y + \frac{1}{c}(\vec{v} \times \vec{H})_y]$$

(7.57)

$$E_z' = \gamma[E_z + \frac{1}{c}(\vec{v} \times \vec{H})_z]$$

which are the desired relations.

Using (7.37), (7.38) and (7.39), we finally obtain Newton's Equations for a charged particle in the form

$$\frac{d}{dt}(\gamma m_o \vec{v}) = e[\vec{E} + \frac{1}{c}(\vec{v} \times \vec{H})]$$

(7.58)

Thus, starting with the notion of 4-current and then using the transformation properties of the associated 4-potential, we have arrived at Equation (7.58), the right-hand side of which is the applied force experienced by a particle moving

with velocity v in frame S, but at rest in the frame S'
(comoving with the particle) and feeling therefore only an
electric field E' in that frame. The force in (7.58) is the
ordinary force, developed originally from Coulomb's Law and
Ampere's Laws, but appearing here automatically from the
basic two assumptions of special relativity, together with
the existence of a 4-potential. If this did not exist, the
force expression in (7.58) would be different in form, and
obtainable by transformation from S', where the functional
dependence of the force on x', y', z', t' is given.

II. NON-INERTIAL FRAMES (GENERAL RELATIVITY)

   When an observer is in a non-inertial frame of reference,
such as any system which is accelerated with respect to an
inertial frame, the description of events is usually rather
complicated. In an elevator which is accelerating, the force
felt by the occupant is different from that exerted by an
elevator in uniform motion. Familiar effects arising because
a system rotates are centrifugal and Coriolis forces.
Einstein considered a system with gravitational forces to be
a non-inertial frame, and showed how one links up such a
frame with an inertial one, thus generalizing the Lorentz
transformation of special relativity. We now proceed to do
this, using a rotating frame of reference as a convenient
example.

Equations of Motion: Free Particles

   Let us start with some inertial system S with Cartesian
coordinates x, y, z, ct, and suppose that there are no forces
acting on the particle. We have already seen that Newton's
Equations must be corrected, when a particle moves fast, to

the form

$$\frac{d}{dt} (m_o \gamma \frac{dx}{dt}) = X, \text{ etc.}$$

The x-component of momentum is

$$p_x = m_o \gamma \, dx/dt = m_o \, dx/d\tau$$

$$= \partial L/\partial (dx/dt)$$

where the Lagrangian L has for its kinetic part (see 1.24)

$$T^* \equiv L \text{ (kin)} = - m_o c^2 \{1 - (v^2/c^2)\}^{1/2} \quad (7.59)$$

It is highly desirable, especially when finding the equations of motion in a new coordinate system, to take as the independent variable the proper time τ, because this is invariant and a standard, rather than the time t which changes whenever we transform. For the Lorentz transformation, both x and t change, but τ does not. With this in mind, then, we can write (7.59) as

$$L \text{ (kin)} = - m_o c \, ds/dt = - m_o c \, (ds/d\tau) \, (d\tau/dt)$$
$$(7.60)$$

where

$$ds^2 = c^2 \, dt^2 - dx^2 - dy^2 - dz^2 = c^2 \, d\tau^2$$
$$(7.61)$$

or

$$\dot{s}^2 = c^2 \, \dot{t}^2 - \dot{x}^2 - \dot{y}^2 - \dot{z}^2 = c^2 \quad (7.62)$$

$$= (c^2 - u^2) \, \dot{t}^2, \text{ and } \dot{t} = \gamma,$$

where dots now denote differentiation re the proper time τ.

Now (7.60) is lop-sided in favoring t, so let us introduce a new Lagrangian $L^* = L\dot{t}$. Then

$$L^* \text{ (kin)} = - m_o c \, ds/d\tau$$
$$= - m_o c \{c^2 \, \dot{t}^2 - \dot{x}^2 - \dot{y}^2 - \dot{z}^2\}^{1/2}$$

and

$$\frac{\partial L^*}{\partial \dot{x}} = m_o \, \dot{x} = p_x$$

The original equations can be put in the form (just multi-plying by $\dot{t}$), if the field is conservative,

$$\frac{d}{d\tau} \left(\frac{\partial L^*}{\partial \dot{x}} \right) = \frac{\partial L}{\partial x} \, \dot{t} = \frac{\partial L^*}{\partial x}$$

so that $L^* = L\dot{t}$ is a satisfactory Lagrangian, and we shall have for the equations of motion in any coordinate system

$$\frac{d}{d\tau} \frac{\partial L^*}{\partial \dot{q}} = \frac{\partial L^*}{\partial q} \qquad (7.63)$$

The <u>significance</u> of the <u>line element</u> (7.61), apart from it being equal to $c^2 \, d\tau^2$, is that it <u>gives directly an invariant kinetic part of a Lagrangian</u> L^* <u>from which one can immediately derive the equations of motion</u>. In general, one may write it as

$$ds^2 = g_{ij} \, dx^i \, dx^j \qquad (7.64)$$

and the g_{ij}'s will be found later to contain within themselves "inertial" and/or "gravitational" terms.

In Cartesian coordinates (7.63) gives, since L^* does not depend on x, y, z, or t,

$$m_o \dot{t} = \text{const}, \; m_o \dot{x} = \text{const, etc.}$$

The ratio $\dot{x} : \dot{y} : \dot{z} : \dot{t}$ is constant, hence dx/dt = const., dy/dt = const., dz/dt = const., and therefore the particle goes in a straight line in (x, y, z) space, with constant velocity. Also $\dot{t} = \gamma$, and $m_o \dot{t} = m$ contains a part $m - m_o$ which is the constant kinetic energy divided by c^2.

Rotating Axes

Transformation to a rotating frame is accomplished by setting

$$x = r \cos(\theta + \omega t), \quad y = r \sin(\theta + \omega t)$$

Then

$$\frac{dx}{dt} = -r \sin(\theta + \omega t)\left(\frac{d\theta}{dt} + \omega\right) + \frac{dr}{dt} \cos(\theta + \omega t)$$

$$\frac{dy}{dt} = r \cos(\theta + \omega t)\left(\frac{d\theta}{dt} + \omega\right) + \frac{dr}{dt} \sin(\theta + \omega t)$$

from which

$$\left(\frac{dx}{dt}\right)^2 + \left(\frac{dy}{dt}\right)^2 = r^2 \left(\frac{d\theta}{dt} + \omega\right)^2 + \left(\frac{dr}{dt}\right)^2 = v^2 \tag{7.65}$$

and

$$x \frac{dy}{dt} - y \frac{dx}{dt} = r^2 \left(\frac{d\theta}{dt} + \omega\right) = P_\theta \tag{7.66}$$

Equation (7.65) expresses the relationship between angular velocities and the linear velocity v (constant in the inertial system). Equation (7.66) is an expression for the angular momentum, constant in the inertial system. Alternative expressions, involving the proper time, are

$$r^2 (\dot{\theta} + \omega \dot{t})^2 + (\dot{r})^2 = v^2 t^2 = v^2 \gamma^2 \tag{7.65a}$$

$$r^2 (\dot{\theta} + \omega \dot{t}) = P_\theta \dot{t} \equiv p_\theta \tag{7.66a}$$

Lagrangian Formulation – Rotating Axes

These equations, which were obtained by direct transformation, can also be found from the Lagrangian, as is to be expected. This will now be demonstrated, preparatory to the use of this method for a general non-inertial system.

The line element is now given by

$$ds^2 = c^2\, dt^2 - dr^2 - dz^2 - r^2\, (d\theta + \omega\, dt)^2$$

$$= (c^2 - r^2\, \omega^2)\, dt^2 - dr^2 - r^2\, d\theta^2 - 2\omega r^2 d\theta dt - dz^2$$

The coefficients are functions of the coordinates, and there is a cross term in $d\theta dt$. The differential of the Lagrangian is

$$dL^* = -\frac{m_o c}{2\dot{s}}\, d\, (\dot{s})^2 = -\frac{m_o}{2}\, d(\dot{s})^2 \qquad (7.67)$$

since $\dot{s} = ds/d\tau = c$. Letting $m_o = 1$, the partials are

$$\partial L^*/\partial \dot{r} = \dot{r}, \quad \partial L^*/\partial \dot{\theta} = r^2\, (\dot{\theta} + \omega \dot{t}) = p_\theta$$

$$-\partial L^*/\partial \dot{t} = (c^2 - r^2\, \omega^2)\, \dot{t} - \omega\, r^2\, \dot{\theta} = c^2\, \dot{t} - \omega p_\theta$$

$$\partial L^*/\partial r = r\, (\dot{\theta} + \omega \dot{t})^2 = p_\theta^2/r^3$$

and the equations of motion are

$$\ddot{r} = p_\theta^2/r^3, \quad p_\theta = \text{const}, \quad \dot{z} = \text{const}.$$

$$c^2\, \dot{t} - \omega p_\theta = \text{const.}, \quad \dot{t} = \text{const.} = \gamma$$

The first integral of the radial equation is

$$\dot{r}^2 + (p_\theta^2/r^2) = \text{const.} = v^2\gamma^2$$

which can also be obtained by combining (7.65a) and (7.66a).

If the particle is momentarily at rest in the rotating system, then $\dot{\theta} = 0$ and $\ddot{r} = r\, \omega^2\, \dot{t}^2$, which converts to the familiar form, $d^2r/dt^2 = r\omega^2$, the centrifugal force per unit mass.

Motion in General Non-Inertial Frame

More generally, for a non-inertial system more complicated than the rotating system, the expression for ds^2 will be a quadratic function of the dx^i, and may be written

$$ds^2 = g_{ij} \, dx^i \, dx^j \qquad (7.64)$$

Then, using (7.67)

$$\frac{\partial L^*}{\partial \dot{x}_i} = - m_o \, g_{ij} \, \dot{x}_j$$

$$\frac{\partial L}{\partial x_i} = - \frac{1}{2} \, m_o \, \frac{\partial g_{kj}}{\partial x_i} \, \dot{x}_k \dot{x}_j$$

and the equations of motion are

$$\frac{d}{d\tau} \, (g_{ij} \, \dot{x}_j) = \frac{1}{2} \frac{\partial g_{kj}}{\partial x_i} \, \dot{x}_k \dot{x}_j \quad (i, \, j, \, k = 0, \, 1, \, 2, \, 3)$$
$$(7.68)$$

These will now be specialized to the case where a particle is momentarily at rest ($\dot{x}_k = 0$, $k = 1, 2, 3$). The result will show how the inertial effects enter in the form of potentials (such as the centrifugal potential).

Splitting off the term involving the time, with $x_o = ct$,

$$c^2 = g_{oo} \, (\dot{x}_o)^2 + 2g_{io} \, \dot{x}_i \, \dot{x}_o + g_{ik} \, \dot{x}_i \, \dot{x}_k \quad (i, \, k = 1, \, 2, \, 3)$$

$$= \frac{1}{g_{oo}} \, (g_{oo} \, \dot{x}_o + g_{io} \, \dot{x}_i)^2 + \gamma_{ik} \, \dot{x}_i \, \dot{x}_k \qquad (7.69)$$

where

$$\gamma_{ik} = g_{ik} - \gamma_i \, \gamma_k$$

and

$$\gamma_i = g_{io} \, / \, g_{oo}^{\,1/2} \qquad (7.69a)$$

Straightforward differentiation, which is necessary for the simplification of the equations (7.69) leads (see Addendum, end of chapter) to the <u>equations of motion</u>

$$\gamma_{ik} \frac{d^2 x_k}{dt^2} = \frac{1}{2} c^2 \frac{\partial g_{oo}}{\partial x_i} - c\, g_{oo}^{1/2} \frac{\partial \gamma_i}{\partial t} \qquad (7.70)$$

The left-hand side contains the accelerations in the non-inertial system, and the right-hand side the forcing terms which arise because we have transformed from an inertial system. They may be called the inertial forces, or, since inertial mass is observed to be equivalent to gravitational mass, the <u>gravitational</u> forces peculiar to the non-inertial system which we consider.

For the rotating frame, $g_{oo} = 1 - (r^2\omega^2/c^2)$, $g_{11} = g_{33} = -1$, $g_{22} = -r^2$, $g_{20} = -\omega r^2/c$, $\gamma_1 = \gamma_3 = 0$, $\gamma_2 = -\omega r^2/(c^2-\omega^2 r^2)^{1/2}$, and $\gamma_{22} = -c^2 r^2/(c^2 - r^2\omega^2)$. The equations of motion are then

$$\frac{d^2 r}{dt^2} = \omega^2\, r, \quad \frac{d^2\theta}{dt^2} = 0, \quad \frac{d^2 z}{dt^2} = 0.$$

The centrifugal term $r\omega^2$ is the same as that derived just above, and is the negative gradient of a scalar (inertial) potential $\chi = -(r^2\omega^2)/2$. The coefficient g_{oo} contains this, namely $g_{oo} = 1 + (2\chi/c^2)$.   (7.71)

The right-hand side of (7.70) is similar in form to the expression for the electric force on a charged particle, namely

$$eE_i = -e\frac{\partial\Phi}{\partial x^i} - \frac{e}{c}\frac{\partial A^i}{\partial t}$$

Consequently, it seems natural to call the function χ in (7.71) the scalar (gravitational) potential, and γ_i the ith

component of a vector (gravitational) potential. For a
particle outside a spherical body with mass M, Equation (7.71)
becomes

$$g_{oo} = 1 - (2m/r), \qquad (7.72)$$

where m is proportional to M.

To recapitulate, if the metric coefficients g_{ij} are given,
then the equations of motion for a particle (not subject to
non-gravitational forces) are expressed by (7.68). The
quantities g_{oo} and g_{oi} are connected with the scalar and
vector gravitational potentials, through (7.69a) and (7.72).
The accelerations of a particle momentarily at rest are given
by (7.70), and suggest (7.72) as the manner in which g_{oo}
depends on the scalar potential.

However, g_{oo} is just one component of a covariant tensor of
second rank, and the other components need to be determined
so that D'Alembert's equations will be satisfied for weak
fields. That is, we want to find equations which are general-
izations of $\nabla^2 \chi - \dfrac{1}{c^2} \dfrac{\partial^2 \chi}{\partial t^2} = - 4\pi\rho$ (where ρ = mass density) and
of its associates.

A solution to this problem was proposed by Einstein whereby
the field equations are

$$R_{ij} - \frac{1}{2} R\, g_{ij} = - \kappa T_{ij} \qquad (7.73)$$

with R_{ij} and R complicated functions of g_{ij} and first and
second order derivatives thereof, while T_{ij} is the energy-
momentum tensor with component T_{oo} proportional to ρ. (For
details, see Møller "Theory of Relativity").

Thus we have the equations of motion (7.68) determining the
motion from the metric coefficients, and these coefficients
being themselves determined from the distribution (position
and velocity) of the matter.

These interrelations are very complicated, non-linear, and outside the scope of the present treatment.

Motion of a Planet Outside of a Spherical Mass

When the particle is outside a spherically symmetrical body of mass M, the field equations (7.73) simplify, $T_{ij} = 0$, and the expression (7.64) is found to be

$$ds^2 = (1 - \frac{\alpha}{r})\ c^2\ dt^2 - r^2\ (d\theta^2 + \sin^2\theta d\phi^2) - \frac{dr^2}{1 - \frac{\alpha}{r}}$$

$$(7.74)$$

where $\alpha = 2kM/c^2$ and k = gravitational constant. For the Sun, $\alpha/r = 4.2 \times 10^{-6}$, which is very small. (The value of r is the radius of the Sun.)

Since the field is symmetrical, the plane of the orbit can be taken as $\theta = \pi/2$. The Lagrange partials with $m_o = 1$ are, from (7.67)

$$\frac{\partial L^*}{\partial \dot\phi} = r^2\ \dot\phi = p_\phi, \quad \frac{\partial L^*}{\partial \dot r} = \frac{\dot r}{1 - \frac{\alpha}{r}} \qquad (7.75)$$

$$\frac{\partial L^*}{\partial \dot t} = - (1 - \frac{\alpha}{r})\ c^2\ \dot t = - h^*$$

where p_ϕ is the (proper) angular momentum and h^* is the energy constant, equal to $c^2 \gamma$ for $\alpha = 0$.

Solving (7.74) for $c\dot t$,

$$(1 - \frac{\alpha}{r})^{1/2}\ c\dot t = (c^2 + r^2\dot\phi^2 + \frac{\dot r^2}{1 - \frac{\alpha}{r}})^{1/2}$$

Expanding and substituting (7.75),

$$\frac{h^*}{c^2}\ (1 + \frac{\alpha}{2r}) = 1 + \frac{1}{2c^2}\ (r^2\ \dot\phi^2 + \frac{\dot r^2}{1 - \frac{\alpha}{r}})$$

or

$$h \equiv h^* - c^2 = \frac{1}{2} \left(r^2 \dot{\phi}^2 + \frac{\dot{r}^2}{1 - \frac{\alpha}{r}} \right) - \frac{kM}{r}$$

For small velocities, $t \to \tau$ and this is the Newtonian expression for <u>conservation of energy</u>.

When the <u>angular momentum is zero</u>, substitution of (7.75) into (7.74) yields

$$c^2 = \frac{(h^*)^2}{c^2 (1 - \frac{\alpha}{r})} - \frac{\dot{r}^2}{1 - \frac{\alpha}{r}} \qquad (7.76)$$

or

$$\dot{r}^2 = \frac{(h^*)^2}{c^2} - (1 - \frac{\alpha}{r}) c^2$$

When $r \to \alpha$, the second term of the r.h. side approaches zero and can be neglected, so that

$$\dot{r} \cong - h^*/c = - (1 - \frac{\alpha}{r}) c\dot{t} \qquad (7.77)$$

for inward motion. From this,

$$\frac{dr}{dt} \cong - c \, \frac{r - \alpha}{\alpha} \qquad (7.78)$$

The constant α is called the <u>Schwarzschild</u> radius. If a particle could be at rest there, the proper time would be small relative to the time t measured by a distant observer. Whereas the radial velocity $dr/d\tau$ in the system where the particle is momentarily at rest is constant, from (7.77), the radial velocity dr/dt goes to zero. A distant observer will say that it takes the particle an infinite time to reach the radius α from greater values of r.

According to Oppenheimer and coworkers, gravitational forces operating on a massive star will eventually cause collapse and the material becomes superdense. Once the radius

of the star gets to be less than α, there is no way by which
a particle could leave in a finite time. Hence there can be
no radiation and the star has become a black hole. Since the
ratio α/r for the Sun is very small, the Schwarzschild radius
is not important here, but it may very well become signifi-
cant in late stages of stellar evolution.

When the angular momentum of the particle is not zero,
there is an extra term added to (7.76), and

$$\dot{r}^2 = \frac{(h^*)^2}{c^2} - (1 - \frac{\alpha}{r})\, c^2 - (1 - \frac{\alpha}{r})\, r^2\, \dot{\phi}^2$$

Dividing this by $\dot{\phi}^2 = p_\phi^2/r^4$, we obtain

$$(\frac{dr}{d\phi})^2 = \frac{r^4}{p_\phi^2}\, [\frac{(h^*)^2}{c^2} - (1 - \frac{\alpha}{r})\, c^2] - r^2\, (1 - \frac{\alpha}{r})$$

Substituting $u = 1/r$,

$$dr = -\, r^2\, du,$$

$$(\frac{du}{d\phi})^2 = -\, u^2\, (1 - \alpha u) + (1/p_\phi^2)\, [(h^*)^2/c^2) - (1 - \alpha u)\, c^2]$$

$$= \alpha u^3 - u^2 + (1/p_\phi^2)\, [(h^*)^2/c^2) - (1 - \alpha u)\, c^2]$$

$$= \alpha\, (u - u_1)\, (u - u_2)\, (u - u_3) \qquad\qquad (7.79)$$

where u_1, u_2, and u_3 are the roots of the cubic and satisfy
the relation

$$u_1 + u_2 + u_3 = 1/\alpha \qquad\qquad (7.80)$$

In applying this theory to the motion of Mercury around
the Sun, we start with the knowledge that the orbit of the
planet is approximately an ellipse with small eccentricity.
The actual value of αu is about 5×10^{-8}, so that $\alpha u \ll 1$. If
this neglection is made in (7.79), and we write

$(h^*)^2 - c^4 = (h^* - c^2)(h^* + c^2) \cong 2c^2h$, then there results

$$(\frac{du}{d\phi})^2 \cong -u^2 + \frac{1}{p_\phi^2}(2h + \alpha c^2 u) = (u_1 - u)(u - u_2)$$

This agrees with (4.5) if we identify p_ϕ with J and $\alpha c^2 = 2kM$ with -2κ. The motion is along the ellipse (note that $\ell \equiv a$)

$$u = \frac{1}{r} \cong \frac{1}{a}(1 + e \cos \phi)$$

with semimajor axis a and eccentricity e. The roots of the quadratic are $u_1 \cong \frac{1}{a}(1 - e)$, $u_2 \cong \frac{1}{a}(1 + e)$, from which

$$u_1 + u_2 = 2/a$$

$$u_2 - u_1 = 2e/a.$$

The motion oscillates between u_1 and u_2 (aphelion and perihelion) in a sinusoidal manner, and the mean value is

$$\bar{u} = (u_1 + u_2)/2 = 1/a$$

Returning now to the exact equation (7.79), let us make the substitutions (see 3.15, 3.16, 3.18 and 3.20)

$$x = (\phi/2)[\alpha (u_3 - u_1)]^{1/2}$$

$$y^2 = (u - u_1) / (u_2 - u_1)$$

$$k^2 = (u_2 - u_1) / (u_3 - u_1) \cong \alpha (u_2 - u_1)$$

Since $\alpha u << 1$, then $k^2 << 1$.

This results in the standard elliptic equation

$$(\frac{dy}{dx})^2 = (1-y^2)(1-k^2y^2)$$

with solution $y = \text{sn } (x + \delta)$. From this,

$$u - u_1 = (u_2 - u_1) \text{ sn}^2 (x + \delta)$$

which has period (in x) of

$$2K = 2 \int_0^1 [(1-y^2)(1-k^2y^2)^{-1/2} \, dy$$

$$\cong 2 \int_0^1 (1-y^2)^{-1/2} (1 + \frac{1}{2} k^2y^2) \, dy$$

$$= \pi (1 + \frac{1}{4} k^2)$$

The corresponding change in ϕ is, using (7.80),

$$\Delta\phi = 2 \pi [1 + \frac{\alpha}{4} (u_2 - u_1)][1 + \frac{\alpha}{2} (2 u_1 + u_2)]$$

$$\cong 2 \pi [1 + \frac{3\alpha}{4} (u_1 + u_2)]$$

$$= 2 \pi [1 + \frac{3\alpha}{2a}]$$

From this, we see that there is for Mercury an advance of the perihelion each period by an amount

$$\Delta\phi - 2 \pi = 3 \alpha\pi/a$$

This amounts to about 43" per century. Comparison with observation seems to indicate agreement, but the other planets give rise to a calculated effect amounting to 5557", while the observed advance is roughly 5600". The difference is practically the same as that predicted from the Einstein field equations, but there has been a certain skepticism concerning the accurary of the observations, as well as the inclusion of all relevant factors in the calculations.

Deflection of Light by the Sun

   For light rays, the paths are described by ds = 0. We can
use $d\tau$ as a parameter and then pass to the limit $d\tau = 0$. The
terms in $\dot{t}$, $\dot{\phi}$, and $\dot{r}$ all become large, so that c^2 can be neglected. Then (7.79) becomes

$$\left(\frac{du}{d\phi}\right)^2 = \alpha u^3 - u^2 + \frac{(h^*)^2}{c^2 p_\phi^2} \qquad (7.81)$$

If we neglect the term αu^3, the integral is

$$u = u_m \sin \phi \text{ or } r\sin \phi = r_m \qquad (7.82)$$

where $r_m = cp_\phi/h^*$ and (7.82) describes a straight line passing
the origin at a distance r_m from it.

   The roots of the cubic are obtained from (7.80) and (7.81),
since $u_2 \cong u_m$, $u_1 \cong - u_m$, and

$$u_2^2 = u_m^2 + \alpha u_2^3 \cong u_m^2 (1 + \alpha u_m)$$

$$u_1^2 = u_m^2 + \alpha u_1^3 \cong u_m^2 (1 - \alpha u_m)$$

from which $u_2 \cong u_m (1 + \frac{1}{2} \alpha u_m)$, $u_1 \cong - u_m (1 - \frac{1}{2} \alpha u_m)$
and

$$u_3 = (1/\alpha) - u_1 - u_2$$

$$= (1/\alpha) - \alpha u_m^2$$

   The change in ϕ as u varies from zero ($r = \infty$) to u_2
(closest approach) is

$$\Delta\phi = \int_o^{u_2} \frac{du}{[\alpha(u_1 - u)(u_2 - u)(u_3 - u)]^{1/2}}$$

But

$$u - u_1 \cong u + u_2 - \alpha u_2^2$$

and

$$u_3 - u \cong (1/\alpha) - \alpha u_2^2 - u$$

so that, expanding and neglecting small terms,

$$\Delta\phi = \int_0^{u_2} \frac{du}{(u_2^2 - u^2)^{1/2} (1 - \dfrac{\alpha u_2^2}{u + u_2})^{1/2} (1 - \alpha u - \alpha^2 u_2^2)^{1/2}}$$

$$\cong \int_0^{u_2} \frac{du}{(u_2^2 - u^2)^{1/2}} (1 + \frac{1}{2}\alpha u + \frac{1}{2}\frac{\alpha u_2^2}{u + u_2})$$

$$= \frac{\pi}{2} + \frac{\alpha u_2}{2} + \frac{\alpha u_2}{2} = \frac{\pi}{2} + \alpha u_2$$

as may be verified by elementary integration.

The total deflection of a ray which just grazes a body of mass M and radius r is

$$2 \Delta \phi - \pi = 2\alpha/r = 4kM/c^2 r.$$

The deflection of a ray which just grazes the Sun is $2 \alpha/r_m = 1.75''$, which seems to be consistent with observations on stars during total eclipses. However, this is just within the limit of experimental error, so that this test of the metric is not regarded as conclusive.

Retardation of Light by the Sun

A massive body can influence the speed of a light ray, as measured by a distant observer. To see this, set ds = 0 in (7.74) and suppose the ray proceeds radially. Then

$$\frac{dr}{dt} = \pm c (1 - \frac{\alpha}{r})$$

As $r \to \alpha$, the light velocity goes to zero, so that it would take light an infinite time to arrive at, or leave from, a black hole.

However, until there is more evidence for the existence of black holes, it is in order to look for an experimental check on the speed of light passing the Sun. We shall assume grazing incidence and shall ignore the small deflection. Letting the ray proceed (Figure 7.9) in a straight line, the x-direction, we shall recast (7.74) into a more suitable form.

Introduce the substitution

$$r = r' \left(1 + \frac{\alpha}{4r'}\right)^2$$

$$dr = dr' \left(1 + \frac{\alpha}{4r'}\right) \left(1 - \frac{\alpha}{4r'}\right)$$

$$1 - \frac{\alpha}{r} = \frac{\left(1 - \frac{\alpha}{4r'}\right)^2}{\left(1 + \frac{\alpha}{4r'}\right)^2}$$

and obtain

$$ds^2 = - \left(1 + \frac{\alpha}{4r'}\right)^4 [(dr')^2 + (r')^2 \, d\phi^2] + \frac{\left(1 - \frac{\alpha}{4r'}\right)^2}{\left(1 + \frac{\alpha}{4r'}\right)^2} c^2 \, dt^2$$

$$\equiv \left(1 - \frac{\alpha}{r'}\right) c^2 \, dt^2 - \left(1 + \frac{\alpha}{r'}\right) (dx^2 + dy^2)$$

$$(7.83)$$

when the field is weak. Here $x = r' \cos \phi$, $y - r' \sin\phi$.

If the ray is in the x-direction, then (7.83) with ds = 0 gives

$$\frac{dx}{dt} = c \left(1 - \frac{\alpha}{r'}\right)$$

The second term on the right-hand side represents the slowing due to the Sun. If we ignore it momentarily, the trajectory is, with c = 1,

$$x(t) = x_o(\tau) + (t - \tau) \qquad (7.84)$$

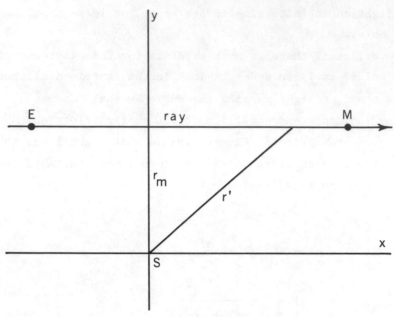

Figure 7.9

Radar experiment, ray from Earth grazes Sun on way to Mercury.

The delay will be, if r_m is the distance of closest approach

$$- \alpha \int_{\tau}^{t} \frac{dT}{\{[x_o(\tau) + (T - \tau)]^2 + r_m^2\}^{1/2}}$$

$$= - \alpha \int_{x_o}^{x} \frac{dx}{(x^2 + r_m^2)^{1/2}} = - \alpha \ln \frac{r' + x}{r' + x_o}$$

For a signal to make a round trip from the earth to Venus or Mercury (on the other side of the Sun), the total delay has been calculated (Tausner, 1966) to be of the order of 10^{-4} sec. Experimental results agree (Shapiro et al., 1971) with theory to within about 1%.

ADDENDUM--Derivation of (7.70)

Equation (7.69) can be put in the form

$$c^2 = g_{oo} \, (\dot{x}_o)^2 \, A + \gamma_{ik} \, \dot{x}_i \, \dot{x}_k \tag{1}$$

with

$$A = \left(1 + \frac{\gamma_i \dot{x}_i}{\sqrt{g}_{oo} \, \dot{x}_o} \right)^2$$

When the particle is __momentarily at rest__, $\dot{x}_i = 0$, and

$$g_{oo}^{1/2} \, \dot{x}_o = cA^{-1/2}, \quad \dot{t} = g_{oo}^{-1/2} \tag{2}$$

$$A = (1 + \frac{\gamma_i \dot{x}_i}{c})^2 \rightarrow 1, \quad \dot{A} = 2\gamma_k \, \ddot{x}_k/c \tag{3}$$

$$\dot{\gamma}_i = \dot{x}_o \, (\partial\gamma_i/\partial x_o) = g_{oo}^{-1/2} \, (\partial\gamma_i/\partial t) \tag{4}$$

$$\frac{d^2 x_k}{d\tau^2} = \frac{1}{\sqrt{g}_{oo}} \frac{d}{dt} \left(\frac{1}{\sqrt{g}_{oo}} \frac{dx_k}{dt} \right) = \frac{1}{g_{oo}} \frac{d^2 x_k}{dt^2} \tag{5}$$

Using (7.69a) and (2), we have

$$g_{io} \, \dot{x}_o = \gamma_i \, g_{oo}^{1/2} \, \dot{x}_o = \gamma_i \, cA^{-1/2} \tag{6}$$

and, with (3) and (4)

$$\frac{d}{d\tau} \, (g_{io} \, \dot{x}_o) = c\dot{\gamma}_i - \frac{1}{2} \, c\gamma_i \, \dot{A}$$

$$= cg_{oo}^{-1/2} \, (\partial\gamma_i/\partial t) - \gamma_i \, \gamma_k \, \ddot{x}_k \tag{7}$$

But equation (7.68) reduces to

$$g_{ik} \, \ddot{x}_k + (d/d\tau) \, (g_{io} \, \dot{x}_o) = \frac{1}{2} \, (\partial g_{oo}/\partial x_i) \, \dot{x}_o^{\,2}$$

which with (2), (5) and (7) yields

$$\gamma_{ik} \frac{d^2 x_k}{dt^2} = \frac{c^2}{2} \frac{\partial g_{oo}}{\partial x_i} - cg_{oo}^{1/2} \frac{\partial\gamma_i}{\partial t} \tag{7.70}$$

which was to have been derived.

VIII

Beams of Particles
(Transformations and Invariants)

The general picture of the behavior of a dynamical system will
be much clearer if it encompases many trajectories at once,
instead of just a single one. For instance, particles of the
same energy may issue from a point in various directions.
Alternatively, particles of the same energy may cross a certain surface normally. One may want to know where such particles can go, whether or not there are forbidden regions,
whether the beam can be focussed, and whether the motion stays
bounded. This way of posing the problem was the one adopted
by Hamilton, who stressed the parallelism between geometrical
optics and dynamics. It has become all the more important for
the design of modern apparatus involving motion of charged
particles, as for example cathode-ray tubes and synchrotrons.
We shall now be concerned with the way in which wave fronts
move with time.

   It is instructive to trace the similarity between wave
optics and wave mechanics, to make the approximation that the
wave length is small relative to the path length, and to find
that the mathematics applicable to optics is applicable to
mechanics as well. In the limit of small wave length, wave
optics is replaced by geometrical optics and wave mechanics
by geometrical (classical) mechanics.

The equation for a wave traveling in the x-direction with phase velocity w is

$$\frac{\partial^2 \psi}{\partial x^2} = \frac{1}{w^2}\frac{\partial^2 \psi}{\partial t^2}$$ (8.1)

If the wave be monochromatic, i.e. $\psi = ue^{-i\omega t}$ where $\omega = 2\pi\nu$, then

$$\frac{d^2 u}{dx^2} + k^2 u = 0, \text{ where } k = \omega/w$$ (8.2)

But $k = 2\pi/\lambda$, with λ = wave length, so that

$$\frac{d^2 u}{dx^2} + \frac{4\pi^2}{\lambda^2} u = 0$$ (8.3)

The phase velocity w is given by c/n, where n is the index of refraction and may vary with position. If λ_o is the wave length in vacuo, and $k_o = 2\pi/\lambda_o$, then $k_o = \omega/c$ and $k = nk_o$. Equation (8.2) now becomes

$$\frac{d^2 u}{dx^2} + n^2 k_o^2 u = 0$$ (8.4)

The wave equation for a particle of momentum p is obtained by noting that the particle has a (De Broglie) wave length $\lambda = h/p$, where h is Planck's constant. Then Equation (8.3) becomes

$$\frac{d^2 u}{dx^2} + \frac{4\pi^2 p^2}{h^2} u = 0$$

or

$$\frac{d^2 u}{dx^2} + \frac{8\pi^2 m}{h^2} (E - V) u = 0$$ (8.5)

where V is some (potential) function of x. This is known as the <u>Schrödinger equation</u>.

When the index of refraction n is a constant, a solution of (8.4) is

$$u = A e^{ikx} \tag{8.6}$$

The function $\psi = A e^{i(kx - \omega t)}$ represents a monochromatic wave advancing in the direction of positive x with phase velocity $w = \omega/k$. The phase is measured by kx.

A typical optical problem is to determine what happens to a beam of parallel light rays which comes say from $x = -\infty$, $n = 1$, and is incident upon a region where the index of refraction varies with position. The rays will be bent while in this region and on emerging will go off in various directions.

The analogous wave mechanical problem is to let a beam of particles, at first parallel to each other and of energy E, fall on a spatial region where the potential varies, and to find out how the wave front changes as it advances.

The connection between the various disciplines is as shown in the following diagram:

Wave Optics $\xrightarrow{\lambda \to 0}$ Geometrical Optics

Wave Mechanics $\xrightarrow{\lambda \to 0}$ Classical Mechanics

Except for a constant factor, the transition from optics to mechanics is made by replacing the index of refraction n by the momentum p.

Transition from Wave Optics to Geometrical Optics

Now let us consider Equation (8.4) for the case where the wave length is small and k_o, the wave number, is large. The solution will be taken so as to reduce to (8.6) for n constant, namely

$$u = A(x) e^{ik_o L(x)} \qquad (8.7)$$

$$k_o L = \text{phase}$$

where A and L are assumed to be real functions. A represents the amplitude of the wave and $L = \omega t + \text{const.}$ is the equation of a surface (if motion is in 3 dimensions) of constant phase, or <u>wave front</u>, at time t.

Differentiating (8.7)

$$u' = (A' + ik_o L' A)e^{ik_o L}$$

and

$$u'' = [A'' + 2ik_o L'A' + ik_o L''A - k_o^2 (L')^2 A]e^{ik_o L}$$

Substituting in (8.4)

$$A'' - k_o^2 (L')^2 A + n^2 k_o^2 A + ik_o (L''A + 2L'A') = 0$$
$$(8.8)$$

Equating the imaginary part to zero and integrating,

$$L' A^2 = \text{const.} \qquad (8.9a)$$

Equating the real part of (8.8) to zero,

$$(L')^2 = n^2 + (A''/k_o^2 A)$$

$$= n^2 + \frac{1}{k_o^2} [\frac{3}{4} (\frac{L''}{L'})^2 - \frac{1}{2} \frac{L'''}{L'}] \qquad (8.9b)$$

where we have differentiated the expression $A = \text{const.} (L')^{-1/2}$ twice to obtain the equivalent of A''/A.

If the term in brackets is not large and k_o is large (or λ is small) then (8.9b) becomes approximately (see justification below)

$$\left(\frac{dL}{dx}\right)^2 = n^2 \qquad\qquad (8.10)$$

This is the _eiconal_ equation. The quantity $L = \int n\,dx$ is the generalization of the distance traversed in vacuo and is called the _optical length_. Combining (8.7), (8.9a) and (8.10), the solution is

$$u = \frac{const}{\sqrt{k}}\, e^{i\int k\,dx} \qquad\qquad (8.11)$$

Now using the relation $k = 2\pi p/h$, and defining the _action_ $S = \int p\,dx$, the equations corresponding to (8.10) and (8.11) are

$$\left(\frac{dS}{dx}\right)^2 = p^2 \qquad\qquad (8.12)$$

and

$$u = \frac{const}{\sqrt{p}}\, e^{(2\pi i/h)\int p\,dx}$$

$$= (const/p^{1/2})\, e^{2\pi i S/h} \qquad\qquad (8.13)$$

The generalization to 3 dimensions replaces (8.12) by

$$\left(\frac{\partial S}{\partial x}\right)^2 + \left(\frac{\partial S}{\partial y}\right)^2 + \left(\frac{\partial S}{\partial z}\right)^2 = p^2 \qquad\qquad (8.14)$$

or

$$|\nabla S| = p \qquad\qquad (8.15)$$

The approximation represented by (8.11) or (8.13) was introduced by Jeffreys, and later rediscovered and popularized by Wentzel, Kramers, and Brillouin, and is usually known as the W-K-B approximation. For a fairly simple treatment of Equation (8.5) by this method, an article by Langer should be consulted.

Validity of The Approximation (8.10)

One method of justifying the approximation (8.10) for small wave lengths (large k_o) is to expand L in a power series in $1/k_o^2$ and to equate coefficients of like powers when the expansion is substituted into (8.9b). (This expansion is an asymptotic one (divergent) but does give good results if broken off after a finite number of terms, with k_o large.) Carrying out the substitution of

$$L = L_o + L_1/k_o^2 + \cdots , \quad L_o' = n,$$

we obtain

$$2L_o' L_1' = \frac{3}{4} \left(\frac{L_o''}{L_o'} \right)^2 - \frac{1}{2} \frac{L_o'''}{L_o'}$$

or

$$2\lambda_o L_1' = \frac{1}{2} \lambda'' - \frac{1}{4} \frac{\lambda'^2}{\lambda}$$

Integrating,

$$2\pi \frac{L_1}{k_o} = \frac{1}{4} \lambda' - \frac{1}{8} \int^x \frac{(\lambda')^2}{\lambda} \, dx$$

If, now the rate of change of wave length with distance is small, i.e. $\lambda' < < 2\pi$, then $L_1/k_o < < 1$ and therefore the multiplicative factor $\exp (iL_1/k_o)$ may be dropped from consideration in (8.7). The approximation (8.10) is therefore valid when k_o is large and $\lambda' < < 2\pi$.

Contact Transformations

From the foregoing discussion, it is apparent that one can treat a beam of particles by the same methods which have been

fruitful in geometrical optics. However, the exact connection with the equations of motion is not immediately evident,
and so we shall study the propagation of a wave front.

Confining our discussion to one dimension, suppose that at
time $t = 0$ we have a collection of particles with momentum a
smooth function p_o of position x_o. At a small time Δt later
they will be described by a changed function p of position x.
Let $S_o = \int_a^{x_o} p_o \, dx_o$ and $S = \int_b^x p \, dx$, where a and b are
fixed. Then an arbitrary variation δ, with time constant,
gives

$$\delta S_o = p_o \delta x_o \quad \text{and} \quad \delta S = p \delta x.$$

The actual motion is such that x_o goes to x, and $x_o + \delta x_o$
to $x + \delta x$, and is governed by the relations

$$x = x_o + \dot{x}_o \Delta t$$

$$\text{(8.16)}$$

$$p = p_o + \dot{p}_o \Delta t$$

Neglecting higher powers of Δt,

$$p \, \delta x = p_o \, \delta x_o + (\dot{p}_o \, \delta x_o + p_o \, \delta \dot{x}_o) \, \Delta t$$

$$= p_o \, \delta x_o + [(\partial L/\partial x_o) \, \delta x_o + (\partial L/\partial \dot{x}_o) \, \delta \dot{x}_o] \, \Delta t$$

$$= p_o \, \delta x_o + \delta L \, \Delta t$$

But the variation of $\Delta S = S - S_o$ is just $\delta \Delta S = p \delta x - p_o \delta x_o$,
which results in

$$\delta \, \Delta S = \delta L \, \Delta t \qquad \text{(8.17)}$$

Extension of the above to three dimensions involves
replacing the action integrals by the expressions

$$S_o = \int (p_{ox} \, dx_o + p_{oy} \, dy_o + p_{oz} \, dz_o)$$

and
$$S = \int (p_x \, dx + p_y \, dy + p_z \, dz)$$

with resulting

$$\delta \Delta S = \sum_{xyz} (p_x \delta x - p_{ox} \delta x_o)$$

$$= \delta L \, \Delta t \qquad\qquad (8.17a)$$

The sum is an exact differential of $L \Delta t = \Delta S$, which will be useful in the further work. The action increment is a function of both initial and final coordinates, i.e.,

$$\Delta S = \Delta S \, (x_o, y_o, z_o, \, x, y, z).$$

If there are two functions $S(x)$ and $S_o(x_o)$ such that $p = dS/dx$ and $p_o = dS_o/dx_o$, or if

$$p \, dx - dS = 0 \text{ and } p_o \, dx_o - dS_o = 0 \qquad (8.18)$$

hold simultaneously and if $\Delta S = S - S_o$, then (8.17) will follow. But (8.18) is the condition for a <u>contact transformation</u> (see below), and (8.17) is a typical representation of one. The derivation of (8.17) was for the infinitesimal transformation (8.16), but it holds for finite transformations as well, since they can be constructed from many infinitesimal ones, for each of which (8.17) holds. In the following, we shall, unless otherwise specified, set $S_o = 0$ and $\Delta S = S$. This will be convenient for treating trajectories which issue from a point.

Geometry of a Contact Transformation

Whereas a transformation $X = X(x,y)$, $Y = Y(x,y)$ can map a point (x,y) into another point (X,Y) and is called a <u>point transformation</u>, a <u>contact transformation</u> is a more involved mapping which has to do with the mapping of one <u>line element</u>

into another. A <u>line element</u> is defined as a point (x,y)
plus a line with slope p through it. We then have to
consider mappings of (x,y,p) into (X,Y,P), defined by
functional equations

$$X = X(x,y,p), \quad Y = Y(x,y,p), \quad P = P(x,y,p)$$

(8.19)

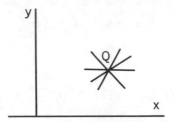

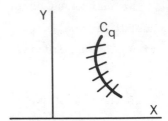

Figure 8.1   Figure 8.2

Lines through a point Q   Map of lines through Q into
fixed in the (x,y) plane.   (X,Y) plane.

If Q(x,y) be held fixed (Figure 8.1) and p be allowed to
vary, the map will be a succession of points C_q: (X,Y) with
a line of slope P through each (Figure 8.2). This by itself
is not particularly useful. But if we are given an arbitrary
curve C: y = y(x) and demand that a point Q plus the tangent
to this curve at Q is mapped into a point Q' plus the tangent
to a corresponding curve C': Y = Y(X), then the result is a
<u>contact transformation</u>, which means that if there is contact
(tangency) originally, there will be finally.

   The restrictive condition on the mapping is now: Whenever
$(dy/dx)_C$ = p, then $(dY/dX)_{C'}$ = P or

$$P \, dX - dY = 0, \quad p \, dx - dy = 0$$

must hold simultaneously.

   Now a point Q_1 on C will be mapped into a curve C_1 and a
neighboring point Q_2 on C into a neighboring curve C_2

(Figures 8.3 and 8.4). If Q_2 is allowed to approach Q_1, then
the line Q_1Q_2 approaches the tangent to C at Q_1, with
$P_1 = (dy/dx)_1$, while the curves C_1 and C_2 become tangent to
an envelope C', and our restriction on the mapping is that
at the point of tangency $P = (dY/dX)_{C'}$.

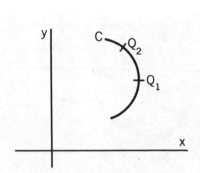

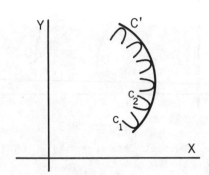

Figure 8.3   Figure 8.4

Curve C in (x,y) Plane   Contact Transformation of C
                                  into (X,Y) Plane. The curve
                                  C' is the envelope of wave-
                                  lets C_1, C_2.

A simple example of a contact transformation is described
by

$$X = x + a \cos \theta$$

$$Y = y + a \sin \theta \qquad (8.20)$$

$$P = p = \tan (\theta + \frac{\pi}{2}) = - \cot \theta$$

with a = constant. Otherwise expressed,

$$(X - x)^2 + (Y - y)^2 = a^2$$

so that the new curve C' (envelope) (Figure 8.5) is at distance a away from the old curve. Since

$$X - x = a \cos \theta$$

and

$$Y - y = a \sin \theta$$

$$P \, dX - p \, dx = p \, d \, (X - x) = a \cos \theta \, d\theta = d(Y-y)$$

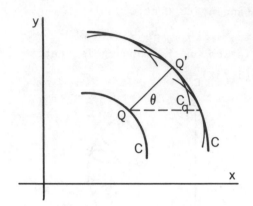

If p is the slope of the tangent to C, the normal has slope $(\tan\theta)_n = - 1/p$, and the map C_q of Q is tangent to the envelope C' with slope P = p at a point Q' on the normal to C.

Figure 8.5

Contact Transformation (8.20)

Motion in Two Dimensions

For 2 dimensions, the contact transformation (C.T.) is generalized (abandoning zero subscripts) to

$$P_x \, dX + P_y \, dY - p_x \, dx - p_y \, dy = dS(X,Y,x,y,).$$

(8.21)

The initial positions are denoted by (x,y) and the final ones by X,Y.

The momentum components are

$$P_x = \partial S/\partial X \quad \text{and} \quad P_y = \partial S/\partial Y \qquad (8.22)$$

while

$$p^2 = |\nabla S|^2 \qquad (8.23)$$

For simple systems where $P_x = m\dot{x}$ and $P_y = m\dot{y}$, the motion will be in the direction $\vec{\nabla S}$. Equation (8.23) is of the same form as (8.14), but we now have the additional information that the particle moves normal to a surface of constant action.

Huygens' Construction

If, then, we construct a succession of curves (2-dimensional case) on which the action is constant, the normals will provide the trajectories. This construction is easy when the Hamiltonian does not depend explicitly on the time and the particles all have the same "total" energy h. Assume that the particles are initially moving perpendicular to some surface, or emanating from a point, at which the action S can be taken as constant. Then calculate $p^2 = 2m(h - U)$ as a function of x and y. Let ΔS be a conveniently small constant, and let n denote the direction of motion (normal to the surface). Then, from (8.23), the distance traversed by the particle will be $\Delta n = \Delta S/p$, which can be calculated for each initial point. Around each such point, draw an arc of a circle (Huygens wavelet) with radius Δn and center at the point. The envelope of these arcs will be a curve on which the action equals the constant

$$S + \Delta S \text{ (or } S - \Delta S).$$

Proceeding step by step in this manner, we can obtain the desired family and hence all the trajectories at once. This is a powerful method for studying where particles can go, and whether or not there are forbidden regions.

During the time Δt, the increase in action, from (8.17a), is

$$\Delta S = L\Delta t = (h - 2U)\Delta t.$$

Since U is not in general constant, the increase of action during Δt will not be constant, so that the above curves are not such that the time is constant on them. However, this is of no importance in determining how the particles move in space.

Imaging Relations

For optical systems, it is important to know those properties of object space which are preserved in the image space. Since the same mathematics applies to dynamical systems, it is likewise important here to discover what invariant properties may exist. The discussion will begin with a treatment of an ordinary second order differential equation, where the invariance is easily demonstrated, and will then be generalized to Hamiltonian systems and their Lagrange invariants.

Suppose that a particle of mass m would ordinarily move in a straight line (as along the axis of a cathode-ray tube) with velocity v, so that the distance x = vt, and that a transverse force - qy acts upon it, where y is the sideways displacement and q is some continuous function of x. Then

$$m\ddot{y} = - qy \qquad (8.24)$$

Choosing units so that v = 1,

$$m \, (d^2y/dx^2) + qy = 0 \qquad (8.25)$$

If y_1 and y_2 are two independent solutions of (8.25),

$$y_1 \, (my_2')' - y_2 \, (my_1')' = 0$$

which has the integral

$$y_1 \, my_2' - y_2 \, my_1' = \text{const} \qquad (8.26)$$

or

$$m \, W \, (y_1, y_2) = \text{const} \qquad (8.27)$$

where W, the Wronskian, equals $y_1 \, y_2' - y_2 \, y_1'$.

Equation (8.27) is true for any equation of the form

$$(my')' + qy = 0$$

where m is a continuous function of x.

The transverse momentum is p_y = m dy/dx (in these units). Equation (8.26) can be rewritten as

$$y_1 \, p_{y_2} - y_2 \, p_{y_1} = \text{const.} \qquad (8.28)$$

The function $y_1(x)$ is a displacement, or a _variation_, from the zero solution and it will be convenient to use the notation dy for it, and correspondingly dp_y for p_{y_1}. Similarly for the independent variation y_2, we shall use δy for it and δp_y for p_{y_2}. In this notation (8.28) becomes

$$dy \, \delta p_y - \delta y \, dp_y = \text{const.} \qquad (8.29)$$

Now specialize the solutions as shown in Figure 8.6. Let the initial values be

$$dy = 1 \quad dp_y = 0$$
$$\delta y = 0 \quad \delta p_y = m \qquad (8.30)$$

Suppose that δy again becomes equal to zero and that then dy = - M. Then, from (8.29)

$$(dy \, \delta p_y)_i = (dy \, \delta p_y)_f$$

or

$$M(\delta p_y)_f = -m \qquad (8.31)$$

or

$$(\delta y')_f = - 1/M \qquad (8.32)$$

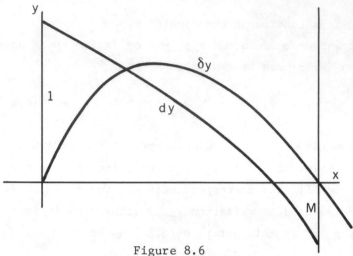

Figure 8.6

Varied Motion, dy and δy, vs. x

Equation (8.32) shows the relation between the linear magnification M and the slope of δy, which represents the angular magnification, since the initial slope of δy is unity.

Area Preservation in Phase Plane (Lagrange Invariant)

Let us now obtain the imaging relations for a Hamiltonian
system, characterized by the contact transformation

$$P \ dX - p \ dx = dS \qquad\qquad (8.18')$$

The variations will be taken to be differentials denoted
by d and δ as above. The initial state of motion will be
represented by coordinates in the (x, p) phase plane, and the
subsequent state by the coordinates (X, P). The variations
are shown in Figure 8.7 as the differential vectors $\vec{d}$ and $\vec{\delta}$,
and x + dx + δ(x + dx) = x + δx + d(x + δx) so that $\delta dx = d\delta x$.

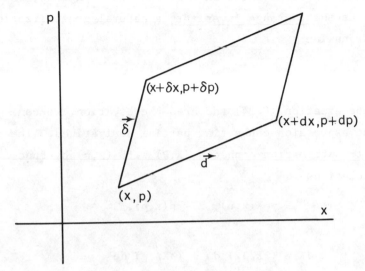

Figure 8.7

Variations d and δ in Phase Plane (x,p)

The area of the parallelogram is $|d \times \delta|$ = dx δp − δx dp. Variations of dS = PdX − pdx and δS = PδX − pδx give

$$\delta P\ dX + P\ \delta dX - \delta p\ dx - p\ \delta dx = \delta dS$$
$$= dP\ \delta X + P\ d\delta X - dp\ \delta x - p\ d\delta x = d\delta S \qquad (8.33)$$

But the differentials commute with each other, so that

$$dX\ \delta P - \delta X\ dP = dx\ \delta p - \delta x\ dp \qquad (8.34)$$

Equation (8.34), which is just (8.29) with different notation, states that <u>area in the phase plane</u> is preserved under a contact transformation and <u>is a constant of the motion</u>.

In more than one dimension, if the coordinates in phase space are $(x_1 \ldots \ldots x_n,\ p_1 \ldots \ldots \ldots p_n)$, representable by the vectors $\vec{x}$ and $\vec{p}$, equation (8.34) is replaced by the assertion of the time invariance of $d\vec{x} \cdot \delta\vec{p} - \delta\vec{x} \cdot d\vec{p}$, which is known as the <u>Lagrange invariant</u>, a natural generalization of the Wronskian.

Poisson Brackets

From the equation (8.34) for area preservation, we can obtain an expression connecting partial derivatives. The contact transformation connects (X,P) with (x,p) by functional relations

$$X = X(x,p), \quad P = p(x,p),$$

so that

$$dX = (\partial X/\partial x)\ dx + (\partial X/\partial p)\ dp$$

$$dP = (\partial P/\partial x)\ dx + (\partial P/\partial p)\ dp$$

Inserting these in (8.34) and comparing coefficients,

$$\frac{\partial X}{\partial x}\frac{\partial P}{\partial p} - \frac{\partial X}{\partial p}\frac{\partial P}{\partial x} = 1 \qquad (8.35)$$

The left hand side is known as a Poisson Bracket and is written $[X,P]$. Equation (8.35) is the condition that the transformation shall be a contact one, of form (8.17). For more variables, there will be more conditions, but these can be obtained by the same procedure as gave (8.35) from the Lagrange invariant.

Transformation of Hamiltonian Systems

If it is desired to transform from one coordinate system to another, or to use some variable other than time as the independent one, the most convenient and powerful attack is that

which uses associated differential forms and the ideas which
are connected with contact transformations.

Start with some differential form

$$\sum_i X_i \; dx_i = f_d \qquad\qquad (8.36)$$

and calculate

$$\delta f_d - df_\delta = \sum_i (\delta X_i \; dx_i - dX_i \; \delta x_i)$$

$$= \sum_{i,j} \left(\frac{\partial X_i}{\partial x_j} - \frac{\partial X_j}{\partial x_i} \right) \delta x_j \; dx_i$$

$$= \sum_{ij} a_{ij} \; dx_i \; \delta x_j$$

This expression is known as the <u>bilinear covariant</u> (B.C.)
of the form, and vanishes if the form is an exact differential.
If a transformation is made to variables y_i and the form
becomes $\sum_i Y_i \; dy_i$, the bilinear covariant will not change in
value, it being immaterial what variables are used to express
it. Its new expression will be $\sum_{ij} b_{ij} \; dy_i \; \delta y_j$, where $b_{ij} =$
$(\partial Y_i / \partial y_j) - (\partial Y_j / \partial y_i)$.

The condition for a contact transformation is that

$$\sum_i (P_i \; dX_i - p_i \; dx_i) = dS$$

where dS is an exact differential. But the B.C. of an exact
differential equals zero, so that

$$\text{B.C.} \left(\sum_i P_i \; dX_i \right) = \text{B.C.} \left(\sum_i p_i \; dx_i \right)$$

or

$$\sum_i (\delta P_i \; dX_i - dP_i \; \delta x_i) = \sum_i (\delta p_i \, dx_i - dp_i \, \delta x_i)$$

or

$$\delta \vec{P} \cdot d\vec{X} - d\vec{P} \cdot \delta \vec{X} = \delta p \cdot dx - dp \cdot \delta x$$

Thus, the motion of a Hamiltonian system, which is characterized by the unfolding of a contact transformation, has an
associated <u>Lagrange invariant</u> $\delta\vec{p}\cdot\vec{dx} - \vec{dp}\cdot\delta\vec{x}$.

The B.C. vanishes not only if $a_{ij} = 0$, but also if the
variations δx_j are independent and arbitrary and the following equations connecting the dx_i hold:

$$\sum_i a_{ij} \, dx_i = 0 \qquad\qquad (8.37)$$

This set of equations is known as the <u>First Pfaff's System</u>
for the form (8.36).

When the change to the above variables y_i is made, the new
first Pfaff's system is

$$\sum_i b_{ij} \, dy_i = 0 \qquad\qquad (8.38)$$

since the variations δy_j are independent and the B.C.
vanishes. Equations (8.38) are completely equivalent to
(8.37)

<u>Hamilton's canonical equations are a consequence of the</u>
<u>vanishing of the B.C. of the differential form</u>

$$\sum_r p_r \, dq_r - H \, dt. \qquad\qquad (8.39)$$

The B.C. is $\sum_r (\delta p_r \, dq_r - dp_r \, \delta q_r) - (\delta H dt - dH\delta t)$. Setting
the coefficients of δq_r, δp_r and δt equal to zero,

$$- dp_r - \frac{\partial H}{\partial q_r} \, dt = 0$$

$$dq_r - \frac{\partial H}{\partial p_r} \, dt = 0$$

$$dH - \frac{\partial H}{\partial t} \, dt = 0$$

These are just Hamilton's Canonical Equations

$$\dot{q}_r = \frac{\partial H}{\partial p_r} \qquad \dot{p}_r = -\frac{\partial H}{\partial q_r} \tag{8.40}$$

plus a third equation which follows from these.

If the differential form be $X_1\, dx_1 + X_2\, dx_2 + X_3\, dx_3$

$$\tag{8.41}$$

then

$$\frac{dx_1}{dx_3} = -\frac{\partial X_3}{\partial X_1}, \quad \frac{dX_1}{dx_3} = \frac{\partial X_3}{\partial x_1} \tag{8.42}$$

are the canonical equations for x_1 and X_1. So, to find the equations of motion in a new system of coordinates, just transform the differential form (8.39) and find a new form such as (8.41), and then write down the canonical equations such as (8.42) immediately. This is a very neat method and saves much time.

Any contact transformation will preserve the form of the canonical equations and can therefore be called a canonical transformation (i.e. one which leaves such equations canonical). The proof of this utilizes the results just obtained.

Let the transformation be from (q_r, p_r, t) to (Q_r, P_r, t) such that

$$\sum_r P_r\, dQ_r - \sum_r p_r\, dq_r = dS = dW - \frac{\partial W}{\partial t}\, dt$$

where W is a function which can depend on time as well as on the q_r's and Q_r's. Then

$$\Sigma p_r\, dq_r - H\, dt + dW = \Sigma P_r\, dQ_r - K\, dt \tag{8.43}$$

where

$$K = H - \frac{\partial W}{\partial t} \tag{8.44}$$

The canonical equations in the new system are then

$$\dot{Q}_r = \frac{\partial K}{\partial P_r}, \quad \dot{P}_r = -\frac{\partial K}{\partial Q_r} \qquad (8.45)$$

since the exact differential dW does not affect the B.C. and
the differential form (8.43) is determining. As advertised,
the form of the canonical equations is preserved by the
contact transformation (8.42).

The value of $\partial W/\partial t$ can be chosen rather arbitrarily. If
$\partial W/\partial t = 0$, then $K = H$, i.e. the new Hamiltonian equals the
old one. If $\partial W/\partial t = E$, then $K = H - E = 0$ for the actual
motion. In this case $\dot{Q}_r = 0$ and $\dot{P}_r = 0$, so that the transformation is to a set of variables (Q_r, P_r) which are constant
during the motion.

This may be illustrated for the <u>harmonic oscillator</u> with
unit mass and frequency. The Hamiltonian is

$$H = \frac{1}{2}(p^2 + q^2) = E = a^2/2, \text{ say.} \qquad (8.46)$$

We shall try to determine a function $S(q, E)$ describing a
contact transformation $dS = p\, dq + \phi\, dE$, where ϕ and E will
be the new canonically conjugate variables. That is,
$p = \partial S/\partial q$ and $\phi = \partial S/\partial E$. Integrating, and using (8.46)

$$S = \int p\, dq = \int (a^2-q^2)^{1/2}\, dq = \frac{1}{2}[q(a^2-q^2)^{1/2} + a^2 \sin^{-1} q/a] \qquad (8.47)$$

$$\phi = \int (\partial p/\partial E)\, dq = \int dq/(a^2-q^2)^{1/2} = \sin^{-1}q/a = \tan^{-1} q/p \qquad (8.48)$$

Combining (8.47) and (8.48)

$$S = \frac{1}{2}q^2 \cot \phi + E\phi, \quad p = q \cot \phi$$

If $\partial W/\partial t = 0$, then $K = H = E$, and

$$\dot{\phi} = \partial H/\partial E = 1, \quad \dot{E} = -\partial H/\partial \phi = 0 \qquad (8.49)$$

so that $\phi = t + \alpha$, $E = $ const, and

$$q = (2E)^{1/2} \sin (t + \alpha) \qquad (8.50)$$

which gives q explicitly in terms of t and thus describes the motion completely.

If $\partial W/\partial t = E$, then $K = H - E = 0$, and

$$\dot{Q} = \frac{\partial K}{\partial E} = \frac{\partial (H - E)}{\partial E} = \dot{\phi} - 1 \qquad (8.51)$$

Hence $Q = \phi - t = \alpha$, so that the coordinate Q is just the integration constant used above. Writing the canonical equations in the form (8.51) rather than (8.49) is largely a matter of taste.

For the harmonic oscillator, it would of course be easier to proceed directly from the equation $dq/dt = \dot{q} = p = (2E - q^2)^{1/2}$ and to integrate to obtain $t - t_o = \int dq/(2E-q^2)^{1/2}$. But in more complicated cases it can be simpler to use (8.48) in the form $t - t_o = \int (\partial p/\partial E) \, dq$.

Reduction of the order of a Hamiltonian system with the aid of the energy integral.

The knowledge that an energy integral exists may, as we now show, enable us to reduce the number of simultaneous equations which it is necessary to solve. This may have some practical value, but with the availability of modern computers it is not really imperative to worry about the order of the system of equations. In fact, the reduction to a lower order is likely to introduce cusps in the solutions, and these will be difficult to handle by computers, whereas, if the reduction be not made, the original system can be integrated readily. For general interest, however, we indicate here how the reduction may be accomplished.

Consider a dynamical system with n degrees of freedom, and where $\partial H/\partial t = 0$. Energy is conserved, and we have

$$H\ (q_1\ q_2 \cdots q_n,\ p_1,\ p_2 \cdots p_n) = E$$

This equation may be solved for p_1, with the result

$$p_1 = -\ K\ (q_2 \cdots q_n,\ p_2 \cdots p_n,\ q_1,\ E)$$

In the original system, the time was chosen as a parameter, and the problem was to find functions $q_1\ (t)$, $q_2\ (t) \cdots$, $p_1\ (t)$, $p_2\ (t) \cdots$ which would represent the motion. One can equally well choose q_1 to be the parameter, and this we now do.

The differential form $\sum\limits_{i=1}^{n} p_i\ dq_i - H\ dt$ becomes

$$\sum\limits_{i=2}^{n} p_i dq_i - E\ dt - K\ (q_2 \cdots q_n,\ p_2 \cdots p_n,\ q_1,\ E)\ dq_1$$

The function K now takes the place of H originally, and it follows from (8.42) that

$$\frac{dq_r}{dq_1} = \frac{\partial K}{\partial p_r}, \ \frac{dp_r}{dq_1} = -\ \frac{\partial K}{\partial q_r} \quad (r = 2, 3, \cdots n) \qquad (8.52)$$

$$\frac{dt}{dq_1} = \frac{\partial K}{\partial E}, \ \frac{dE}{dq_1} = 0 \qquad\qquad (8.53)$$

Equations (8.52) may be used to determine where the particles go (i.e. their trajectories), but do not furnish information on the time required. This may be found from Equation (8.53).

It should be noted that the derivative $\partial K/\partial p_r$ is taken with all the other p's and all the q's, including q_1, kept constant. This corresponds to the canonical equations, where the time derivatives of q and p are equal to partial derivatives of H, t being held constant along with the other q's and p's.

Example of the use of the reduced canonical equations

The Slit Problem: What is the differential equation for the trajectory of a charged particle moving in the field of two infinite conductors separated by a slit of negligible width? One conductor is at potential V_1, the other at potential V_2, as shown in Figure 8.8.

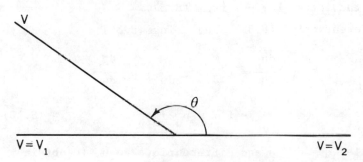

Figure 8.8

Potentials in the Slit Problem

It is convenient to make the conformal transformation $u = \log r$, $v = \theta$. Let the charge be unity.

The kinetic energy is

$$T = \frac{m}{2} (\dot{r}^2 + r^2 \dot{\theta}^2) = \frac{m}{2} e^{2u} (\dot{u}^2 + \dot{v}^2)$$

The potential energy is $V = V_2 + (V_1 - V_2) (\theta/\pi)$.

The momenta are

$$p_u = me^{2u} \dot{u} \text{ and } p_v = me^{2u} \dot{v} \qquad (8.54)$$

and the kinetic energy satisfies the equation

$$p_u^2 + p_v^2 = 2fT = 2f(E - V) \qquad (8.55)$$

where

$$f = me^{2u} \text{ and } f' = 2f \qquad (8.56)$$

The reduced equations of motion are, from (8.54), (8.55) and (8.52)

$$\frac{dv}{du} = \frac{p_v}{p_u} \qquad (8.57)$$

$$\frac{dp_v}{du} = \frac{\partial p_u}{\partial v} = \frac{f}{p_u} \frac{dT}{dv} \qquad (8.58)$$

$$= \frac{f}{p_v} \frac{dT}{du} \qquad (8.59)$$

where the latter follows from (8.57).

Differentiating (8.55), and using (8.59)

$$p_u \frac{dp_u}{du} = f'T + f \frac{dT}{du} - p_v \frac{dp_v}{du}$$

$$= f'T = p_u^2 + p_v^2 \qquad (8.60)$$

Now let $q = dv/du$ and differentiate re u, to obtain

$$\frac{dq}{du} = \frac{d^2v}{du^2} = q \frac{dq}{dv} = \frac{1}{p_u} \frac{dp_v}{du} - \frac{p_v}{p_u^2} \frac{dp_u}{du}$$

$$= \frac{1}{p_u^2} (f \frac{dT}{dv} - q f'T)$$

$$= (1 + q^2) [- q - (a/2T)] \qquad (8.61)$$

where $T = E - V_2 - av$ and $a = (V_1 - V_2)/\pi$.

This equation will enable us, if we are given initial values of v and $q (= \frac{dv}{du})$, to trace out the trajectory in the v, q plane, so that we find a curve $\frac{du}{dv} = f (v)$. A quadrature (simple integration) will then give u as a function of v, which is the desired solution.

If now, $q = 0$ (motion radial), we see that $(d^2v/du^2) = -a/2T$. If the motion is confined to the region between $v = \frac{3}{4} \pi$ and $v = \pi$, then $V_1 < E < V_2$, so that a is negative. The curvature is positive, which is to be expected when the force is toward the plate with potential V_1.

If the radial Lagrange equation is written in terms of r and θ, it is $\ddot{r} = r\dot{\theta}^2$. This shows that the acceleration is always in the direction of increasing r.

It is permissible, then, for r to have a minimum, and particles sent toward the slit will have such trajectories if $V_1 < E < V_2$ and the plate $\theta = \pi$ is not hit first.

Adiabatic Invariants

If for a plane pendulum the length of the suspending thread be changed very slowly, then the quantity h/ω remains constant, where h is the energy (maximum kinetic energy) and ω is the angular frequency. This result will now be demonstrated as a special case of a more general theorem: If a particle librate in a slowly varying potential well described by V(x), with a single minimum, then the increment of action over a period $\oint p \, dx$ remains constant, or is an <u>adiabatic invariant</u>.

We shall suppose that the system is described by a Hamiltonian $H(x,p,\lambda)$ where λ is a parameter which varies only slightly during the period T of the libration, i.e.

$$T \, d\lambda/dt \ll \lambda.$$

During one period, the Hamiltonian can be taken to be
constant, $H = h$.

The change of energy with time, averaged over a period,
will be

$$< \frac{dh}{dt} > = \frac{1}{T} \oint \frac{\partial H}{\partial t} \, dt = \frac{1}{T} \frac{d\lambda}{dt} \oint \frac{\partial H}{\partial \lambda} \, dt$$

Since λ varies slowly, $\dot{\lambda}$ does too, and may be taken as
constant. For a constant λ, we set the momentum $p = p \, (x,h,\lambda)$
and differentiate $H(x,p,\lambda) = h$ re λ. Then

$$\frac{\partial H}{\partial \lambda} + \frac{\partial H}{\partial p} \frac{\partial p}{\partial \lambda} = 0 \text{ and } \dot{x} = - \frac{\partial H}{\partial p}$$

Substituting in the above equation,

$$< \frac{dh}{dt} > = - \frac{d\lambda}{dt} \frac{\oint (dp/d\lambda) \, dx}{\oint (dp/dh) \, dx}$$

or

$$\oint (\frac{dp}{dh} \frac{d<h>}{dt} + \frac{\partial p}{\partial \lambda} \frac{d\lambda}{dt}) \, dx = 0$$

or

$$< \frac{d}{dt} \oint p\,dx > = 0$$

This shows that the action increment remains approximately
constant as the parameter λ is varied slowly.

The integral $\oint$ p dx represents the area enclosed by the
path in the phase plane and is an ellipse for a harmonic
oscillator. The semi-axes are $(2h/m\omega^2)^{1/2}$ and $(2mh)^{1/2}$, with
enclosed area equal to $2\pi(h/\omega)$. This substantiates the
beginning statement, that for a slow change of the parameter

the energy is proportional to the frequency. The change
causes the semi-axes to change in such a way that the area
is kept constant.

Examples of some adiabatic invariants in a plasma are given
by Schmidt, pp 27–29. One such is the magnetic moment of a
particle in the "guiding-center" system (Schmidt, p 21).

Exercises

1. Particles are shot off, with the same speed but in all
directions, from a point on the ground, and are assumed to
be subject only to a constant gravity field (assume the earth
is flat and that Coriolis forces may be neglected). Using
the Huygens' method, construct the surfaces of constant
action and show how they are reflected from a parabolic
envelope, at which they develop cusps.

2. Referring to Figure 8.8 for the slit problem, let
particles be shot off, all with the same energy, from a point
at $\theta = 150°$, and can only reach at most to $\theta = 120°$. Construct
the surfaces of constant action and the envelope of the
trajectories.

3. If the Lagrangian depends explicitly on the time, so
that there is no energy integral, the simple Huygens'
construction of the text is not possible. Is there, under
these circumstances, any practical and useful way to study
the propagation of a surface of constant action (wave front)?
(How is (8.14), the Hamilton-Jacobi equation for constant
energy, to be generalized?)

General Autonomous Systems

INTRODUCTION

We now proceed to a general study of autonomous systems, namely systems of equations which do not explicitly involve the time. Specifically, the mass will be taken to be constant, and the acceleration, or force per unit mass will be assumed to depend upon both the position and velocity, as follows:

$$\ddot{x} = -\ 2p(x,\dot{x})\ \dot{x}\ -\ q(x,\dot{x})\ x \qquad (9.1)$$

In the very special, but familiar, case of the damped harmonic oscillator, the quantity 2p is a positive constant called the damping coefficient, while q is another positive constant, and the combination $q - p^2$ is ω^2, where ω is the angular frequency.

Equation (9.1) thus has a damping coefficient which can
        upon position and velocity, and we shall allow it to be both positive and negative. The frequency coefficient q will be a similar function of position and velocity. After we have discussed the general theory, we shall take as a specific example the Van der Pol equation, where $2p = -\ \varepsilon(1-x^2)$ and $q = 1$, with ε a positive constant.

The general qualitative nature of solutions of (9.1) may be pictured once we have (a) located the points of equilibrium, (b) determined whether the equilibrium is stable or unstable there, (c) found any periodic orbits which may exist, and

(d) determined whether or not the orbits are stable (for definition see below).

If p and q are positive constants, we know that the amplitude of the oscillations dies down continually and with it the kinetic energy. A system of first order equations equivalent to (9.1) is

$$\dot{x} = y, \ \dot{y} = -2py - qx \qquad (9.2)$$

These may be compared with Hamilton's canonical equations

$$\dot{x} = \partial H/\partial y, \ \dot{y} = - \ \partial H/\partial x$$

from which

$$(\partial \dot{x}/\partial x) + (\partial \dot{y}/\partial y) = 0 \qquad (9.3)$$

Condition (9.3) is not satisfied for the system (9.2), so that this is not a Hamiltonian system, and there is no conservation of energy.

However, systems such as (9.2) are of great importance, and we need an appropriate theory to handle them. A very effective method of procedure is to eliminate the time and to find out how the trajectories in the (x,y) phase plane look. For the system (9.2), we have

$$\frac{dy}{dx} = - \ \frac{qx + 2py}{y} \qquad (9.4)$$

or more generally, if $\dot{x} = X(x,y)$ and $\dot{y} = Y(x,y)$,

$$\frac{dy}{dx} = \frac{Y(x, \ y)}{X(x,y)} = f(x,y) \qquad (9.5)$$

where $X(x,y)$ and $Y(x,y)$ are some non-linear functions of x and y.

Now as long as $f(x,y)$ can be determined for the point (x,y) this gives the slope of the trajectory through the point, and so one can easily construct a graphical solution giving dx/dt as a function of x, which leads to the desired solution t(x).

However, this procedure breaks down when $f(x,y)$ is indeterminate, which occurs at points of equilibrium. In this case, there may be many trajectories through such a point, or there may be none. We shall see below what to expect, depending on the particular equation of motion.

The system will be at rest, or in equilibrium, when $\dot{x} = 0$ and $\dot{y} = 0$, or when $X = 0$ and $Y = 0$ simultaneously. Let us assume that the origin is an equilibrium point, and that in its neighborhood the functions X and Y are expressible as ascending power series in x and y, starting with linear terms. For the moment, let us neglect the higher order terms, and take

$$\dot{x} = X = ax + by, \quad \dot{y} = Y = cx + dy \qquad (9.6)$$

[The case where X and Y are homogeneous polynomials of degree m has been studied in considerable detail, but the results from solving Equation (9.6) will serve to indicate what will happen in the more general case.]

Equation (9.5) is now

$$\frac{dy}{dx} = \frac{cx + dy}{ax + by} = f\ (x,y) \qquad (9.7)$$

Since both numerator and denominator equal zero at $x = 0$, $y = 0$, $f\ (x,y)$ is indeterminate and $x = 0$ $y = 0$ is called a singular point (as opposed to an ordinary point, through which there is just one trajectory, according to the existence theorem, Appendix A). Further specification of a, b, c, and d is needed, as we shall now see, to ascertain how many trajectories do go through the origin when it is a singular point, as here.

LINEAR SYSTEMS

The motion near a singular point will depend upon the
coefficients of the linear transformation (9.6). Accordingly,
we shall devote this section to finding out what this dependence signifies in terms of geometry. It will be convenient
to change notation by setting

$$x_1 = x, \; x_2 = y, \; y_1 = \dot{x}_1, \; y_2 = \dot{x}_2$$

$$a_{11} = a, \; a_{12} = b, \; a_{21} = c, \; a_{22} = d. \tag{9.8}$$

Then (9.6) becomes

$$y_1 \equiv \dot{x}_1 = a_{11} \, x_1 + a_{12} \, x_2$$

$$y_2 \equiv \dot{x}_2 = a_{21} \, x_1 + a_{22} \, x_2 \tag{9.9}$$

The transformation (9.9) converts the vector (x_1, x_2) into
another vector (y_1, y_2).

It is possible, as we shall show, to find two linear combinations of x_1 and x_2, which we shall designate as x_1' and x_2',
respectively, such that

$$y_1' \equiv \dot{x}_1' = s_1 x_1'$$

$$y_2' \equiv \dot{x}_2' = s_2 x_2' \tag{9.10}$$

The constants s_1 and s_2 are the roots of the quadratic
equation

$$s^2 - Ts + a = 0 \tag{9.18}$$

with $T = a_{11} + a_{22}$ and $a = a_{11}a_{22} - a_{12}a_{21}$.

Explicitly,

$$s_1 = (T/2) + [(T/2)^2 - a]^{1/2}$$

$$\tag{9.19}$$

$$s_2 = (T/2) - [(T/2)^2 - a]^{1/2}$$

These roots may be real or conjugate complex, their product
equals a, and they will have the same sign (if real) when
a > 0 and opposite signs when a < 0.

Equations (9.10) show that the <u>motion</u> of x_1' is <u>independent</u>
of that of x_2'. Furthermore, these equations can be integrated
immediately to give

$$x_1' = A e^{s_1 t}$$

$$\tag{9.11}$$

$$x_2' = B e^{s_2 t}$$

where A and B are the values of x_1' and x_2' at t = 0. The
constants s_1 and s_2 are called <u>characteristic exponents.</u> A
consequence of (9.11) is that

$$x_2' = B (x_1'/A)^{\sigma} \tag{9.12}$$

with

$$\sigma = s_2/s_1$$

Equations (9.11) show how the coordinates x_1' and x_2' change
with time, while (9.12) shows how they are interrelated. If
they are real, then this equation describes the curves traced
out in the (x_1', x_2') plane. Eventually, we need to find out
what the motions in the original (x_1, x_2) plane are, and for
that will have to determine the linear transformation from
the one plane to the other. This will be done below. For the
present, we shall borrow the result that real values of
s_1 and s_2 lead to real coefficients of the transformation.
Hence, the qualitative nature of the motion in the (x_1, x_2)
plane is the same as that in the (x_1', x_2') plane, the curves

just being distorted somewhat by the transformation. When
(x_1', x_2') are complex quantities, one can still perform the
algebraic operations, but the motion is to be represented by
curves in the (x_1, x_2) plane only.

General Nature of Motion - Real Exponents

Let us consider first the case where A, B, s_1 and s_2 are
real.

Inspection of (9.11) shows that, if $t \to \infty$, then $x_1' \to 0$ if
$s_1 < 0$ and $x_1' \to \infty$ if $s_1 > 0$. A similar statement can be
made about x_2'.

The motion in the plane, as $t \to \infty$, is:

     (a) If $s_1 < 0$, $s_2 < 0$, then $x_1' \to 0$, $x_2' \to 0$;

     (b) If $s_1 > 0$, $s_2 > 0$, then $x_1' \to \infty$, $x_2' \to \infty$;

     (c) If $s_1 < 0$, $s_2 > 0$, then $x_1' \to 0$, $x_2' \to \infty$;

     (d) If $s_1 > 0$, $s_2 < 0$, then $x_1' \to \infty$, $x_2' \to 0$.

According to one definition of stability, a point is <u>stable</u>
if a particle stays near it or moves toward it as $t \to \infty$, and
<u>unstable</u> otherwise. In this sense, then, the origin will be
stable only for case (a).

The actual curve in the (x_1', x_2') plane is given by (9.12)
and its shape is determined by the value of σ. If $\sigma < 0$
(s_1 and s_2 opposite in sign) then the curves are like <u>hyper-</u>
<u>bolas</u>, (see Figure 9.1), with the x_1' and x_2' - axes as asymp-
totes. The origin is then a <u>saddle-point or col or hyperbolic</u>
<u>fixed point</u>. If $\sigma > 0$ (s_1 and s_2 have the same sign) the
curves are like <u>parabolas</u> through the origin, which is called
a <u>node</u>. For $\sigma < 1$, the curves are tangent to the x_2' - axis

at the origin; for $\sigma > 1$, they will be tangent to the
x_1' - axis. Figure 9.2 shows the various kinds of nodes for
$\sigma < 1$ and for $\sigma > 1$. For $\sigma = 1$, a special treatment is
necessary (see below) and the curves are shown in Figure 9.3.

   This is about as far as we can proceed without discussing
the details of the transformation from (x_1, x_2) to (x_1', x_2'),
so those will be presented next, and then applications to
special cases will be made.

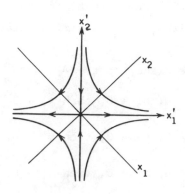

Figure 9.1

Hyperbolic Point $(\sigma < 0)$

Transformation to Independent Coordinates

   As mentioned above, there are linear combinations of x_1 and
x_2, namely x_1' and x_2', which execute motions independent of
each other and specified by (9.10) and (9.11). In this
section, we show in detail how to find these linear combinations and the values of s. There follows (see end of section)
a different version of this treatment, which is a compaction
into matrix language. To begin, introduce a transformation,
with non-zero determinant, from (x_1', x_2') to (x_1, x_2) as
follows: $x_1 = c_{11} x_1' + c_{12} x_2'$, $x_2 = c_{21} x_1' + c_{22} x_2'$. (9.13)

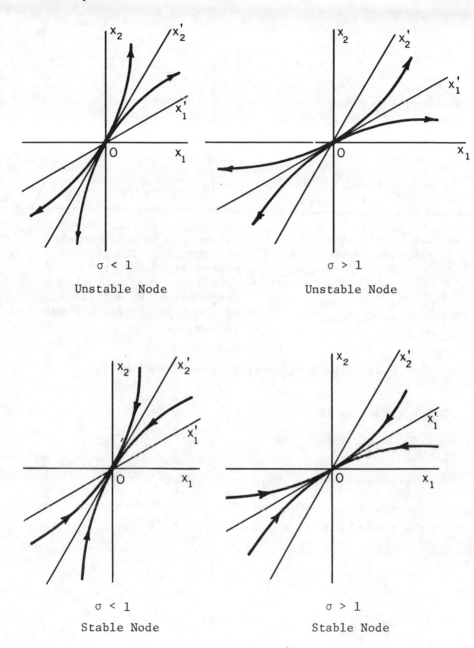

Figure 9.2

Nodal Points

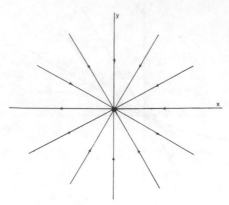

a. The origin is a   b. The origin is a degenerate
    singular stable node.   stable node.   The curves
                                 represent (9.32), for x > 0.
                                 (Similar curves, not shown,
                                 are obtained by substituting
                                 -x for x and -y for y simul-
                                 taneously.)

Figure 9.3

Nodes with Real Equal Eigenvalues ($\sigma = 1$)

Differentiating (9.13) re t, and using (9.10)

$$\dot{x}_1 = c_{11} \, \dot{x}_1' + c_{12} \, \dot{x}_2' = c_{11} \, s_1 \, x_1' + c_{12} \, s_2 \, x_2'$$

$$(9.14)$$

$$\dot{x}_2 = c_{21} \, \dot{x}_1' + c_{22} \, \dot{x}_2' = c_{21} \, s_1 \, x_1' + c_{22} \, s_2 \, x_2'$$

On the other hand, if we substitute (9.13) into (9.9), we
find

$$\dot{x}_1 = (a_{11} \, c_{11} + a_{12} \, c_{21} \,)x_1' + (a_{11} \, c_{12} + a_{12} \, c_{22} \,)x_2'$$

$$(9.15)$$

$$\dot{x}_2 = (a_{21} \, c_{11} + a_{22} \, c_{21} \,)x_1' + (a_{21} \, c_{12} + a_{22} \, c_{22} \,)x_2'$$

Equating (9.14) with (9.15),

$$a_{11} \, c_{11} + a_{12} \, c_{21} = s_1 \, c_{11}$$

$$a_{21} \, c_{11} + a_{22} \, c_{21} = s_1 \, c_{21}$$

$$a_{11} \, c_{12} + a_{12} \, c_{22} = s_2 \, c_{12} \qquad\qquad (9.16)$$

$$a_{21} \, c_{12} + a_{22} \, c_{22} = s_2 \, c_{22}$$

From (9.16), it is apparent that s_1 and s_2 are the roots of
the determinantal equation

$$\begin{vmatrix} a_{11} - s & a_{12} \\ a_{21} & a_{22} - s \end{vmatrix} = 0 \qquad\qquad (9.17)$$

which is quadratic in s, namely

$$s^2 - Ts + a = 0 \qquad\qquad (9.18)$$

with

$$T = a_{11} + a_{22} \quad \text{and} \quad a = \begin{vmatrix} a_{11} & a_{12} \\ a_{21} & a_{22} \end{vmatrix}$$

The roots are

$$s_1 = (T/2) + \{(T/2)^2 - a\}^{1/2}$$

$$\qquad\qquad (9.19)$$

$$s_2 = (T/2) - \{(T/2)^2 - a\}^{1/2}$$

and may be real or conjugate complex. Once they have been
found, one can substitute back into (9.16) to obtain the
ratios

$$c_{11} : c_{21} = - a_{12} : a_{11} - s_1$$

$$\qquad\qquad (9.20)$$

$$c_{12} : c_{22} = - a_{12} : a_{11} - s_2$$

If the roots are real and distinct, the ratios in (9.20) are real and distinct. The corresponding linear combinations of x_1 and x_2 are found by solving (9.13), and are given by

$$c \, x_1' = c_{22} \, x_1 - c_{12} \, x_2$$

$$c \, x_2' = - \, c_{21} \, x_1 + c_{11} \, x_2$$

(9.21)

where

$$c = \begin{vmatrix} c_{11} & c_{12} \\ c_{21} & c_{22} \end{vmatrix}$$

Motions in the Phase Plane Near Equilibrium

We shall now consider, for various linear transformations (9.9), what the types of motion are which can occur near an equilibrium point.

For real exponents, this is an elaboration and justification of the previous treatment; for complex exponents, we show explicitly how to describe the motion in the (x_1, x_2) plane.

When the exponents are real then the ratios c_{21}/c_{11} and c_{22}/c_{12} in (9.20) will be, and if we take c_{11} and c_{12} to be real (which is permissible since they are arbitrary) then the new coordinates x_1' and x_2' will be real, by (9.21), and may be specified by (9.22). The axes $x_1' = 0$ and $x_2' = 0$ will in general be non-orthogonal (skew) to each other. Let us now look at the geometrical relations involved, for a general skew system.

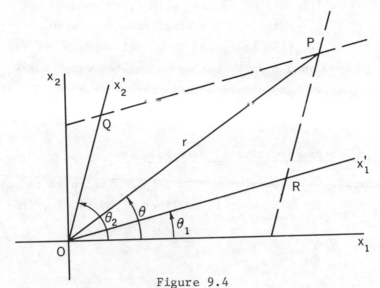

Figure 9.4

Skew Coordinates

Transformation to Skew Coordinates

If we introduce new coordinates x_1' and x_2' and denote by θ_1 and θ_2 the angles from the x_1 - axis to the x_1' - axis and x_2' - axis, respectively, then from Figure 9.4 with $x_1' = OR$ and $x_2' = OQ$, we have

$$x_1' \sin (\theta_2 - \theta_1) = r \sin (\theta_2 - \theta)$$
$$= x_1 \sin \theta_2 - x_2 \cos \theta_2 \qquad (9.22)$$
$$x_2' \sin (\theta_2 - \theta_1) = r \sin (\theta - \theta_1)$$
$$= -x_1 \sin \theta_1 + x_2 \cos \theta_1$$

This is a special case of (9.21), with $c_{11} = \cos\theta_1$, $c_{12} = \cos\theta_2$, $c_{21} = \sin\theta_1$, $c_{22} = \sin\theta_2$ and $c = \sin(\theta_2 - \theta_1)$. The ratios $(c_{21}/c_{11}) = \tan\theta_1$ and $(c_{22}/c_{12}) = \tan\theta_2$ are real if the roots s_1 and s_2 are, but can otherwise take on any values. Once we have obtained these ratios from (9.20), the angles θ_1 and θ_2 are thereby fixed, and hence (9.22).

Condition for Preservation of Orthogonality

In general, the line $x_1' = 0$ will not be orthogonal to the line $x_2' = 0$. The slopes of these lines, from (9.21), are:

$$\text{for } x_1' = 0, \quad \frac{x_2}{x_1} = \frac{c_{22}}{c_{12}} = \frac{a_{11} - s_2}{-a_{12}}$$

$$\text{for } x_2' = 0, \quad \frac{x_2}{x_1} = \frac{c_{21}}{c_{11}} = \frac{a_{11} - s_1}{-a_{12}}$$

For orthogonality, one slope must be the negative reciprocal of the other. That is,

$$(a_{11} - s_2)(a_{11} - s_1) = -a_{12}^2 \qquad (9.23)$$

But

$$s_1 + s_2 = a_{11} + a_{22}$$

and

$$s_1 s_2 = a_{11} a_{22} - a_{12} a_{21}$$

Substituting these values in (9.23) <u>the condition that the new axes be orthogonal is that</u> $a_{21} = a_{12}$. $\qquad$ (9.24)

Examples of Orthogonal Transformations

Assume the initial equations to be

$$\dot{x}_1 = ax_1 + bx_2$$
$$\dot{x}_2 = bx_1 + ax_2$$

(9.25)

where a and b are real constants. By (9.24), the new coordinates x_1' and x_2' will be orthogonal.

Eliminating x_2, we find

$$\ddot{x}_1 - 2a\dot{x}_1 + (a^2 - b^2)\, x_1 = 0 \qquad (9.26)$$

Equation (9.26) is that of a damped harmonic oscillator, (non-oscillatory condition), with negative damping if a > 0, positive damping if a < 0. Its general solution is

$$x_1 = A\, e^{(a+b)t} + Be^{(a-b)t} \qquad (9.27)$$

with A and B arbitrary constants. The variable x_2 also obeys (9.26). In terms of the above formalism, $s_1 = a+b$, $s_2 = a-b$, and from (9.20), $c_{21}/c_{11} = 1$ and $c_{22}/c_{12} = -1$, so that $\tan\theta_1 = 1$ and $\tan\theta_2 = -1$. Then (9.22) becomes

$$x_1' = (1/2)^{1/2}\, (x_1 + x_2)$$
$$x_2' = (1/2)^{1/2}\, (-x_1 + x_2)$$

The coordinates x_1' and x_2' are orthogonal to each other and rotated $\pi/4$ from the original x_1 and x_2.

If a>b>0, then s_1 and s_2 both have the same sign and the origin is an unstable node, as is evident from (9.27). The

motion is as shown in Figure 9.5, outward from the origin.
If $a^2 < b^2$, then s_1 and s_2 must have opposite signs, since
$s_1 s_2 = a^2 - b^2 < 0$. The origin is then a hyperbolic point,
and Figure 9.1 shows the nature of the motion if $x_2' \to 0$ as
$t \to \infty$, which will occur if $b > a > 0$.   [If, on the other hand,
$b < a < 0$, then $x_1' \to 0$ as $t \to \infty$].

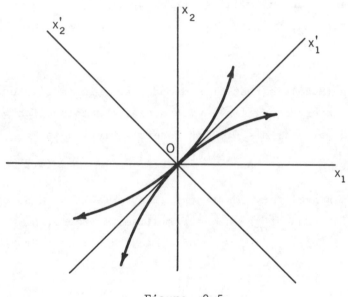

Figure   9.5

Unstable Node

There is a very simple case of (9.26) when a = 0, namely

$$\ddot{x} - b^2 x = 0$$

This is the approximate equation for the angle of deflection of a pendulum from its straight up position. This is an
equilibrium position, for if $\dot{x} = 0$ there, x = 0 and the
particle will stay there, since $\ddot{x} = 0$. The first integral is

$$\dot{x}^2 - b^2 x^2 = h$$

described by hyperbolas near the origin, which is a hyperbolic point.

Now our system of first order equations can be written in
two ways:

$$\text{(a)} \quad \dot{x} = y, \; \dot{y} = b^2 x$$

or

$$\text{(b)} \quad \dot{x} = by, \; \dot{y} = bx$$

In case (b), the hyperbolas are represented by $y^2 - x^2 =$
constant, and their asymptotes have slopes ± 1, so that the
x_1' and x_2' axes are orthogonal. In case (a), when one just
plots $\dot{x}$ vs x, the asymptotes have slopes $\pm b$.

Motion for Real Equal Roots

Having shown that real unequal roots result in either a
hyperbolic point ($\sigma < 0$) or a node ($\sigma > 0$), we now turn our
attention to the case of equal roots. Two examples will
serve to indicate what may be expected. First, the system
$\dot{x} = sx, \; \dot{y} = sy$ has for a solution

$$x = x_o e^{st}, \; y = y_o e^{st}$$

If $s < 0$, then the motion will approach the origin as
$t \to +\infty$, from both x_o, y_o and from $-x_o$, $-y_o$. Every straight
line thru the origin represents two trajectories, and the
origin is called a __singular node__. The motion is as shown
in Figure 9.3a.

Secondly, the system

$$\dot{x} = a_{11}x; \; \dot{y} - a_{11}y = a_{21}x \tag{9.28}$$

has for its solution, with $s = a_{11}$

$$x = x_o e^{st}; \; y = a_{21} x_o t e^{st} + y_o e^{st} \tag{9.29}$$

Therefore

$$\frac{y}{x} = \frac{y_o}{x_o} + a_{21} t \qquad (9.30)$$

and

$$\frac{dy}{dx} = \frac{a_{11} y + a_{21} x}{a_{11} x} = \frac{a_{21}}{a_{11}} + \frac{y_o}{x_o} + a_{21} t \qquad (9.31)$$

If $s < 0$, motion is toward the origin as $t \to \infty$, from Equation
(9.29). If $a_{21} > 0$ the slope is positively infinite there,
from (9.31), but it steadily decreases to negative infinity
as $t \to -\infty$, where, from (9.30), $\frac{y}{x}$ is negative. The equation
of the curves is

$$\frac{y}{x} = \frac{y_o}{x_o} + \frac{a_{21}}{s} \ln \frac{x}{x_o} \qquad (9.32)$$

The origin is here a <u>degenerate node</u>. The motion in the
first quadrant is shown in Figure 9.3b. Similar motion in
the third quadrant is obtained by reflection of this in the
origin.

(For a very thorough discussion of cases of this kind,
where solutions of the type in Equation (9.29) occur, see
N. G. Chetayev, "The Stability of Motion," Pergamon, 1961).

Motion when Exponents are Complex

Let us now examine the motion in a phase plane when the
exponents s_1 and s_2 are conjugate complex roots of (9.18).
If we eliminate x_2 from (9.9), we have for $x = x_1$ the
equation

$$\ddot{x} - T\dot{x} + ax = 0 \qquad (9.33)$$

This is the equation for the damped harmonic oscillator, with
positive damping if $T < 0$. Setting $T = -2b$ and

$$\omega^2 = a - (T/2)^2 = a - b^2,$$

we have the general solution from (9.11) and (9.19) as

$$x = x_o e^{-bt} \cos (\omega t + \delta). \qquad (9.34)$$

Instead of proceeding as we did with real roots, it is more enlightening to construct a representation in terms of polar coordinates. To do this, introduce as a new variable

$$u = \dot{x} + bx \qquad (9.35)$$

Then, from (9.33)

$$\dot{u} = -b(u-bx) - ax$$
$$\qquad (9.36)$$
$$= -bu - \omega^2 x$$

Going to polar coordinates, set

$$\omega x = r \cos 0, \ u = r \sin \theta \qquad (9.37)$$

Differentiating re t

$$\omega \dot{x} = \dot{r} \cos \theta - r \sin \theta \ \dot{\theta}$$
$$\dot{u} = \dot{r} \sin \theta + r \cos \theta \ \dot{\theta}$$

Solving for $\dot{r}$ and $\dot{\theta}$

$$\dot{r} = \omega \dot{x} \cos \theta + \dot{u} \sin \theta$$
$$\qquad (9.38)$$
$$r\dot{\theta} = - \omega \dot{x} \sin \theta + \dot{u} \cos \theta$$

Substituting (9.35) and (9.36) in (9.38),

$$\dot{r} = - br$$

$$\dot{\theta} = - \omega$$

Eliminating the time,

$$\frac{dr}{r} = \frac{b}{\omega} \theta$$

Integrating

$$r = r_o \, e^{b\theta/\omega}$$

which represents a logarithmic spiral in the plane whose
Cartesian coordinates are ωx and $\dot{x} + bx$, and which we take
as our phase plane here. The origin is stable if $b > 0$,
unstable if $b < 0$. In both cases it is encircled an infinite
number of times, and called a <u>focus</u>.

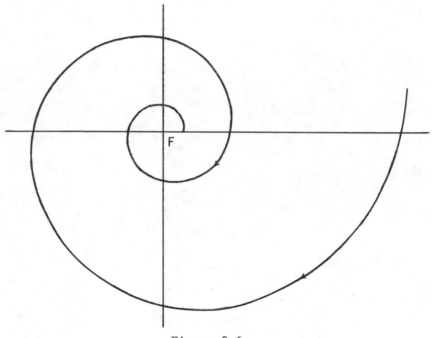

Figure 9.6
Stable Focus

Figure 9.6 shows how, for a stable focus, the spiral is swept
out clockwise with r becoming ever smaller, being diminished
by a fraction exp ($- 2\pi$ b/ω) each time around. For an
unstable focus, r becomes ever larger with time.

Motion when Exponents are Imaginary

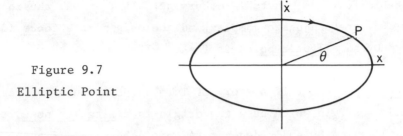

Figure 9.7

Elliptic Point

When the damping is zero, i.e. $T = 0$, Equation (9.33) for $a > 0$ takes the form

$$\ddot{x} + \omega^2 x = 0$$

This integrates to

$$\dot{x}^2 + \omega^2 x^2 = 2h$$

and the motion is along an ellipse around the origin, which is called an __elliptic fixed point__, and is stable because the motion stays around. The explicit solution is of the form

$$x = A \cos (\omega t + \delta)$$

where A and δ are constants.

Summary of Types of Motion Near Equilibrium

Summing up, the origin is a singular point for the linear system (9.6), which is equivalent to the second order differential equation (9.33), with constant coefficients. The solutions are exponential in time, with the characteristic exponents s_1 and s_2 the roots of (9.18), given explicity by

(9.19). Motion in the phase plane depends upon the quanti-
ties $T = a_{11} + a_{22}$ and $a = a_{11} a_{22} - a_{12} a_{21}$. The origin is
stable if $T \leq 0$, unstable otherwise. If $T^2 < 4a$, the roots
s_1, s_2 are conjugate complex and the origin is a <u>focus</u> if
$T \neq 0$, an <u>elliptic point</u> if $T = 0$. If $T^2 > 4a$, the roots are
real and distinct, and the origin is a <u>node</u> if $a > 0$, a
<u>hyperbolic</u> point if $a < 0$. If $T^2 = 4a$, the roots are real
and equal, in which case the origin is a <u>singular node</u> if
$a_{12} = a_{21} = 0$, and a <u>degenerate node</u> otherwise. Figures
9.1 - 9.3 and 9.5 - 9.7 show the various types of motion
which may arise. Since (9.33) describes the damped harmonic
oscillator, further treatment is presented in the next
section, with the aim of correlating the time behavior with
the phase plane behavior.

The system (9.9) will be a <u>Hamiltonian</u> one, from (9.3),
if $a_{11} = -a_{22}$, or $T = 0$. If $q > 0$, the system will oscillate
and the origin is an elliptic point; if $q = -b^2$, then the
origin is a <u>hyperbolic point</u>, which will be approached by
the solution $x = x_o \exp (-bt)$ and left by the solution
$x = x_o \exp bt$, $(b > 0)$, the exponents being real.

Matrix Formulation of Transformation

Although the preceeding discussion is sufficient to
characterize the motion in the phase plane, it is useful to
show how the treatment can be made more compact by adopting
matrix notation.

We start with the transformation (9.9) in the form

$$y \equiv \dot{x} = Ax \qquad \qquad (9.9a)$$

where

$$A = \begin{pmatrix} a_{11} & a_{12} \\ a_{21} & a_{22} \end{pmatrix}, \quad x = \begin{pmatrix} x_1 \\ x_2 \end{pmatrix}, \quad \dot{x} = \begin{pmatrix} \dot{x}_1 \\ \dot{x}_2 \end{pmatrix},$$

and

$$y = \begin{pmatrix} y_1 \\ y_2 \end{pmatrix} \text{ are matrices.}$$

A transformation from x' to x is introduced, namely

$$x = Cx', \quad C = \begin{pmatrix} c_{11} & c_{12} \\ c_{21} & c_{22} \end{pmatrix} \qquad (9.13a)$$

or

$$x' = C^{-1}x \qquad (9.21a)$$

The problem is now to determine C such that

$$y' = \dot{x}' = Sx' \qquad (9.10a)$$

where

$$S = \begin{pmatrix} s_1 & 0 \\ 0 & s_2 \end{pmatrix}, \text{ a diagonal matrix.}$$

Differentiating (9.13a) re t, and using (9.10a)

$$\dot{x} = C\dot{x}' = CSx' \qquad (9.14a)$$

But if we combine (9.13a) with (9.9a), we have

$$\dot{x} = Ax = ACx' \qquad (9.15a)$$

Equating (9.14a) with (9.15a), since x' is arbitrary,

$$AC = CS \qquad (9.16a)$$

or

$$C^{-1} AC = S \qquad (9.16b)$$

The values of s_1 and s_2, the diagonal elements of S, are found by solving the determinantal equation (9.17). These values are called the eigenvalues of the matrix A. Corresponding to s_1, there is an associated eigenvector with components (c_{11}, c_{21}) and to s_2 there is the eigenvector with components (c_{12}, c_{22}). These eigenvectors are represented by the columns of the matrix C. It is thus possible to determine a matrix C which will reduce A to diagonal form S, and the reduction is expressed by (9.16b). A transformation of the form $C^{-1}A\,C$ is called a similarity transformation. The trace (sum of the diagonal elements) remain unaltered by such a transformation, for $s_1 + s_2 = T = a_{11} + a_{22}$, by (9.19).

THE DAMPED HARMONIC OSCILLATOR

(MOTION IN THE PHASE PLANE)

We can now apply the preceding considerations to the damped harmonic oscillator. Set q = 1 (in 9.1), which amounts to taking the unit of time so that an undamped oscillator would have unit angular frequency $\omega = 1$. Then, from (9.2), (9.6) and (9.8), $a_{11} = 0$, $a_{21} = -1$, $a_{22} = -2p$, $a_{12} = 1$.

Equation (9.18) now becomes

$$s^2 + 2ps + 1 = 0 \qquad\qquad (9.39)$$

For the moment, assume p > 0, which corresponds to positive damping. Similar arguments can be advanced for negative damping, where p < 0.

Case I p > 1 (a) Motion near the origin

If p is greater than unity, let p = ch ψ. Then the solutions of (9.39) are $s_1 = -\,$ch $\psi + $ sh ψ, $s_2 = -\,$ch $\psi - $ sh ψ

which are both negative. Also

$$\sigma \equiv s_2/s_1 > 1.$$

The normal coordinates obey (9.11), and the inverse transformation is given by (9.13). We have, if A_1 and A_2 are constants,

$$x_1' = A_1 \, e^{s_1 t}, \; x_2' = A_2 \, e^{s_2 t} \qquad (9.40)$$

from which

$$x_2' = k \, (x_1')^{\sigma}$$

This represents a family of parabolas tangent to the x_1' axis at the origin, and the motion is inward as $t \to \infty$, so that the origin is a stable node. (See Figures 9.2 and 9.9.)

The original coordinates, from (9.27), are

$$x \equiv x_1 = x_1' + x_2' = A_1 \, e^{s_1 t} + A_2 \, e^{s_2 t} \qquad (9.41)$$

$$\dot{x} = y \equiv x_2 = s_1 \, x_1' + s_2 \, x_2' = A_1 \, s_1 \, e^{s_1 t} + A_2 \, s_2 \, e^{s_2 t}$$

Now, as t becomes greater, the term $\exp s_2 t$ damps out faster, since $|s_2|$ is greater. So at $t = -\infty$ this term will be dominant, while at $t = +\infty$ the term with $\exp s_1 t$ will dominate. Neglecting one term or the other, we have the limiting cases, from (9.41) with $v = \dot{x}/x$,

$$v = \dot{x}/x \to s_2 \text{ as } t \to -\infty$$
$$\qquad (9.42)$$
$$v = \dot{x}/x \to s_1 \text{ as } t \to +\infty$$

If $A_2 = 0$, then the motion stays along the line $\dot{x} = s_1 x$; if $A_1 = 0$, the motion is along $\dot{x} = s_2 x$; in the general case $A_1 \neq 0$, $A_2 \neq 0$ the asymptotic relations (9.42) hold. As a function of t, the motion is of two kinds, one if A_1 and A_2

have the same sign, and one if A_1 and A_2 have opposite signs.
These are shown in Figures (9.8a) and (9.8b), respectively.

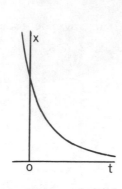

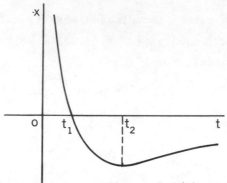

<table>
<tr><td align="center">Figure 9.8(a)</td><td align="center">Figure 9.8(b)</td></tr>
<tr><td align="center">Damped Harmonic Oscillator
p > 1
Steady Decrease
of Amplitude</td><td align="center">Damped Harmonic Oscillator
p > 1
Amplitude Changes Sign</td></tr>
</table>

When A_1 and A_2 have the same sign, the amplitude of x
continually decreases as t increases, but does not become
zero for a finite value of t.

On the other hand, if A_1 and A_2 have opposite signs, then
$x = 0$ at $t = t_1$, where from (9.41)

$$e^{(s_2 - s_1)t_1} = - (A_1/A_2).\qquad(9.43)$$

At this time

$$\dot{x} = \dot{x}_1 = A_2\, e^{s_2 t_1}\, (s_2 - s_1)\qquad(9.44)$$

which is not zero, since $s_2 \neq s_1$. Actually, $\dot{x} < 0$ if $A_2 > 0$.

The amplitude of the motion thus changes sign at $t = t_1$,
the slope (velocity) being finite. As t increases, the
motion with exponent s_2 damps out faster, and then the motion
approaches that with exponent s_1. If x starts out positive

it takes on its minimum, and $\dot{x} = 0$, at $t = t_2$, where

$$e^{(s_2 - s_1)t_2} = - A_1 s_1 / A_2 s_2 \qquad (9.45)$$

The corresponding value of x is

$$x = x_2 = [-A_2(s_2/s_1) + A_2]\, e^{s_2 t_2}$$

$$\qquad (9.46)$$

$$= (s_1-s_2)\,(A_2/s_1)e^{s_2 t_2}$$

Trajectories in the Phase Plane (p>1)

To obtain a qualitative picture of the motion in the phase plane, it is useful to examine the ratio $v \equiv \dot{x}/x$. We have, from (9.1) with $q = 1$,

$$\ddot{x} = \dot{x}\,(d\dot{x}/dx) = \dot{v}x + v\dot{x} = -2p\dot{x} - x \qquad (9.47)$$

Two equations follow, namely (dividing by x and $\dot{x}$)

$$\dot{v} = -v^2 - 2pv - 1 = -(v-s_1)(v-s_2) \qquad (9.48)$$

$$\frac{d\dot{x}}{dx} = -2p - \frac{x}{\dot{x}} = -2p - \frac{1}{v} \qquad (9.49)$$

The slope at $x = 0$, $\dot{x} \neq 0$, is $-2p$ and is negative. The second derivative is

$$\frac{d^2\dot{x}}{dx^2} = \frac{1}{v^2}\frac{dv}{dx} = \frac{1}{v^2}\frac{\dot{v}}{\dot{x}} = \frac{1}{v^3}\frac{\dot{v}}{x}$$

$$= -(v-s_1)(v-s_2)/v^3 x \qquad (9.50)$$

There are two types of curves, as we have seen in Figures (9.8a) and (9.8b). For the former, the slope v varies

between s_2 at $t = - \infty$ to s_1 at $t = + \infty$, while x goes from
$+ \infty$ to 0. As we see from (9.48), for $s_2 < v < s_1$, the slope
v increases with t, which is consistent with the previous
statement.

For a curve of type 9.8b, we start at $t = - \infty$, $x = \infty$, and
let v decrease from s_2. At $t = t_1$, given by (9.43), the
negative $\dot{x}$-axis is crossed at $\dot{x} = \dot{x}_1$: (9.44), with slope
$d\dot{x}/dx = - 2p = s_1 + s_2$ (9.49). At $t = t_2$, given by (9.45),
the negative x-axis is crossed at $x = x_2$: (9.46), with
infinite slope, since $\dot{x} = 0$: (9.49).

Now let us see how the second derivative behaves, from
(9.50) and (9.48). If the ratio $d^2\dot{x}/dx^2$: $\dot{x}$ is negative,
the trajectory is concave toward the x-axis; if the ratio
is positive, the curve is convex toward the x-axis. Now,
from (9.50) this ratio has the same sign as

$$\dot{v} = - (v-s_1)(v-s_2) \qquad\qquad (9.48)$$

But our region consists of two parts: (1) $v < s_2 < s_1$
when $x > 0$, and (2) $v > s_1 > s_2$ when $x < 0$. The sign of $\dot{v}$
is negative in both parts, and so the trajectory is <u>concave
to the x-axis</u> from $t = - \infty$ to $t = + \infty$ (when it is asymptotic
to the line $v = s_1$.

Trajectories of both kinds are shown in Figure 9.9 (a,b).
It is seen that the origin is a stable node, as we had
already deduced in (9.40), with the trajectories tangent to
the line $v = s_1$ (x_1' - axis).

(b) Motion At Infinity

To complete the picture in the whole phase plane, we need
to describe the motion at infinity. The standard way is to
set $z = 1/x$, to write the equations of motion in terms of

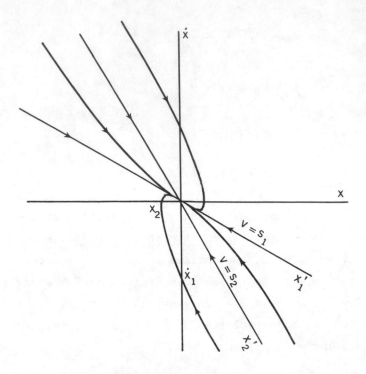

Figure 9.9(a)

Phase Plane Trajectories, p > 1, in Finite Plane

z and v and to find the singular points.   The equations are

$$\dot{z} = -\dot{x}/x^2 = -vx/x^2 = -zv \qquad (9.51)$$

$$\dot{v} = -(v-s_1)(v-s_2) \qquad (9.48)$$

The singular points are at

$\quad$ z = 0, v = s_1; z = 0, v = s_2, with 0 > s_1 > s_2.

Along v = s_1, the solution is

$$|z| = |A_1|^{-1} e^{-s_1 t} \qquad (9.52)$$

Dividing $\dot{x}$ by x in (9.41), rearranging, and setting a = $-A_2/A_1$,

we have

$$\frac{v-s_1}{v-s_2} = a\,e^{(s_2-s_1)t} \tag{9.54}$$

Using (9.52) we have near $v = s_1$

$$v - s_1 \cong (s_1-s_2)\,a\,e^{(s_2 - s_1)t}$$

$$= - (s_1-s_2)(A_2/A_1)\,\left|A_1 z\right|^{1-\sigma}$$

This equation defines a family of hyperbolas at $z = 0$, $v = s_1$, which is therefore a __hyperbolic point__. For a definite value of z, the smaller A_2 is, the closer is the point to $v = s_1$. The value $A_2 = 0$ gives motion along the line $\dot{x}/x = s_1$. As $t \to \infty$, in general, $x \to 0$ and $v \to s_1$.

The corresponding solution at $z = 0$, $v = s_2$ is, from (9.41) and (9.54), $z = (1/A_2)\,e^{-s_2 t}$ and

$$v - s_2 \cong (s_2-s_1)\,a^{-1}e^{(s_1-s_2)t}$$

$$= (A_1/A_2)(s_1-s_2)\,e^{(s_1-s_2)t}$$

$$= (A_1/A_2)(s_1-s_2)(A_2 z)^{(\sigma-1)/\sigma}, \quad \sigma>1$$

This defines a family of parabolas tangent to the v axis at $z = 0$, $v = s_2$, so that this point is a __node__. The motion is such that $\left|z\right|$ increases with t, since s_2 is negative, and the node is therefore __unstable__. When v is __near__ s_1, then $\dot{v} \cong - (v - s_1)(s_1 - s_2)$ so that the __motion is always__ towards s_1. When v is near s_2, then

$$\dot{v} \cong + (s_1 - s_2)(v - s_2)$$

and the <u>motion is away from</u> s_2. For $v < s_2$, $\dot{v} < 0$, so that
v continually decreases toward $-\infty$, which it attains when
$x = 0$.

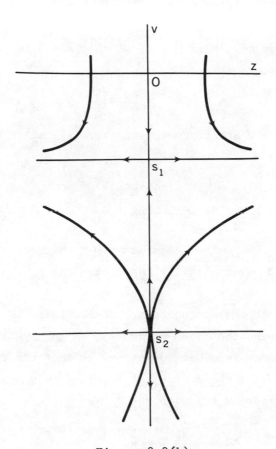

Figure 9.9(b)

Phase Plane Trajectories, $p > 1$, at Infinity

The motion near the singularities at infinity is shown in
Figure 9.9b. This, in combination with Figure 9.9a, gives
the general qualitative picture of the motion for the whole
phase plane ($p > 1$).

Case II p = 1 (a) Motion near the origin

The critical case can be regarded as the limit when the 2 real roots come together, as they do for p = 1.

One solution is readily obtained by setting s = -1 in Equation (9.40), giving $x_1 = ae^{-t}$. If $x = x_1 w$ is a second solution, then, since x_1 satisfies (9.1) with q = 1, and $\ddot{x}_1 + x_1 = 0$, we must have $\ddot{w} = 0$, so that w = t is a solution. The general solution is thus

$$x = a\,e^{-t} + bt\,e^{-t}$$

where a and b are arbitrary constants. ($a = x_o$).   From equation (9.2)

$$\dot{x} = e^{-t}(-a\,-bt\,+b)$$

$$= -x + be^{-t} \qquad\qquad (9.48)$$

At t = 0, $\dot{x}_o + x_o = b$.   Also, as t → ∞ the motion is toward the origin and asymptotically approaches the line $\dot{x} = -x$, or v = -1, where $v = \dot{x}/x$.

Now, since the two exponents are equal, there is naturally no motion between the lines s_1 and s_2, which have coalesced. Now $(1/\dot{x})\,d^2\dot{x}/dx^2$ is always negative, so that the trajectory will be concave to the x axis, as before (see 9.50).

The slope, from (9.49) with p = 1, is

$$\frac{d\dot{x}}{dx} = x\frac{dv}{dx} + v = -2 - \frac{1}{v} \qquad\qquad (9.49a)$$

This can be rewritten as

$$[\frac{1}{v+1} - \frac{1}{(v+1)^2}]\;\frac{dv}{dx} = -\frac{1}{x}$$

which integrates to

$$\ln\,(v+1) + \frac{1}{v+1} = -\ln\,x + \text{const.} \qquad (9.32a)$$

This is the same type of equation as (9.32), and the origin is a <u>stable degenerate node</u>, with all the trajectories tangent at the origin to the line $\dot{x} = -x$. (To see the connection with (9.29), set $x' = y$ and $\dot{x}' + x' = a_{21}x$, $a = y_o$, $b = a_{21}x_o$. Then set $\dot{x}' = vx'$ and drop primes in (9.32) to obtain (9.32a)). The $\dot{x}$ axis is intersected at $t = -a/b$, and the x-axis at $t = 1 - (a/b)$.

p = 1 (b) Motion at Infinity

To obtain the motion at infinity, set $z = 1/x$, and then (9.32a) becomes

$$\ln (v+1) + \frac{1}{v + 1} = \ln z + \text{const.} \qquad (9.32b)$$

The direction $v = -1$ is singular, and we have an <u>unstable improper node</u> at $z = 0$, which has come about by the coalescence of a stable node and a hyperbolic point $(s_2 \rightarrow s_1)$.

Figure (9.9) gives the general idea of the motion, both at the origin and at infinity, if we make $s = -1$ and draw the motion arrows correctly. (Figure 9.3b, also, shows a degenerate node when the y-axis is singular.)

Case III 0<p<1.

The motion in the phase plane near the origin has already been discussed for $0<p<1$ and shown to be a spiral, with the origin a stable focus.

At infinity, set $z = 1/x$ and obtain

$$\dot{z} = -vz$$
$$\dot{v} = -(v^2 + 2pv + 1) \qquad (9.48)$$

whence

$$\frac{dz}{dv} = \frac{vz}{v^2 + 2pv + 1}$$

which has no singularities when p < 1 because the denominator
does not vanish for real values of v. The point at infinity
is thus ordinary.

NON-LINEAR SYSTEMS

General Remarks

A linear system is usually much simpler than a non-linear
one. The motion described by (9.9) or its equivalent (9.33)
is such that the variable x either goes to zero, becomes
infinite, or executes a periodic motion, as time increases.
In the phase plane, the singular points are at $r = 0$ and at
$r = \infty$. If a particle leaves the origin, it will keep going
all the way to $r = \infty$, and not approach a periodic orbit "at"
some finite value of r. There is a periodic orbit through
each point $(x, \dot{x})$ when $T = 0$, so that any disturbance from
one periodic orbit will just result in another periodic orbit.

The non-linear situation may be illustrated by the example

$$\ddot{x} = -2p \, \dot{x} - qx \qquad\qquad (9.1)$$

where

$$2p = - \, \varepsilon(1-x^2) \text{ and } q = 1, \, (\varepsilon > 0).$$

The origin is unstable and a particle will leave its vicinity.
However, when $|x|$ becomes greater than unity, p becomes
positive and the acceleration negative (for $\dot{x} > 0$). One
might guess, and this is borne out by closer study, that a
periodic orbit in the phase plane is possible, and that it
will be approached asymptotically from points either outside
or inside it, as $t \to \infty$. When such is the case, the orbit is
said to be asymptotically stable. So, for a non-linear
system, we need to locate not only the singular points
(equilibrium points) and to determine the nature of the

stability there, but we must also find where the periodic
orbits are and whether or not they are stable, in order to
construct a general qualitative picture of the motion.

Stability of Motion

The motion of a particle is said to be stable if it returns
repeatedly to the vicinity of a given point, unstable if it
does not. For one dimension, it is convenient to represent
the motion by successive points in the $(x, \dot{x})$ plane. If P
denotes the initial point, then after a time τ the particle
will have moved to a new position $Q = TP$, where T is some
transformation of the phase plane. The point P is mapped
into the point Q. If there is a point P_o such that $TP_o = P_o$,
then the motion is periodic with period τ and P_o is a fixed
point of the mapping. The basic question of stability is
this: Does there exist a neighborhood of a fixed point P_o
such that all points within this neighborhood do not leave
it when the transformation T^n is applied, where
$n = \pm 1, \pm 2, \ldots\ldots$? This criterion for stability, where
n is allowed to assume both positive and negative values, is
more restrictive than our criterion used hitherto $(n > 0)$.

To obtain some idea about possible stability, we can
discuss stability in the first approximation, or first order
stability. This just means that, if we start from a periodic
orbit, only first powers of the variation from the standard
orbit will be included in the discussion. In this way, we
can find that some motions are unstable in first order and
that the higher order terms are very unlikely to bring in
enough corrections to make the motion stable. On the other
hand, even though a motion may be stable to first order, one
can by no means assume that it will remain so if the higher
order terms are included.

First Order Stability

Bearing in mind that higher order terms may in some cases have a decisive influence, let us for the moment neglect them and restrict our considerations to terms which are linear in the variations from a standard orbit.

Let the equations of motion be

$$dx_i/dt = X_i \quad (i = 1.....n)$$

where the X_i's are functions of $(x_1...x_n)$ and perhaps t, having a period τ in t. Suppose that a periodic solution $x_1^o(t)$, $x_2^o(t)$,...$x_n^o(t)$ has been found, and consider adjacent solutions of form

$$x_i(t) = x_i^o(t) + \Delta x_i(t)$$

where Δx_i represents a small variation in x_i.

The variational equations, to first order, are

$$d \ \Delta x_i/dt = \sum_j a_{ij} \Delta x_j \qquad (9.V)$$

where

$$a_{ij} = \partial X_i / \partial x_j .$$

If (a) the reference configuration is that of equilibrium, the a_{ij}'s will be constants, but if (b) the basic solution is a set of functions x_i which are periodic in time, then the coefficients a_{ij} will likewise be periodic. In the first case, we shall talk about stability at an equilibrium point, and the equations are of type (9.9), for which the theory has been presented. The second case concerns stability of a periodic solution, or orbital stability, and the theory, which is well known, will be given below. (Chapter X, Sec. A)

Effect of Non-linear Terms on Stability

Although one can often decide about stability by considering
first order variations, this method is not always sufficient
and we must then examine the higher order terms in the
expansion of the forces X_i. To know when this is in order,
let us quote certain theorems due to Liapounov, for autonomous systems.

1. If all the characteristic exponents have negative real
parts, the point of equilibrium is asymptotically stable.

2. If there is at least one characteristic exponent with
a positive real part, the point of equilibrium is unstable.

Even if we have a non-linear mapping

$$\dot{x} = ax + by + \varepsilon(x,y)$$

$$\dot{y} = cx + dy + \eta(x,y)$$

where $ad - bc \neq 0$ and $\varepsilon/r \to 0$, $\eta/r \to 0$ as $r \to 0$, the stability
of the motion, except at an elliptic point, is determined by
the linear terms alone. (For a detailed proof, see Pars,
pp 364-372.)

Practically, then, special investigations are in order when
the real parts of some of the exponents are zero. We shall
consider the cases: (1) when one exponent equals zero and
the others have negative r.p.'s and (2) when there are just
2 conjugate imaginary exponents, the other r.p.'s being < 0.

Case I: One Zero Exponent

Consider, as a typical example when one exponent equals
zero, the motion in 2 dimensions

$$\dot{x} = ax^2 + bxy + cy^2$$
$$\dot{y} = -y + kx + \ell x^2 + mxy + ny^2 \qquad (9.55)$$

For $\dot{y} = 0$, we find

$$y = kx + B_2 x^2 + B_3 x^3 + \ldots$$

where $B_2 = \ell + mk + nk^2$ and $B_3 = (m + 2nk)B_2$.

Then

$$\dot{x} = A_2 x^2 + A_3 x^3 + A_4 x^4 + \ldots$$

with

$$A_2 = a + bk + ck^2$$

$$A_3 = (b + 2ck)\ B_2$$

$$A_4 = (b + 2ck)\ B_3 + cB_2^2$$

For stability, we must have $A_2 = 0$ because otherwise an x
can be found such that $\dot{x}$ has the same sign. If $B_2 = 0$, then
all subsequent $A_i = 0$ and there is <u>stability</u>. If $B_2 \neq 0$ and
$b + 2ck \neq 0$, then there will be stability if A_3 is negative
and instability if A_3 is positive. If $b + 2ck = 0$, then
$A_4 \neq 0$ (and the system will be unstable) unless $c = 0$, when
it follows from $A_2 = 0$ and $b + ack = 0$ that $a = 0$, $b = 0$ and
the equilibrium point is stable.

From (9.55), if we set $x = 0$ and make y small enough, the
y-motion is dominated by $\dot{y} = -y$ and is hence stable. However,
the x-motion with y stationary ($\dot{y} = 0$) is stable or unstable
depending upon what the lowest order term in x is. If the
quadratic term is not zero, then there is instability toward
positive or negative x. If it is zero, and the cubic term
dominates, the sign of its coefficient determines the
stability. And it may be proved that the stability of the
motion is the same in general as with y stationary
(Chetaev, Ch. 7).

Case II: Two Conjugate Imaginary Exponents (Phase Plane)

Let us now consider motion in the phase plane, and suppose
that the characteristic exponents are conjugate imaginaries.
We then have an elliptic point in the linear case, for which
the general transformation will be

$$\dot{x} = -y + \epsilon(x,y)$$

$$\dot{y} = x + \eta(x,y)$$

from which

$$r\dot{r} = x\epsilon + y\eta, \quad \dot{\theta} = 1 + (x\eta - y\epsilon)/r^2 \qquad (9.56)$$

We suppose as before than ϵ/r and η/r both go to zero as
$r \rightarrow 0$. A wide variety of motions can be obtained just by
setting $x\eta - y\epsilon = 0$, so let us do this.

Put

$$\epsilon = xf(r), \quad \eta = yf(r)$$

then, from (9.56)

$$\frac{dt}{dr} = \frac{1}{rf(r)} \qquad (9.57)$$

(a) If $f(r) = r$, or $\dot{r} = r^2$, this integrates to

$$t = \frac{1}{r_o} - \frac{1}{r}$$

The plot of this is shown in Figure 9.10, from
which it is evident that the particle, starting from
$r = r_o$ at $t = 0$, will reach $r = \infty$ at $t = 1/r_o$. The
origin is thus unstable.

(b) If $f(r) = -r$, or $\dot{r} = -r^2$, the integral of (9.57) is

$$t = \frac{1}{r} - \frac{1}{r_o} .$$

The slope of the curve, shown in Figure (9.11), is negative,
and the particle approaches the origin, i.e. $r \to 0$ as $t \to \infty$.
The origin is then <u>asymptotically stable</u>.

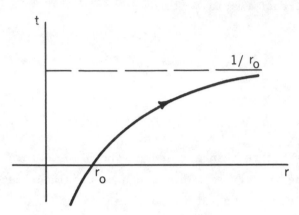

Figure 9.10

Motion to Infinity in Finite Time

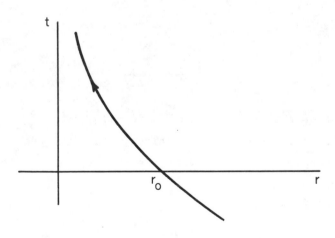

Figure 9.11

Asymptotically Stable Origin

(c) Slightly more complicated is the case $f(r) = r(a - r)$. Here

$$\dot{r} = r^2 (a - r)$$

which vanishes if $r = a$. The orbit will then be a circle, $r = a$, $\dot{\theta} = 1$.

If $r_o < a$, then the particle will spiral outward to approach $r = a$ as $t \to \infty$. If $r_o > a$, the particle will spiral inward, approaching $r = a$ as $t \to \infty$. These relationships are shown in Figure (9.12). Near $r = a$, we have approximately

$$\frac{dt}{dr} = \frac{1}{a^2 (a-r)}$$

which integrates to

$$a^2 t = -\ln |a - r|$$

The periodic orbit $r = a$ which is thus approached is called a limit cycle.

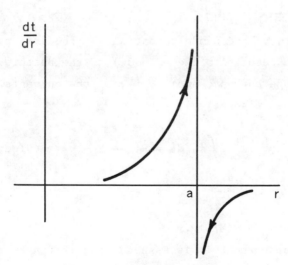

Figure 9.12

Limit Cycle at r = a

We thus see that, <u>when the first order stability indicates</u> <u>a vortex point</u>, <u>the actual stability</u>, <u>or even its existence,</u> <u>can be affected crucially by the higher-order terms of the</u> <u>transformation.</u>

THE VAN DER POL OSCILLATOR

Triode Oscillator with Feedback

Having treated an oscillator with constant damping coefficient, we now turn to the case where this coefficient is variable, which is met with in the example of a triode oscillator coupled both inductively and directly to an R-L-C circuit as in Figure 9.13.

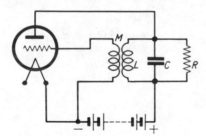

Figure 9.13

(After Van der Pol, B., 1934, Proc. I.R.E. 22, 1051)

Triode Oscillator
with Feedback

Denote by v_a the anode potential, v_g the grid potential (both with respect to the cathode), E the battery voltage, i_a the anode current, and i_L, i_C, i_R the currents in the LCR branches respectively.

Then

$$L \frac{di_L}{dt} = \frac{1}{C} \int i_C \, dt = Ri_R = E - v_a = -v \,(\text{say})$$

$$M \frac{di_L}{dt} = v_g, \quad i_a = i_L + i_R + i_R$$

from which $v_g = -(M/L)v$.

The anode current i_a is assumed to be a function ϕ of the single variable $v_a + \mu v_g$, where μ is the amplification factor of the tube, so that

$$i_a = \phi(v_a + \mu v_g)$$

In the steady (unstable) state, $v_{a_o} = E$, and

$$i_{a_o} = \phi(E)$$

The current variation from the steady state will be denoted by $i = i_a - i_{a_o}$, and obeys the relation

$$\frac{di}{dt} = -\{\frac{v}{L} + \frac{1}{R}\frac{dv}{dt} + C\frac{d^2v}{dt^2}\}$$

The function ϕ is nearly linear over a sizeable range of $v_a + \mu v_g$, so that

$$i_a = (1/R_i) [v_a + \mu v_g]$$

$$= (1/R_i) [E + v - \mu(M/L)v]$$

Substitution in the preceding equation gives

$$C\frac{d^2v}{dt^2} + \{\frac{1}{R} - \frac{1}{R_i} (\mu \frac{M}{L} - 1)\} \frac{dv}{dt} + \frac{v}{L} = 0$$

If there is regenerative feedback, i.e. if M is large enough that

$$\mu \frac{M}{L} - 1 > \frac{R_i}{R}$$

then the damping coefficient is negative and the quantity v is an exponentially increasing oscillation. This will even-tually be limited because the function ϕ actually has higher-order, non-linear, terms in it. For simplicity, we shall take E at the inflection point of the characteristic, and now approximate ϕ by a cubic polynomial, i.e.

$$i = \phi(v_a + \mu v_g) - \phi(E)$$

$$= - \alpha v + \gamma v^3$$

with

$$\alpha > 0, \gamma > 0$$

The equation of motion now becomes

$$C \frac{d^2 v}{dt^2} + \{(\frac{1}{R} - \alpha) + 3 \gamma v^2\} \frac{dv}{dt} + \frac{v}{L} = 0$$

This is the basic equation of Van der Pol, and it is seen that
the coefficient of dv/dt depends on the amplitude v.   For
$\alpha > 1/R$, the coefficient is negative near v = 0, but becomes
positive when v gets large enough.

The substitutions

$$\frac{1}{C} (\alpha - \frac{1}{R}) = \alpha', \quad \frac{3\gamma}{C} = \gamma', \quad \frac{1}{LC} = \omega_o^2$$

$$\omega_o t = t', \quad \frac{\alpha'}{\omega_o} = \varepsilon, \quad \frac{\gamma' v^2}{\alpha'} = x^2$$

are used to bring the Van der Pol equation to standard form
(dropping the prime from t') i.e.

$$\ddot{x} - \varepsilon(1 - x^2) \dot{x} + x = 0 \qquad\qquad (9.58)$$

If $x^2 \ll 1$, then the motion is that of an oscillator with
negative damping, so that the amplitude builds up as $\exp(\varepsilon t/2)$,
the rate of increase being small if ε is small.   If x is large,
so that $x^2 \gg 1$, the damping coefficient is approximately εx^2
and the amplitude of the oscillation will die down very
rapidly, say as $\exp(-\varepsilon x^2 t/2)$.   So it would seem that there
will be some amplitude for which the oscillation will maintain

the same form as time goes on. It will be convenient to use
the phase plane representation and to take ε as small, to
begin with. The equations are

$$\dot{x} = y, \; \dot{y} = \varepsilon(1 - x^2) \, y - x \qquad (9.59)$$

Substituting $x = r \cos \theta$, $y = r \sin \theta$, we have

$$x\dot{x} + y\dot{y} = r\dot{r} = \varepsilon y^2 (1 - x^2) \qquad (9.60)$$

$$x\dot{y} - y\dot{x} = r^2\dot{\theta} = \varepsilon \, xy(1 - x^2) - r^2 \qquad (9.61)$$

If ε is small, and r is not too large, the r.h. side of (9.61)
reduces to

$$\dot{\theta} = -1. \qquad (9.62)$$

Dividing (9.60) by (9.62), we find that

$$r \, \frac{dr}{d\theta} = - \, \varepsilon \, r^2 \sin^2 \theta (1 - r^2 \cos^2 \theta)$$

$$= + \, \frac{\varepsilon r^2}{2} \, [-1 + \cos 2\theta + \frac{r^2}{4}(1 - \cos 4\theta)]$$

$$= + \, \frac{\varepsilon r^2}{8} \, [r^2 - 4 - r^2 \cos 4\theta + 4 \cos 2\theta] \quad (9.63)$$

If the variation of r with θ is small, r can be assumed to be
constant in the r.h. side of (9.63). Integrating over θ from
0 to $\frac{\pi}{2}$, the last two terms in the brackets contribute zero
and the average rate of change of r with time is then given
by

$$\dot{r} = - \, (\varepsilon/8) \, r \, (r^2 - 4). \qquad (9.64)$$

The motion will be inward for $r > 2$, outward for $r < 2$, and
periodic for $r = 2$. Equation (9.64) can be written as

$$\dot{r} = - \, (\varepsilon/8) \, r \, (r + 2) \, (r - 2)$$

Near $r = 0$, we have approximately

$$\dot{r} \cong \varepsilon r/2$$

so that the amplitude of the oscillation x = r cos t builds
up as exp (εt/2), as stated above.

Near r = 2, the amplitude changes as

$$\dot{r} \cong - \varepsilon(r-2)$$

which is of the same form $\dot{r}$ = const (r - a) that was discussed
previously and which yields a limit cycle at r = 2.

To terms of first order in ε, the calculation has thus shown
that the circle r = 2 in the phase plane is a <u>stable limit
cycle</u>, which is approached asymptotically from within and
without by spiral trajectories winding on to it. For small
ε, then, the oscillations can build up from small amplitude
to a limiting motion

$$x \cong 2 \cos t.$$

In the neighborhood of the limit cycle, if ε is small but not
zero, the amplitude will fluctuate slightly because it
contains terms proportional to ε and periodic in θ.

For ε = 0.1, the oscillation builds up with time to the
constant amplitude as shown in Figure 9.14a. The winding on
to the limit cycle C in the phase plane is shown in Figure
9.15a.

Van der Pol integrated the system (9.59) in the form

$$\frac{dy}{dx} = \varepsilon(1-x^2) - \frac{x}{y} \qquad\qquad (9.59')$$

numerically for ε = 0.1, ε = 1.0, and ε = 10, and obtained
the curves of Figure (9.15). Heavy lines are trajectories
winding on to the limit cycles and light lines are curves of
constant slope (9.59'), called isoclines. Such isoclines were
of help in the old-style numerical integration, for one could
use them to sketch in various trajectories.

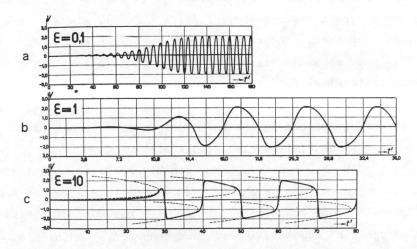

Figure 9.14

(After Van der Pol, B., 1934, loc. cit.) Oscillations, Amplitude vs. Time, for Van der Pol Oscillator.

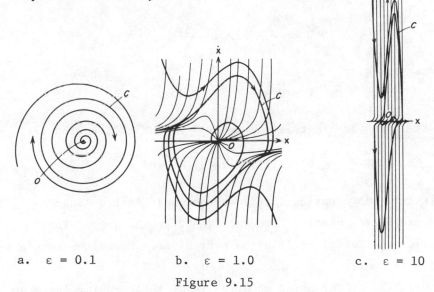

a.   ε = 0.1   b.   ε = 1.0   c.   ε = 10

Figure 9.15

(After Van der Pol, B. 1926, Phil Mag 2, 978) Trajectories in phase plane, for Van der Pol oscillator.

The trajectory in the phase plane gives $\dot{x}$ as a function of x, and a <u>second integration</u>, of $dt/dx = f(x)$, gives x as a function of time. The results, for $\varepsilon = 0.1$, $\varepsilon = 1.0$, and $\varepsilon = 10$, are shown in Figure 9.14. The oscillations for $\varepsilon = 1.0$ reach their final form after about two periods, and this final form has already departed considerably from a sine wave, which is further evidenced by the great deformation from a circle of the phase plane limit cycle, Figure 9.15b.

For $\varepsilon = 10$ the wave form, shown in Figure 9.14c, has lost all resemblance to a sine wave and is more like a square wave, with small slope from $|x| = 2$ to $|x| = 1$ for most of the time, interrupted by almost discontinuous behavior, jumping from 1 to -2 or -1 to +2 at regular intervals, the period being roughly $T = 20 = 2\varepsilon$. In the region where the acceleration is small, we can neglect the first term in (9.58), obtaining for x > 1

$$dt = - \varepsilon (xdx - \frac{dx}{x})$$

Integrating,

$$t + C = \varepsilon (- \frac{x^2}{2} + \ell nx) \qquad (9.65)$$

Putting in the limits x = 2 and x = 1, the half-period is

$$\frac{T}{2} = \varepsilon (1.5) - \ell n \, 2) = .807 \, \varepsilon$$

Curves representing (9.65) are shown as dotted lines in Figure 9.14c, passing through $|x| = 2$ at the jump. They follow the actual oscillation very closely where the acceleration is small.

When the differential equation is of the form involving an effective resistance r and a capacity C only, i.e.

$$r\dot{x} + x/C = 0$$

then signals die out or increase exponentially with a time
constant, or <u>time of relaxation</u>, equal to rC. Since the
oscillations for large ε involve a region which can be
described by such a differential equation, they are called
<u>relaxation oscillations</u>.

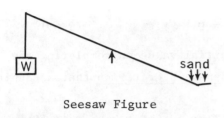

Seesaw Figure

A very simple relaxation
oscillator is a seesaw with
a weight attached at one end
and sand pouring into a
shovel at the other end.
When enough sand has accumu-
lated, the end is low enough so the sand runs out, and the
shovel springs back up, whereupon the process starts over.
Another familiar example is the repetitive discharge of a
neon tube shunted by a capacitance and fed by a battery
through a resistance.

Liénard Phase Plane

In order to describe the motion when ε is large, it is
more convenient to introduce a different set of variables
first used by Liénard. Then both the slow and fast changes
can be described simply. Observe that the second term in
(9.58) can be put in the form $\varepsilon\dot{F}$, where

$$F = - \left(x - \frac{x^3}{3}\right) \qquad (9.66)$$

Then Equation (9.58) itself becomes

$$\ddot{x} + \varepsilon\dot{F} + x = 0$$

Consequently, setting $y = \dot{x} + \varepsilon F$, we have

$$\dot{x} = y - \varepsilon F, \quad \dot{y} = -x \qquad (9.67)$$

Now

$$0 = \dot{x}(\dot{y} + x) = x\dot{x} + y\dot{y} - (y - \dot{x})\dot{y}$$

$$= x\dot{x} + y\dot{y} - \epsilon F \dot{y} \qquad (9.68)$$

If the orbit be a closed one, then integration of (9.68) yields

$$\oint F \, dy = 0 \qquad (9.69)$$

since $x^2 + y^2$ returns to its initial value on completing the circuit. Equation (9.69) is <u>Liénard's criterion</u> that an orbit be periodic.

This is satisfied by the oscillations of Figure 9.14, shown as limit cycles in Figure 9.15 [$(x,\dot{x})$ plane] and in Figure 9.16 [(x,y) plane].

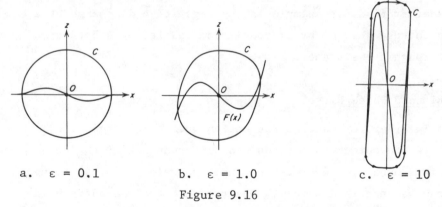

a. $\epsilon = 0.1$ b. $\epsilon = 1.0$ c. $\epsilon = 10$

Figure 9.16

(After Le Corbeiller, Ph., 1931) Van der Pol Trajectories in the Liénard Phase Plane.

For any value of ϵ, the motion can be visualized by referring to Figure 9.17. Equations (9.67) become, if we set $y = \epsilon z$,

$$\dot{x} = \epsilon(z - F); \dot{z} = -x/\epsilon$$

The motion will be toward positive x if $z > F$, vertical if $z = F$, and toward negative x if $z < F$. For positive x,

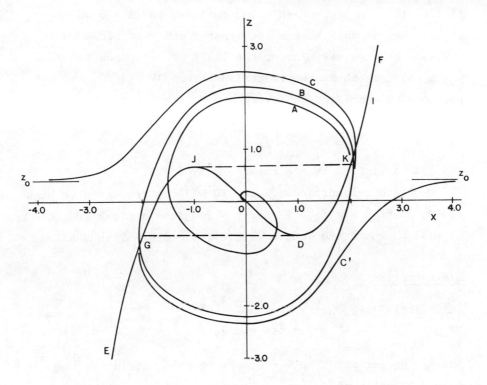

Figure 9.17

Sample Van der Pol Orbits in Liénard Phase Plane

the motion will be downward, for x = 0 horizontal, and for
negative x the motion will be upwards. Consequently, if the
curve z = F be represented by EJDF, a periodic orbit will be
typified by Curve B, which is for ε = 1. Curve C starts at
x = $-\infty$, z = z_0 and winds rapidly onto Curve B, the limit
cycle. Curve C', from x = $+\infty$ and z = z_0, shows a similar
behavior. Curve A starts from the origin 0 and spirals rapidly
outward to wind onto Curve B. The slope of a trajectory is

$$\frac{dz}{dx} = -\frac{x}{\varepsilon^2 (z - F)}$$

If this is to be appreciable and positive for x > 0 and ε
large, then we must have z less than F and very close to it.

When ε becomes very large, the limit cycle approaches JKDG,
which consists of two horizontal portions JK and DG, and two
portions KD and GJ of the curve z = F. The equation
$(z - F)dz + x\, dx/\varepsilon^2 = 0$ goes in the limit to

$$(z - F)\, dz = 0$$

which is satisfied by z = const and by z = F.

The horizontal portions JK and DG will be traversed at high
speed, since $\dot{x} = \varepsilon\,(z - F)$ and $z \neq F$. The portions KD and
GF will be where the motion is slow, since z - F is small.

Singularities

The system

$$\dot{x} = y + \varepsilon\left(x - \frac{x^3}{3}\right); \quad \dot{y} = -x \qquad (9.70)$$

has a singular point at x = 0, y = 0. The matrix for the
linear transformation is $\begin{pmatrix} \varepsilon & 1 \\ -1 & 0 \end{pmatrix}$ with secular equation
$\lambda^2 - \varepsilon\lambda + 1 = 0$. The origin is a focus (unstable) if $\varepsilon < 2$, an
unstable node if $\varepsilon \geq 2$. The trajectories depart from the
origin and spiral onto the limit cycle.

The x-axis

To find the nature of the singular points at infinity, let
us first suppose x $\neq$ 0 and make the substitutions x = 1/η and
y = vx = v/η. Also neglect x re $x^3/3$. Then

$$\frac{dy}{dx} = -\frac{x}{y + \varepsilon\left(x - \frac{x^3}{3}\right)}$$

$$= x\frac{dv}{dx} + v \cong -\frac{x}{y - \frac{\varepsilon x^3}{3}}$$

and

$$\eta \frac{dv}{d\eta} = v + \frac{3\eta^2}{3\eta^2 v - \varepsilon} \qquad (9.71)$$

The second term on the r.h. side of (9.71) goes to zero as η^2 if v is finite, and so may be neglected. The origin $v = 0$, $\eta = 0$ is a <u>singular unstable node</u>, with $v = c/x$ or $y = $ const.

The y-axis

If $x = 0$, substitute $y = 1/w$, $x = uy = u/w$.

This gives

$$\frac{du}{dw} = \frac{u}{w} + \frac{1}{uw} + \varepsilon \left(\frac{1}{w} - \frac{u^2}{3w^3} \right) \qquad (9.72)$$

Since near $u = 0$, $w = 0$, this derivative can range from zero to infinity, it is convenient to introduce a new time variable τ such that $w^2 d\tau = dt$ and such that $du/d\tau$ and $dw/d\tau$ each become zero at $u = 0$, $w = 0$. We shall, for this discussion, now use a dot to denote differentiation re τ. Then (9.72) can be replaced by the system

$$\dot{u} = u^2 w^2 + w^2 + \varepsilon \left(uw^2 - \frac{u^3}{3} \right)$$

$$\dot{w} = uw^3$$

Along the u-axis ($w = 0$), $\dot{u} = -\varepsilon u^3/3$ and the motion is inward (see Figure 9.18), toward the origin. Along the w-axis ($u = 0$), $\dot{w} = 0$ and $\dot{u} = w^2$ so that the motion is at right angles to the w-axis, i.e., in the +u direction.

To learn more about the motion, form

$$r\dot{r} = u\dot{u} + w\dot{w} = (u^3 + \varepsilon u^2 + u)w^2 - \frac{\varepsilon}{3}u^4 + uw^4$$

The radial velocity $\dot{r} = 0$ when $u = 0$ (w-axis) or when

$$w^2 = (\varepsilon u^3/3)/(1 + \varepsilon u + u^2 + w^2) \qquad (9.73)$$

The motion is vertical when $\dot{u} = 0$, i.e.

$$w^2 = (\varepsilon u^3/3)/(1 + \varepsilon u + u^2) \qquad (9.74)$$

It is radial when $u\dot{w} - w\dot{u} = 0$, i.e.

$$w^2 = (\varepsilon u^3/3)/(1 + \varepsilon u) \qquad (9.75)$$

In this event, $r\dot{r} = u^3 w^2 + uw^4$, which is positive when u is.

Alternatively, corresponding statements can be made for the motion in the (x,y) plane. From (9.67) and (9.66), at $x = 0$, we have $dy/dx = -x/y = 0$, so the +y axis is crossed at right angles to it, and toward positive x. The <u>slope will be</u> infinite when $y = \varepsilon(\frac{x^3}{3} - x)$, or $\dot{x} = 0$, which is the same as

$$w^2 = (\varepsilon u^3/3)/(1 + \varepsilon u)$$

the condition that <u>the motion in the (u,w) plane is radial</u>. Finally, <u>the motion in the (x,y) plane will be radial</u> when $x\dot{y} = y\dot{x}$, or

$$y (y - \varepsilon F) = -x^2$$

which reduces to (9.74) in terms of u and w, and corresponds to $\dot{u} = 0$.

For a definite u (>0), comparison of (9.73) and (9.74) shows that $w^2(\dot{r} = 0) < w^2(\dot{u} = 0)$. Near $u = 0$, $w = 0$, the motion is inward along both positive and negative u-axes. There is a type of trajectory which leaves the w-axis normally and crosses $y = \varepsilon F$ radially. At smaller w, finite u, the motion is at first inward, then bends upward and back, with first $\dot{r} = 0$ and then $\dot{u} = 0$. Its trajectory cannot intersect the ones coming from $u = 0$. There is a separatrix which leaves the origin $u = 0$, $w = 0$ and divides the plane into two regions, corresponding to the two types of motion. The origin $u = 0$, $w = 0$ is a <u>saddle point</u>, and the paths to and from it are

tangent to, or coincide with, the u-axis. Thus, just as in
the case of the damped harmonic oscillator, the singular points
at infinity consist of nodes and cols, only here the col is
slightly more complicated in form.

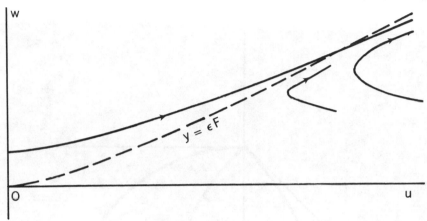

Figure 9.18

Van der Pol Trajectories at Infinity

Trajectories Projected Onto a Sphere

In order to have a comprehensive picture of how all the
motions in the phase plane look, it is extremely useful to
project all the points of this plane onto a sphere with unit
radius, and then to project the points of the sphere onto a
plane parallel to the original phase plane.

Let then the original phase plane be indicated in Figure
9.19 by x0y, and have a unit sphere below it tangent to the
plane at 0. The center of this sphere shall be labeled 0'.
If P(x,y,1) be any point in the plane x0y, draw 0'P and have
it intersect the sphere at P'. (For clarity, just the
octant OAB of the sphere is shown). The direction cosines

of O'P are

$$U = x/(x^2 + y^2 + z^2)^{1/2}$$

$$V = y/(x^2 + y^2 + z^2)^{1/2}$$

$$Z = 1/(x^2 + y^2 + z^2)^{1/2}$$

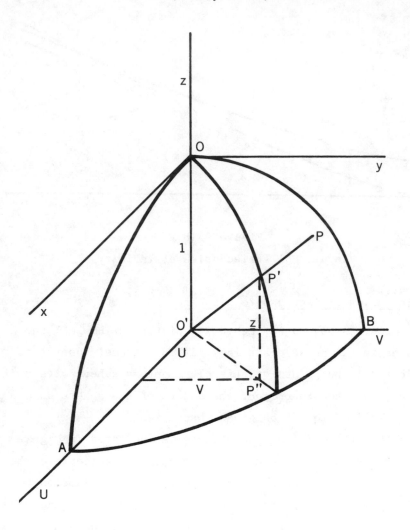

Figure 9.19

Projection onto a Sphere

If $y = 0$ and $x \to \infty$, then $U \to 1$ and $Z \to 0$ so that the point A is approached. If $x = 0$ and $y \to \infty$ then $V \to 1$ and $Z \to 0$ and point B is approached. If we now look down on the sphere, we see the projection of P' onto the (U,V) plane at P''. All trajectories, including those near the singular points at infinity (mapped into A, B, A', B') will be represented by their maps, $P \to P''$ being the map of one point on a trajectory. Typical curves are shown in Figure 9.20.

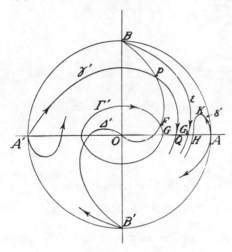

Figure 9.20

(From Lefschetz, S., 1963, "Differential Equations: Geometric Theory," Interscience, with permission) Projection of Van der Pol Trajectories onto the (U,V) Plane.

Curve B'OFPB represents $y = \varepsilon F$, and A'PQ a trajectory crossing it. The curve AKH is a trajectory coming in from $x = \infty$ and ε is the separatrix which divides the region containing trajectories of type AKH from those of type A'PQ. The curve Γ' is the limit cycle, onto which the trajectories wind. It will now be convenient to recapitulate our knowledge of the curves outside this limit cycle.

Paths from Infinity

Let us consider two paths which come from infinity, one (I) from $x = -\infty$, $y = y_0$ and the other (II) from $x = +\infty$, $y = y_0$. From our result that there is a singular unstable node, we know that paths for $|x|$ infinite are approximately along $y = $ const.

Figure 9.21

(From Lefschetz, S., loc.
cit.) Van der Pol Tra-
jectories outside limit
cycle.

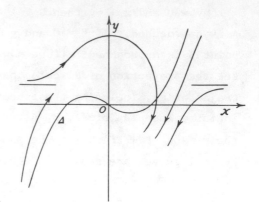

The slope of a trajectory is

$$\frac{dy}{dx} = - \frac{x}{y-\varepsilon F} \qquad (9.76)$$

which for finite y and large x is approximately

$$\frac{dy}{dx} = \frac{3}{\varepsilon x^2}$$

This is positive, so curve I rises from its asymptote $y = y_o$,
while curve II falls (see Figures 9.17 and 9.21). Also,

$$r\dot{r} = x\dot{x} + y\dot{y} = -\varepsilon x^2 \left(\frac{x^2}{3} - 1\right)$$

so that $\dot{r} < 0$ for $x^2 > 3$. In this region, both curves I and
II come closer to the origin.

The curvature is

$$\frac{d^2y}{dx^2} = - \frac{y - \varepsilon F - x(y' - \varepsilon F')}{(y - \varepsilon F)^2}$$

which for $x^2 < 3$ and large y is approximated by $y'' \cong - \frac{1}{y}$.
The curves are therefore concave to the x-axis here, curve I
falling for $x > 0$ (since $\dot{y} = -x$). The line $y = \varepsilon F$ is crossed
vertically by curve I, which has positive slope after the
crossing (according to (9.76)). Curve I cannot intersect
curve II, so must stay between curve II and the origin. The

result is that they both spiral inward onto the limit cycle.
As y_o is increased, curves I and II approach each other and
a common separatrix, which issues from $w = 0$, $u = 0$ ($y \to \infty$,
$u = 0$) and winds around the limit cycle.

X

Non-Autonomous Systems

(Periodic Variation with Time)

Having discussed examples of autonomous systems such as
(1) Hamiltonian systems with no explicit time dependence,
(2) the damped harmonic oscillator and (3) the Van der Pol
oscillator, we still need to look at systems where the time
does enter explicitly, and these are called non-autonomous.
The most general such system will have nothing peculiar about
it and consequently its theoretical treatment will be pedes-
trian and of little interest. However, if the coefficients
in the equation of motion vary periodically with the time,
then such a restriction brings with it simplification of the
theoretical treatment and the possibility of certain kinds of
invariance. In what follows, then, our attention will be
confined to this type of non-autonomous system.

The importance of considering systems with periodic coefficients is evident when one deals with various problems which
occur commonly in physics. For instance, if one system is
known to undergo periodic motion or has a periodicity of some
kind, and if it exerts an influence on another system, then
this second system may be regarded as non-autonomous of the
kind we are discussing. (1) In Chapter 9, the idea of stability of a periodic orbit was introduced, and it was shown that
the variational equations which determine the first order

stability are simultaneous first order linear equations (9.V)
with periodic coefficients. (The theory will be presented in
this chapter.) (2) Again, when an electron moves through a
periodic crystal lattice, it encounters a force field which
varies periodically in space. The linear wave equation has a
periodic coefficient, and solution of the equation shows that
a standing wave system (stationary state) is only possible
for certain bands of the total energy. (3) Finally, an
electron may move in a synchrotron, where the field is a
function of the angle and so is periodic in time. The
transverse force must obey certain conditions if the electron
is not to be lost from the beam (periodic orbit).

A. FIRST ORDER STABILITY: THEORY

In attempting to understand the factors responsible for
stability, it is probably best to begin with motion in one
dimension, specified by $(x, \dot{x})$, and to study that to first
order only, since the equations are then linear. Later, we
shall show how to treat general motion in one dimension. This
is about as far as one can usefully go at the present time,
for with these methods it is possible with various devices to
reach conclusions about the stability of motion of one
particle (in an external force field). However, if the system
should consist of two particles, this would involve six space
coordinates and six momenta, so that the mathematics is rather
opaque to the casual observer and not likely to yield simple
general results in any case, since the degree of coupling of
the particles can be expected to affect the motion strongly.

The stability problem is thus presented to us in two guises.
On the one hand, as in the synchrotron, the force is proportional to the displacement and is periodic in time as expressed
by a coefficient $p(t)$ which has a period τ, i.e. $p(t + \tau) = p(t)$.

The equation of motion is

$$\frac{d^2x}{dt^2} + p(t)\, x = 0 \qquad\qquad (10.1)$$

This is known as the <u>Mathieu-Hill Equation</u>, and its theory is
due to Floquet.

The other version of the problem comes from the variational
equations (9.V). Setting $x_1 = \Delta x$ and $x_2 = \Delta \dot{x}$, these become

$$\frac{dx_1}{dt} = a_{11}(t)\, x_1 + a_{12}(t)\, x_2$$

$$\qquad\qquad (10.2)$$

$$\frac{dx_2}{dt} = a_{21}(t)\, x_1 + a_{22}(t)\, x_2$$

Initially, we are given the displacement x and its time
variation (velocity) $\dot{x}$, (or alternatively x_1 and x_2) and this
can be represented by a point $P(x,\dot{x})$ in the phase plane. In
order to characterize the stability, it is not necessary to
find the detailed motion for all values of the time. It will
suffice to sample it at intervals of one (or more) periods
apart, i.e. to find the maps of P at times τ, 2τ, etc. For
the linear case, (10.1) or (10.2), the mapping for just one
period is sufficient to determine the stability for all time.
In the general case, where the higher order terms are important, one can find the maps at $t = \tau$, 2τ, etc. and from this
determine the trend as $t \to \infty$.

The first thing to do, then, is to integrate Equation (10.1)
or (10.2) for just one period. Let us select the first alternative, just to be definite. Since Equation (10.1) is of
second order, we may find two independent solutions f(t) and
g(t). Let us choose these so that at $t = 0$ the solutions are
specified by   f: $x = 1,\ \dot{x} = 0$

$$\qquad\qquad (10.3)$$

                 g: $x = 0,\ \dot{x} = 1$

A general solution can be written as a linear combination of
these two solutions, namely

$$x(t) = a\,f(t) + b\,g(t)$$
$$\dot{x}(t) = a\,\dot{f}(t) + b\,\dot{g}(t)$$

(10.4)

By setting $t = 0$, we find $a = x(0)$, $b = \dot{x}(0)$.

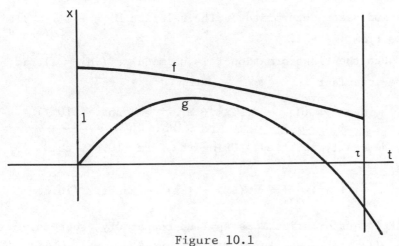

Figure 10.1

Independent Solutions of (10.1) for One Period

Upon integration, we shall have found f, g, $\dot{f}$, and $\dot{g}$ as
functions of time. Denote their values at $t = \tau$ by M_{11}, M_{12},
M_{21}, and M_{22}, respectively. Substituting into (10.4)

$$x(\tau) = M_{11}\,x(0) + M_{12}\,\dot{x}(0)$$
$$\dot{x}(\tau) = M_{21}\,x(0) + M_{22}\,\dot{x}(0)$$

(10.5)

This is a linear transformation T (through one period) of
the phase plane into itself. It transforms the initial point
$(1,0)$ into (M_{11}, M_{21}) and $(0,1)$ into (M_{12}, M_{22}). Since
the Wronskian $f\dot{g} - g\dot{f}$ of the two solutions f and g is constant

and initially unity, the matrix M has determinant 1 and the transformation can be written

$$x(\tau) = (\cos \sigma + \alpha \sin \sigma)x(0) + \sin \sigma \, \beta \dot{x}(0)$$

(10.6)

$$\beta \dot{x}(\tau) = -(1 + \alpha^2)\sin \sigma \, x(0) + (\cos \sigma - \alpha \sin \sigma) \, \beta \dot{x}(0)$$

where α, β, and σ are arbitrary constants. In terms of the M_{ij}'s, $\cos\sigma = (M_{11} + M_{22})/2$, $\alpha = (M_{11} - M_{22})/2 \sin \sigma$, and $\beta = M_{12}/\sin \sigma$. (The sign of $\sin \sigma$ must be specified, and then α and β are determined.) The relation $M_{21} = -(1 + \alpha^2) \sin\sigma$ follows from $|M| = 1$.

Now when the transformation T is unimodular ($|M| = 1$), it has the invariant

$$M_{21}x^2 + (M_{22} - M_{11})x\dot{x} - M_{12}\dot{x}^2 = \text{const.} \quad (10.7)$$

as may be easily verified. Using the form (10.6), this becomes

$$(1 + \alpha^2)x^2 + 2\alpha x \, (\beta \dot{x}) + (\beta \dot{x})^2 = \text{const.} \quad (10.8)$$

If the transformation T is applied repeatedly, successive points will lie on the curve (10.8), which can be an ellipse or an hyperbola.

The formulae become especially simple if $p(-t) = p(t)$ for then Equation (10.1) is invariant if $-t$ is substituted for t. Solutions of (10.1) will then be even or odd in t, i.e.

$$x(-t) = + x(t) \text{ or } x(-t) = - x(t).$$

The inverse of (10.5) yields

$$x(-\tau) = M_{22} \, x(0) - M_{12} \, \dot{x}(0) \qquad (10.9)$$

$$\dot{x}(-\tau) = -M_{21} x(0) + M_{11} \, \dot{x}(0)$$

Putting $x(0) = 1$, $\dot{x}(0) = 0$ and $x(\tau) = x(-\tau)$, we have

$$M_{11} = M_{22} \text{ and thus } \alpha = 0.$$

Then equation (10.6) reduces to

$$x(\tau) = \cos \sigma \ x(0) + \sin \sigma \ \beta\dot{x}(0)$$

$$\text{(10.10)}$$

$$\beta\dot{x}(\tau) = -\sin \sigma \ x(0) + \cos \sigma \ \beta\dot{x}(0)$$

and Equation (10.8) to

$$x^2 + (\beta\dot{x})^2 = \text{const} \qquad \text{(10.11)}$$

Equation (10.10) shows that T just rotates the vector
$(x, \beta\dot{x})$ through an angle σ (if σ is real), the magnitude of
which is determined by p(t) alone and not by the initial
values of x and $\beta\dot{x}$. If we just consider the points t = nτ,
n = integer, the quantity $x^2 + \beta^2 \dot{x}^2$ is constant. When σ is
real, β is real, the locus of the point $(x,\dot{x})$, for t = nτ, is
an ellipse, and the origin is called an <u>elliptic fixed point</u>.
If nσ = 2πm, with m and n integers, the initial P and the
consequent points TP, T^2P, etc. will be fixed under T^n. If
$\sigma/2\pi$ is irrational, repeated application of T generates more
and more points on the ellipse.

When $| \cos \sigma | > 1$, σ is imaginary, and β must be imaginary
also, since M_{12} is real. Put $\sigma = i\psi$ and $i\beta = \gamma$, a real
quantity. Equation (10.10) becomes

$$x(\tau) = x(0) \ \text{ch} \ \psi + \gamma\dot{x}(0) \ \text{sh}\psi$$

$$\text{(10.12)}$$

$$\gamma\dot{x}(\tau) = x(0) \ \text{sh} \ \psi + \gamma\dot{x}(0) \ \text{ch}\psi$$

For the points t = nτ, the quantity $x^2 - \gamma^2\dot{x}^2$ is constant,
and the locus consists of the branches of a hyperbola. For
this case, the origin is called a <u>hyperbolic</u> fixed point.

The asymptotes to the hyperbolas are given by x = $\gamma\dot{x}$ and
x = $- \gamma\dot{x}$. If the initial point P is on the first asymptote,
then the map TP will also be, but the distance from the origin
will have been stretched by the factor λ_1 = ch ψ + sh ψ, as

we see by substitution into (10.12). Similarly, the distance
from the origin to a point on the second asymptote will be
shrunk, the factor now being λ_2 = ch ψ - sh ψ.

The x-axis is mapped into x(τ) = x(0) ch ψ, $\gamma\dot{x}(\tau)$ = x(0) shψ,
from the first of which we can obtain ch ψ, and from the
second we can find γ, assuming that the mapping has been done
numerically. The stretch factors λ_1 and λ_2 follow immediately
from ch ψ.

If $\dot{x}(0)$ = 0, then Equation (10.10) yields cos σ = x(τ)/x(0).
Consequently, if x = 0, $\dot{x}$ = 0 is a fixed point, we can test
whether it is elliptic or hyperbolic by seeing how a point on
the x-axis maps under T. If $|x(\tau)| < |x(0)|$, the fixed point
is elliptic; if $|x(\tau)| > |x(0)|$ the fixed point is hyperbolic.

Matrix Formulation

If in (10.12) we set x_1 = x and x_2 = $\gamma\dot{x}$, then the equations
become

$$x_1(\tau) = x_1(0) \text{ ch } \psi + x_2(0) \text{ sh } \psi$$
$$x_2(\tau) = x_1(0) \text{ sh } \psi + x_2(0) \text{ ch } \psi$$

(10.13)

or in matrix form

$$X(\tau) = AX(0)$$

where

$$A = \begin{pmatrix} \text{ch } \psi & \text{sh } \psi \\ \text{sh } \psi & \text{ch } \psi \end{pmatrix} \text{ and } X = \begin{pmatrix} x_1 \\ x_2 \end{pmatrix}$$

We now ask: Is it possible to find a vector (eigenvector)
such that the transformation stretches it by a factor λ
(called an eigenvalue)? If so, then

$$x_1(0) \text{ ch } \psi + x_2(0) \text{ sh } \psi = \lambda \, x_1(0)$$
$$x_1(0) \text{ sh } \psi + x_2(0) \text{ ch } \psi = \lambda \, x_2(0)$$

(10.14)

An eigenvalue λ must satisfy the equation

$$\lambda^2 - 2\lambda \text{ ch } \psi + 1 = 0 \qquad (10.15)$$

so that there are 2 possible values, namely $\lambda_1 = e^{\psi}$ and $\lambda_2 = e^{-\psi}$. For the value λ_1, the eigenvector has components $x_{21}(0) = x_{11}(0)$ and for λ_2 the components satisfy

$$x_{22}(0) = - x_{12}(0).$$

In other words, the eigenvectors lie along the above asymptotes, and the eigenvalues are the stretch factors already mentioned. In matrix notation,

$$X(\tau) = AX(0) = \Lambda X(0)$$

where

$$\Lambda = \begin{pmatrix} \lambda_1 & 0 \\ 0 & \lambda_2 \end{pmatrix} \text{ and } X = \begin{pmatrix} x_{11} & x_{12} \\ x_{21} & x_{22} \end{pmatrix}$$

If we set $\lambda = e^{s\tau}$, then

$$x(t + \tau) = e^{s\tau} x(t)$$

Multiplying both sides by e^{-st},

$$e^{-s(t + \tau)} x(t + \tau) = e^{-st} x(t) = P(t),$$

a periodic function of t.

The quantities s are called the <u>characteristic exponents</u>. They are real at a hyperbolic point, and imaginary at an elliptic point. Since $\lambda_1 = 1/\lambda_2$, because of area preservation for a Hamiltonian system, it follows that $s_1 = - s_2$. The characteristic exponents thus <u>occur in pairs</u> for a Hamiltonian system.

The motion corresponding to the characteristic exponent s is described by the function $x_s(t) = e^{st} P_s(t)$ where $P_s(t)$ is a function with period τ. If s is a pure imaginary, the motion is oscillatory and neither runs away nor damps to zero. If s has a negative real part, then $x_s(t) \to 0$ as $t \to \infty$ and the motion is said to be <u>asymptotically stable</u>. If s has a positive real part, $|x_s(t)| \to \infty$ as $t \to \infty$ and the motion is termed <u>unstable</u>. In the design of actual apparatus, it is most practical to aim for the oscillatory case (since asymptotic stability requires rather special injection conditions) and this is achieved when $|\cos\sigma| \le 1$. But when $\alpha = 0$, this means that $|M_{11}| \le 1$, or that M_{11} can range between -1 and $+1$. If $p(t) = p_o + p_1(t)$, where p_o is a constant, then it is possible for p_o to vary over a considerable range of values and still keep the motion oscillatory. This range of eigenvalues p_o will be continuous and may be called the <u>stability band</u>. In general, more than one such band can exist.

B. FIRST ORDER STABILITY: ISOENERGETIC VARIATIONS

When there is a simply—periodic motion in a plane, and the Hamiltonian does not involve the time explicitly, the preceding theory can be used to decide whether or not the system is stable to first order. The treatment will be limited to a comparison of simply-periodic orbits which are symmetric about the x-axis and which have the same value of the energy constant E.

If an orbit near to a periodic orbit leaves the x-axis with initial values x_o and $\dot{x}_o$, it will after about one period again cross the x-axis in the same direction with values x_1 and $\dot{x}_1$. The initial point is thus mapped into the final point by a transformation T

$$x_1 = f(x_o, \dot{x}_o, E)$$
$$\dot{x}_1 = g(x_o, \dot{x}_o, E)$$

$$(10.16)$$

Repeated application of this transformation may give a
series of points in the $(x, \dot{x})$ plane which lie near the initial
point. If this continues indefinitely, then the periodic
orbit is __stable__. However, the points $T^n(P)$ may leave the
neighborhood of the initial point, often quite rapidly, and
there may be no intersection of the orbit with the x-axis, in
which case the periodic solution will be termed __unstable__.

Since the system is a Hamiltonian one, area in the $(x, \dot{x})$
plane is preserved, i.e. $\partial(f, g)/\partial(x_o, \dot{x}_o) = 1$. Furthermore,
the equations of motion and the Jacobi integral remain the
same if the signs of x and of t are changed simultaneously.
Therefore, Equation (10.16) has as a consequence

$$x_o = f(x_1, -\dot{x}_1, E)$$
$$-\dot{x}_o = g(x_1, -\dot{x}_1, E) \tag{10.17}$$

(This is because each orbit has a corresponding orbit obtained
by reflection in the x-axis, and traced out in the opposite
direction).

Now if $(x_o, \dot{x}_o)$ refer to the invariant point, the variational
equations will be

$$\Delta x_1 = a\, \Delta x_o + b\, \Delta \dot{x}_o$$
$$\Delta \dot{x}_1 = c\, \Delta x_o + d\, \Delta \dot{x}_o \tag{10.18}$$

where

$$a = f_x, \quad b = f_{\dot{x}}, \quad c = g_x, \quad d = g_{\dot{x}} . \tag{10.19}$$

Since area is preserved, $ad - bc = 1$. From (10.17)

$$\Delta x_o = a\, \Delta x_1 - b\, \Delta \dot{x}_1$$
$$-\Delta \dot{x}_o = c\, \Delta x_1 - d\, \Delta \dot{x}_1 \tag{10.20}$$

But inversion of (10.18), combined with (10.19), gives

$$\Delta x_o = d \, \Delta x_1 - b \, \Delta \dot{x}_1$$

$$\Delta \dot{x}_o = -c \, \Delta x_1 + a \, \Delta \dot{x}_1$$

(10.21)

If the orbit is simply periodic and symmetric about the x-axis, then

$$x_o = x_1, \; \dot{x}_o = \dot{x}_1 = 0 \qquad (10.22)$$

and the coefficients a, b, c, and d in (10.20) and (10.21) are the same, since they refer to the same point. Comparing (10.20) and (10.21), we have

$$a = d. \qquad (10.23)$$

The eigenvectors of the transformation are defined by

$$\Delta x_1 = \lambda \, \Delta x_o, \; \Delta \dot{x}_1 = \lambda \, \Delta \dot{x}_o$$

so that (10.18) will be satisfied when λ is a root of the equation

$$\lambda^2 - 2a \, \lambda + 1 = 0$$

If $|a| > 1$, there are 2 real roots and the fixed point is hyperbolic and unstable.

If $|a| < 1$, the characteristic exponents s defined by $\lambda = e^{st}$, are imaginary and of opposite signs. The fixed point is elliptic, and a special investigation is required to determine whether or not it is stable.

To ascertain whether the fixed point is hyperbolic or elliptic, one need only set $\Delta \dot{x}_o = 0$ and calculate $a = \Delta x_1 / \Delta x_o$. This is exact only if Δx_o is sufficiently small, but in practice one obtains the value of a with satisfactory precision.

C.   MATHIEU-HILL EQUATION: SOLIDS

An early application of the Mathieu-Hill equation was made
by Kronig and Penney, who examined the wave motion of an electron in a one-dimensional periodic potential U.   They concluded that standing waves (stationary states) are possible if
the total energy E is confined to certain ranges of values.
In other words, if the potential energy inside a solid is
periodic, then there exist energy bands.   (The Kronig-Penney
model is a simplified one, because it is a one-dimensional,
rather than a three-dimensional, one, and the actual potential
energy used in their calculations is really only schematic.
Nevertheless, it does indicate the general nature of the
solutions and has led to the introduction of more accurate
models.)

The amplitude u of an electron wave in one dimension, moving
in a periodic potential $U(x)$, is governed by the wave equation

$$\frac{d^2u}{dx^2} + p(x)\, u = 0 \qquad\qquad (10.24)$$

where $p(x) = (2m_o/\hbar^2)\,(E - U)$.   The stipulation is made that
the charge density, which is proportional to u^2, shall be
finite everywhere.

Equation (10.24) has the same form as (10.1), and so the
same kind of analysis is applicable.   The first thing to do is
to integrate the equation for an interval x equal to the
period τ of the potential.   For simplicity, the potential will
be taken to be of a square wave form, for which the function
$p(x)$ assumes alternately two constant values.

The potential energy will be taken as U_o in the range
$-b/2 \le x \le b/2$ and equal to zero from $b/2$ to $a + (b/2)$.   The

total energy is to be denoted by E as usual, and the potential is to be periodic with period $\tau = a + b$.

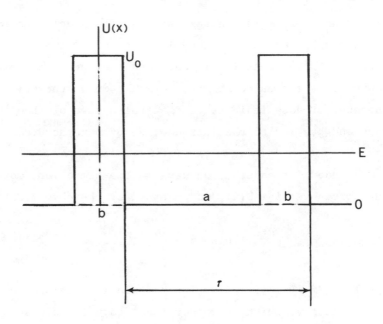

Figure 10.2

Square-Wave Potential: Broad Valleys

The function $p(x)$ is symmetrical about $x = 0$, and is given
by

$$p = (2m_o/\hbar^2)\ (E - U_o) = -m^2 \quad (-\frac{b}{2} < x < \frac{b}{2})$$

$$p = (2m_o/\hbar^2)\ E = n^2 \quad (\frac{b}{2} < x < a + \frac{b}{2})$$

In the region where $U = 0$, the vector $(u, u'/n)$ undergoes a
rotation through the angle na, while for $U = U_o$ there is a

corresponding transformation according to (10.12) with angle
mb.   Setting β = 1/n and γ = 1/m, the transformation from
0 to a + b will be the result of three separate ones, from
0 to (b/2), from (b/2 to (b/2 + a), and from there to a + b.
Thus the total matrix M operating on (u,u') is M = BAB where

$$A = \begin{pmatrix} \cos na & (1/n)\sin na \\ -n \sin na & \cos na \end{pmatrix}$$

$$B = \begin{pmatrix} ch\ mb/2 & (1/m)sh\ mb/2 \\ m\ sh\ mb/2 & ch\ mb/2 \end{pmatrix}$$

Matrix M has determinant equal to unity, since this is so for
each individual matrix.   Its diagonal element is

$$\cos \sigma = ch\ mb \cos na + \frac{m^2 - n^2}{2mn}\ sh\ mb \sin na$$

$$(10.25)$$

The condition for stability, or that the wave function remain
finite at infinity, is that $|\cos \sigma| \leq 1$, as we have seen
above.   This takes a simpler form if we let b go to zero and
U_o go to infinity such that m^2 b remains finite.   If we let
P = lim m^2ab/2, then

$$\cos \sigma = \cos na + P\ (\sin na)\ /\ na. \quad (10.25a)$$

The right-hand side may be plotted against na.   Starting at
P + 1 for na = 0, it decreases to -1 for na = π, and oscillates
with decreasing amplitude.   There are sections of the
horizontal axis for which $|\cos \sigma| \leq 1$, and the corresponding
values of n (or E) give the energy bands. (See Figure 10.3.)

If P = 0, this means that there is no barrier and all values
of the energy are allowed.

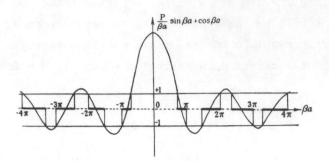

Figure 10.3

(After Kronig & Penney.) The right-hand side of (10.25a)
is plotted as a function of the argument βa (na in the
present notation), for P = 3π/2. The dark segments of
the horizontal axis correspond to stable (non-runaway)
solutions of (10.24)

If P → ∞, the barrier height becomes very great, and the
energy bands shrink down to discrete levels for values of na
which are integral multiples of π, which corresponds to the
wave function vanishing at the steep walls at t = 0, π, 2 π,
etc.

Now let us adopt a somewhat more physical model, that of a
deep well (Figure 10.4) of depth V_o and width b, where
b → 0 and $V_o b$ remains finite while V_o → ∞. Let the energy
E' be zero at the top of the well, so that E' = E − V_o. Then

$$p = (2m_o/\hbar^2) \, E' = -m^2 \quad (-\frac{a}{2} < x < \frac{a}{2})$$

$$p = (2m_o/\hbar^2)(E' + V_o) = n^2 \quad (\frac{a}{2} < x < \frac{a}{2} + b)$$

In the limit, $m^2 \ll n^2$ and $n^2 b/2 = m_o V_o b/\hbar^2 = Q$, say.
Setting σ = kτ, with τ ≅ a, we have, from (10.25)

$$\cos k\tau = \mathrm{ch}\ m\tau - (Q/m)\ \mathrm{sh}\ m\tau$$

$$= \frac{1}{2}\,[e^{m\tau}\,(1 - \frac{Q}{m}) + e^{-m\tau}\,(1 + \frac{Q}{m})]$$

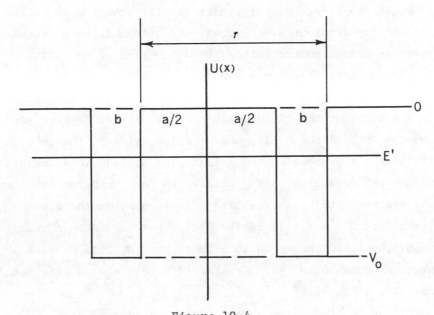

Figure 10.4

Square-Wave Potential: Narrow Valleys

If now $Q\tau$ be made large, then $e^{-Q\tau}$ is small, and for $\cos k\tau$ to be within bounds (for real k) then $Q/m \to 1$. Neglecting the second term in the bracket,

$$1 - \frac{Q}{m} = 2 \cos k\tau \, e^{-Q\iota}$$

or
$$m^2 \cong Q^2 \, [1 + 4e^{-Q\tau} \cos k\tau]$$

The allowable band energies are then given by

$$E' = - E'_o \, [1 + 4e^{-Q\tau} \cos k\tau]$$

$$= - E'_o - 4E'_o e^{-Q\tau} + 2E'_o k^2 \tau^2 \, e^{-Q\tau} \text{ (small k).}$$

The width of the band is proportional to $e^{-Q\tau}$, which is a small quantity if Qb is large.

D. MATHIEU-HILL EQUATION: A. G. SYNCHROTRON

Another application of the Mathieu-Hill equation is to the alternating-gradient synchrotron, which is a circular accelerator in which the field index is a function of the azimuth ϕ, the variation being made so as to ensure that the number of particles lost from the beam is kept to a minimum. Whereas in the betatron the field index $n = - (\rho/B) (dB/d\rho)$ was kept constant around the orbit, now we allow $dB/d\rho$ to change sign frequently (by changing the orientation of individual magnets) whence the term <u>alternating gradient</u>. For the betatron, there was the very stringent stability requirement on the field index that $0 < n < 1$ (Equation 6.35), but now, when n changes sign, the field index can be made large, and this results in great economy in design. We proceed to discuss in some detail how this comes about.

For the betatron, the oscillation equations were:
Radial,

$$\frac{d^2x}{d\phi^2} + (1 - n)\ x = 0 \qquad\qquad (6.36)$$

Vertical,

$$\frac{d^2z}{d\phi^2} + n\ z = 0 \qquad\qquad (6.37)$$

If now, n be made large and alternating between $n_1 > 0$ and $n_2 < 0$, the two equations may both have a net focussing effect, since we have a succession of lenses, convex-concave etc. Positive values of n focus, while negative values defocus. For the vertical oscillations, for instance, let $n = n_1$ for an increment $\Delta\phi = \phi_1$ and $n = - n_2$ for an increment $\Delta\phi = \phi_2$. If the number of alternations in one revolution be N, then $N(\phi_1 + \phi_2) = 2\pi$.

The equation governing the oscillation is now of the form
(10.1), with period $\tau = 2\pi/N$. We have

$$\frac{d^2 z}{d\phi^2} + p(\phi)\, z = 0$$

where

$$p = n_1 \text{ for } 0 < \phi < \phi_1$$

$$p = -\, n_2 \text{ for } \phi_1 < \phi < \phi_1 + \phi_2$$

The analysis then proceeds exactly as in the case for the
energy bands in solids, and we need only set $n_2 = m^2$, $n_1 = n^2$
and apply (10.25) to find the stability region. The criterion
is $|\cos \sigma| \le 1$, where

$$\cos \sigma = \text{ch } m\phi_2 \, \cos n\phi_1 + \frac{m^2 - n^2}{2mn} \text{ sh } m\phi_2 \, \sin n\phi_1$$

$$(10.26)$$

If $n_1 \gg 1$, $n_2 \gg 1$, and $\phi_1 = \phi_2 = \pi/N$, then the region of
stability, according to Courant, Livingston, and Snyder (1952)
is within the "necktie" shown in Figure 10.5. With proper
design, one can easily achieve stable oscillations, both
radial and vertical.

Figure 10.5

(After Courant, Livingston,
   and Snyder)

Regions of stability for
radial and vertical oscillations for a large number of
sectors N, in terms of the
parameters n_1/N^2 and n_2/N^2.

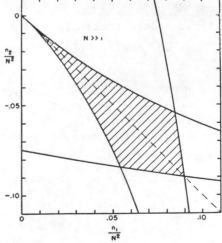

E. NON-LINEAR PERIODIC SYSTEMS

As has already been indicated, the first order variational
equations are not always a reliable or sufficient approximation from which to deduce stability. One can find out where
the hyperbolic fixed points are and then say that the particle
motion tends to avoid these. But as concerns the elliptic
fixed points, the neighborhood in which motion is in an ellipse
is undetermined in magnitude, and this may be a crucial factor
in the design of an accelerator. For these reasons, it is
necessary to study actual non-linear systems, first locating
the periodic orbits and secondly making estimates of the size
of the stable region around each fixed point.

To keep things simple, one can take the motion as one-dimensional and obeying the equation

$$\frac{d^2x}{dt^2} + p(t)\ f(x) = 0 \qquad\qquad (10.27)$$

The function $f(x)$ may be chosen arbitrarily, but it has proved
convenient and instructive to take $f(x) = x^3$ and to study this
case in detail. Periodic solutions have been found, and
factors which seem to be responsible for long-term (not just
first order) stability have been isolated.

Let us consider the differential equation

$$\frac{d^2x}{dt^2} + p(t)\ x^3 = 0 \qquad\qquad (10.28)$$

where $p(t)$ is a periodic square wave function with period τ
and

$$p = p_o \ \text{for} - \tau/4 < t \leq \tau/4, \ p_o > 0$$

$$p = - p_o \ \text{for} \ \tau/4 < t \leq 3\tau/4.$$

Fixed Points (Periodic Solutions)

If the motion starts from any point P in the phase plane,
then after one period the particle will have arrived at a
uniquely determined point P' = TP, or P has undergone a transformation T which moves it to P'. Fixed points of order n are
such that $T^n P = P$, and correspond to periodic solutions of
(10.28), since T refers to the change due to a replacement of
t by t + τ. (The motion is to start in the middle of a sector,
with p > 0.) The fixed points can be located by a systematic
mapping of the points in the phase plane. This was done
(Bartlett 1960, 1967) for fixed points up to n = 14. It was
found that all the fixed points lay either on the x- or ẋ-axis
or were maps under T^k (k < n) of such points. Higher order
fixed points are in general closer to the origin.

Reference to Figure 10.6 shows that, as we go around the
origin, elliptic and hyperbolic points for a given T^n alternate. This is a general property which can be rather intui-
tively understood. For suppose that we have located one elliptic fixed point E such that $T^n E = E$. Other elliptic fixed
points for the same value of n are generated by applying the
transformation T repeatedly to E, i.e. TE, $T^2 E \cdots T^{n-1} E$. (This
corresponds very roughly to rotating the point through a certain
angle each time.) Around each such elliptic fixed point, if we
are close enough to it so that the mapping can be approximated
by a linear one, any point P is mapped by T^n, T^{2n}, etc. into
other points which lie on an ellipse. Now let the distance of
P from the fixed point be gradually increased. A simple-minded
view of what might occur is that the ellipse can become distorted into a oval which swells out to meet its map, a corresponding oval around a neighboring elliptic point, and that the
meeting place is marked by a hyperbolic fixed point.

In practice the "oval" is only approximately a simple closed curve, but its ends do butt up against corresponding ends of neighboring ovals to give hyperbolic trajectories near the hyperbolic point. The alternation of hyperbolic and elliptic fixed points is thus more or less obvious from the topology.

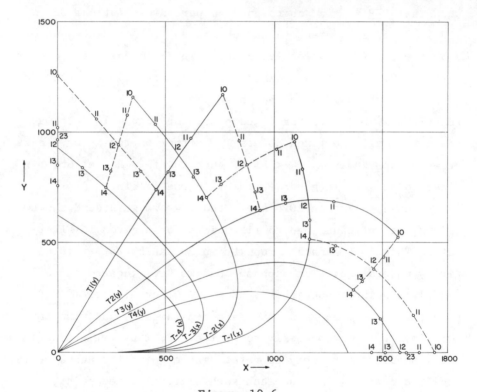

Figure 10.6

Fixed Points for Cubic Equation (10.28), n = 10 to n = 14. Plot of y = $\dot{x}$ vs. x. Vertical Scale: 10^4 $\dot{x}$, Horizontal Scale: 10^3 x. Also mappings $T^{-n}(x)$ and $T^n(y)$ for n = 1 to n = 4.

Stability and Eigencurves

Even though a study of the mapping near a fixed point immediately reveals whether this point is elliptic or hyperbolic, a further investigation is needed to ascertain whether

or not an elliptic point is <u>stable</u>. Such a fixed point is
defined to be stable if there is a simply-connected region
around it such that no particle within this region can get
outside. We shall show below how one can construct such a
region for the mapping in question, if the region exists.

Close to a hyperbolic point $H = T^n H$, the mapping is approximately linear and there are 2 characteristic lines through the

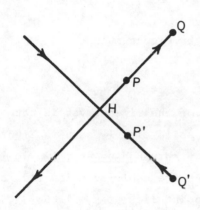

point. Along one line (see
Figure 10.7) any point P will
be mapped by T^n into $Q = T^n(P)$,
the distance HP being stretched
to $HQ = \lambda_1 \times HP$. Along the
other line, a point Q' goes
into a point P' nearer to H,
and the distance $HP' = HQ'/\lambda_1$.

As the distance from the
hyperbolic point is increased,
the mapping usually becomes
non-linear, and these invariant lines are replaced by

Figure 10.7

Mappings along character-
istic lines at hyperbolic
point.

invariant <u>eigencurves</u>, which may be determined numerically by
starting with a short straight part of a characteristic from
the hyperbolic point and finding how it is mapped.

Now consider the situation between 2 consecutive hyperbolic
points H_1 and $H_2 = TH_1$, where $T^n H_1 = H_1$ and $T^n H_2 = H_2$. As
we shall demonstrate below, it is possible to devise a
special mapping T such that an outgoing eigencurve from
H_1 passes through H_2, and such that an outgoing eigencurve
from H_2 passes through H_1, as in Figure 10.8. These
curves enclose a region containing an elliptic fixed point
$E = T^n E$. The motion is rotatory around E, and no points
can cross the eigencurve, so that the interior is completely
stable.

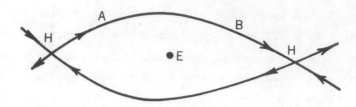

Figure 10.8

Stable Region Bounded by Eigencurves

Oscillations and Instability

When the angle of intersection of the eigencurves is not
zero, they develop oscillations, as is clearly seen in Figure
10.9, which pertains to the region surrounding an elliptic
fixed point E_1 of order $n = 11$ for the cubic equation (10.28).
(For $p_o = 0.037$ and $\tau = 6$, this point lies at $x = 1.666735$,
$x = 0$.) Neighboring hyperbolic points for $n = 11$ are H_1
($x = 1.6395$, $\dot{x} = 0.0169$) and its mirror image, H_2, in the
x-axis, as shown in Figure 10.9. (According to Birkhoff
(1913), an elliptic point of an area-preserving mapping will
have higher order fixed points nearby. Figures 10.10 and
10.11 show such points near E_1 for $n = 33$ and $n = 55$, with h_1
for $n = 33$ located at $x = 1.67382$, $\dot{x} = 2.11 \times 10^{-3}$. These
points will play a part in our further discussion.) Now
halfway between H_1 and H_2, at Y, the outgoing eigencurve
H_1GY intersects the ingoing eigencurve $Y\overline{G}H_2$ at a rather
large angle. (Similarly, at W, the outgoing eigencurve
H_2FW intersects the ingoing eigencurve $W\overline{F}H_1$ at a finite but
smaller angle.) From H_1 to Y the outgoing eigencurve G
does not oscillate, nor does the ingoing eigencurve $\overline{G}$ from
Y to H_2. But G is oscillatory after Y, as is $\overline{G}$ before Y.

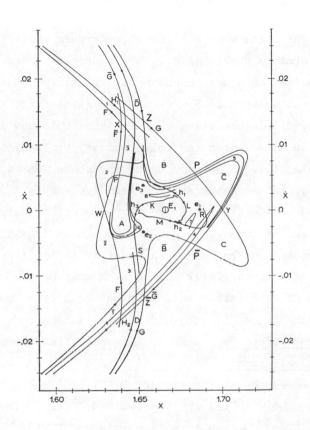

Figure 10.9

Eigencurves of the mapping T^{11} and T^{33}, cubic periodic force. Elliptic points of T^{33}: e_1, e_2, e_3; hyperbolic points of T^{33}: h_1, h_2, h_3. For h_1: $x = 1.67382$, $\dot{x} = 2.11 \times 10^{-3}$.

From inspection, <u>oscillations must occur</u>. The point Y maps into $\bar{Z} = T^n(Y)$ on $\bar{G}$. A point on G just before Y must map into a point just inside $\bar{Z}$ (toward E_1). Therefore the curve G, which has crossed $\bar{G}$ at Y, must cross back between Y and $\bar{Z}$, say at $\bar{P}$.

The region bounded by H_1GY, $Y\bar{G}H_2$, H_2FW, and $W\bar{F}H_1$ will be called a <u>cell</u>, and we shall try to ascertain what fraction of the cell area is stable, i.e., is not mapped outside the cell. The definition of a cell boundary is arbitrary, and we could just as well substitute H_1GP, $P\bar{G}Y\bar{G}H_2$ for the first two curves above, the difference being that the area $\bar{C}$ is now included. However, we have chosen the cell boundary to look as simple as possible, i.e. to be made up of the above non-oscillatory portions of the outgoing and ingoing eigencurves.

Repeated application of T^n to the segment $Y\bar{Z}$ yields smaller segments on the curve $\bar{Z}H_2$, each one of which is smaller than the preceeding one by roughly a constant factor (the "stretch" factor mentioned earlier). Since area must be preserved, the amplitude of oscillation must increase each time to offset the compression against F. The end result is that <u>roughly one half of the points on G</u> near H_1 "escapes" from the cell <u>and the other half remains in the cell</u> and approaches F.

As G proceeds backwards (toward $H_1\bar{F}W$) from Y, it next crosses $\bar{G}$ at $P = T^{-n}(\bar{P})$ and then at $Z = T^{-n}(Y)$. The region B between $\bar{G}$ and G, from Z to P, maps under T^n into a corresponding region C between G and $\bar{G}$ from Y to $\bar{P}$. Thus points inside the cell have gotten outside, or escaped. The complete region of escape (unstable region) is composed of B and the <u>corresponding area 3 between F and $\bar{F}$, plus all of their maps under repeated application of T^{-n}</u>.

If, then, we wish to construct the boundary of the stable region, we can start from Y, proceed along G to P, then

along $\bar{G}$ to Z, and then alternate between G and $\bar{G}$ indefinitely.
The same is to be done starting from W and going toward $Y\bar{G}H_2$,
alternating between F and $\bar{F}$. The composite curve so generated
is enormously complicated, since area 3 maps under T^{-n} successively into 4, 5, 6 .... As we see from the figure, area
5 has a long thin tongue running alongside the boundary of
area B. Area A, mapped under T^{-n}, will give a tongue between
3 and 4. The tips of the tongues appear in many places, and
can on occasion come close to an interior elliptic point
(e.g., e_1, n = 33). Whether they do or not will depend on
the ratio of the <u>fluctuation area</u> $\bar{C}$ to the area of the cell
itself. For small ratios, they stay on the periphery of the
cell, but for large ratios they invade the interior. Referring
to Figure 10.10, we see that the mirror images $\bar{4}$, $\bar{7}$, $\overline{10}$ ...
of 4, 7, 10 ... pile up onto the n = 33 eigencurve h_1L, and
so 4, 7, 10 ... will pile up onto $h_2\bar{L}$. When this happens,
then it becomes in practice impossible to say what points are
stable and what are not, because stable and unstable regions
occur in alternate streaks, outside the interior eigencurves
of n = 33.

On the other hand, Figure 10.11 shows how the interior
eigencurves for n = 33 oscillate very little, with the intersection angle between h_1L and its mirror image $h_2\bar{L}$ very small,
and the ratio of fluctuation area (or loop area) to the cell
area also small. The tongues seem to stay close to the
boundary of the cell, and probably converge on some inner
limiting curve not far from the boundary. If this can be
proved by numerical calculation or otherwise, then it would
definitely show that the region inside the limit is a stable
one, for particles can only escape from inside the tongues,
which would lie outside the limit. Such a calculation has
not yet been made, but it has been shown in cases where the

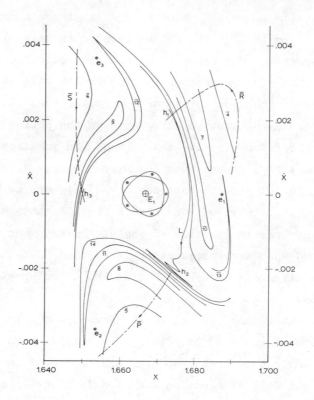

Figure 10.10

Eigencurves, n = 11, 33, 55.   (Cubic periodic force)

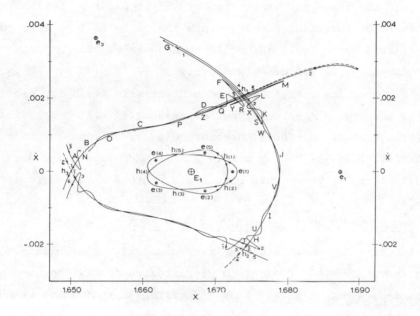

Figure 10.11

Cubic Periodic Force

Eigencurves of T^{33} and T^{55}. Fixed points of T^{33}: e_i and h_i; fixed points of T^{55}: $e(i)$ and $h(i)$.

intersection angle is very small that the probability of
escape from just inside the cell boundary is astronomically
small. In other words, there is good, but as yet not
rigorous, <u>evidence</u>, that <u>there is a region of eternal stability just inside a cell boundary, provided that the intersection angle between forward and backward (from the next
hyperbolic point) eigencurves is imperceptible</u>. (In Figure
10.11, the inner pattern is for n = 55 eigencurves, which
seem to be invariant curves, since the ratio of loop area to
cell area is so small as not to be measurable from a plot
produced by precise computer data.)

The foregoing remarks seem to be generally valid for simple
area-preserving mappings, whether associated with a dynamical
system or not. We shall therefore discuss at some length
what one finds by taking a very simple such mapping, where
there is a bit more control over the factors which enter.

F.   STABILITY OF AN ALGEBRAIC AREA-PRESERVING MAPPING

In order to understand the factors underlying stability, it
is useful to study a simple algebraic area-preserving mapping
which does not originate from a Hamiltonian system but which
does have eigencurves which intersect their companions at a
non-zero angle, and therefore oscillate with increasing
amplitude. This has the advantages (1) that we bypass time-
consuming integration of differential equations and (2) that
we can choose a family of mappings characterized by a parameter, the variation of which can on occasion rapidly destroy
the stability. We take as our example the mapping of Hénon
and Heiles (1964) which is area-preserving and nonlinear, and
hence like those derived from conservative nonlinear dynamical
systems. (It is different in that it is a polynomial mapping,
so that a point can not go to infinity in one step.)

The mapping, T, of the plane onto itself is defined to be:

$$T\ (x,y) = (x',y')$$

where

$$x' = x + a\ (y - y^3)$$

$$y' = y - a\ (x' - x'^3).$$

Its inverse is

$$T^{-1}\ (x'y') = (x,\ y):$$

$$x = x' - a\ (y - y^3)$$

$$y = y' + a\ (x' - x'^3).$$

For a = 1.6, according to Hénon and Heiles, successive points
of the mapping appear to group themselves along invariant
curves. In Figure 10.12 there are such (solid) curves which
enclose the origin, as well as satellite curves enclosing the
fixed points (e_o to e_5) of order n = 6. Since the stable
regions around fixed points with n < 6 are quite limited in
extent, the present investigation deals with stability for
n ≥ 6.

The elliptic fixed points for n = 6 are at (t, 0), (t, -t),
(0, -t), (-t, 0), (-t, t), and (0, t) and are labelled e_o to
e_5 consecutively. The interlacing hyperbolic points, h_o to
h_5, are located at (± r, ∓ s), (± s, ∓ r), and (± s, ∓ s).
Values of r, s, and t as functions of the parameter a are
given in Table I.

The pattern of six elliptic fixed points ringing the origin,
which is an elliptic fixed point, is maintained for values of
the parameter, a, from 1 to 2. For a > 2 the origin becomes
hyperbolic and for a < 1 the satellite sixth order points
disappear. In general, we find that fixed points and invariant
curves of the mapping exhibit symmetry with respect to the line

x + y = 0 or to x - y = 0.

Table I

CONSTANTS FOR FIXED POINTS OF T^6		
a = 1.5	a = 1.6	a = 1.7
r 0.63380654	0.70287083	0.73512227
s 0.27848056	0.28450670	0.28717990
t 0.57735027	0.61237244	0.64168895

   If one plots successive mappings of a few selected points,
then in some cases these successive mappings can be connected
by a closed curve or curves, as is shown in Figure 10.12.
When there is no detectable departure from such a curve after
10^5 mappings, we will assume that a closed invariant curve
has been found. For those shown in Figure 10.12, the central
curve is invariant under T and each of the six 'satellite'
curves is invariant under T^6. They surround the origin and
e_o through e_5 respectively, and the regions inside seem to be
invariant (and therefore stable). For a point sufficiently
distant from an elliptic fixed point, its successive mappings
appear as scattered points outside the closed curve region,
and at first glance they obey no discernible pattern. However,
when one has located the hyperbolic points (not done by Hénon
and Héiles) the manner in which a point will map can become
more evident. Some writers have alluded to the absence of
pattern as evidence that the particle shows an ergodic, or
random, behavior, but this would seem to require proof.
   The existence of closed invariant curves about an elliptic
fixed point has been a debatable issue for some time. But the
calculations show that there are closed curves, to the limits

of accuracy of the computer. There is then for each such
curve an _integral relation_, and so the curves are supposed to
represent an integral of the motion for a Hamiltonian system.
However, _other integrals_ such as the energy integral and the
angular momentum integral _are generally valid_, the former
arising because the Hamiltonian does not involve the time
explicitly, and the latter because the system has no torque
on it. In the present instance, on the other hand, the basic
datum is that area must be preserved, and from it there
follows that, _only in some regions_, _there are closed invariant_
curves.

If we can find the largest closed invariant curve surrounding an elliptic fixed point, then we have bounded the largest
connected invariant (and therefore stable) region associated
with that fixed point. Conversely, a stable region must be
invariant under T^n for some n and by the fixed point theorem
must therefore contain a stable fixed point. This suggests
that the compartmentalization into stable and unstable regions
could be made by finding all elliptic fixed points and the
associated largest closed invariant curves. This, however,
would involve an inordinate amount of computer time, and so
we present below a practical method of approximating the
stability boundary.

Development of Eigencurves

To take advantage of the symmetry about x + y = 0, we have
examined the region between h_o and h_1, where there is a stable
part which includes e_1. The saddle-point nature at h_o means
that there will be one eigencurve along which points are
mapped away (outgoing) from h_o toward h_1, while along another
(incoming) eigencurve points from the vicinity of h_1 are
mapped toward h_o.

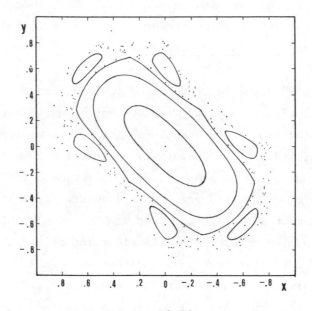

Figure 10.12

Invariant curves of T and T^6, algebraic mapping, a = 1.6.
The scattered points are all successive mappings of one point.

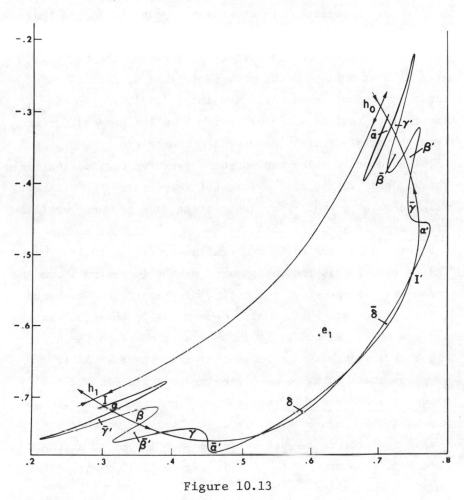

Figure 10.13

Eigencurves for h_o and h_1 for a = 1.6, showing oscillations and how they map.

Figure 10.13 shows a typical eigencurve pattern for a = 1.6
on which the oscillations are apparent on the outer (from the
origin) pair of eigencurves. They also occur on the inner
pair but the curves remain nearly coincident until the fixed
points are approached. In the figure, point I' is the image
under T^6 of I (in this case four cycles of eigencurve oscillation occur between consecutive image nodes of the curves),
area α' is the image of α, etc. Note that α, β, γ, δ and
their antecedents eventually map into α', β', γ', δ' and
thence into regions which may carry them further and further
from the stable regions near e_1 and the origin.

The eventual development of the eigencurve oscillations into
the interior region near e_1 is best illustrated for a larger
value of the parameter, a, because the size of these oscillation loops is greater.

Figures 10.14 and 10.15 are similar to Figure 10.13, except
that a = 1.7, only the inner parts of the loops are shown and
these only on one side of the line of symmetry. The shaded
region represents points which appear stable under 10^4 inter-
actions of T. Figure 10.14 shows a part of the outgoing
eigencurve from h_1 as it is in the early stages. As it
develops under more and more mappings, it becomes wrapped
around the above stable regions, and long narrow tongues
appear as is shown in Figure 10.15. There is an extensive
invasion of the cell by these tongues or loops and at the
same time some area, such as R in Figure 10.15, gets squeezed
out from the cell.

As a comparison of Figure 10.13 and 10.14 will show,
a modest increase in 'a' can produce a large change in the
size of the oscillation loops, and a rapid variation of the
angle of eigencurve intersection. In Figure 10.16 the angle
of intersection of the inner eigencurves of $h_o \rightarrow h_1$ is

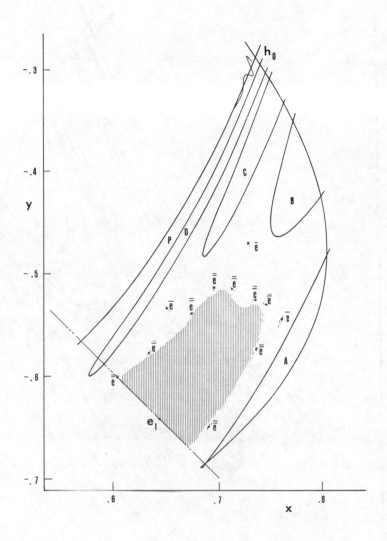

Figure 10.14

Eigencurves for h_o and h_1, for a = 1.7, showing early inner parts of oscillation and region of stability about e_1 as well as the satellite points $\bar{\bar{e}}$ fixed under T^{102}.

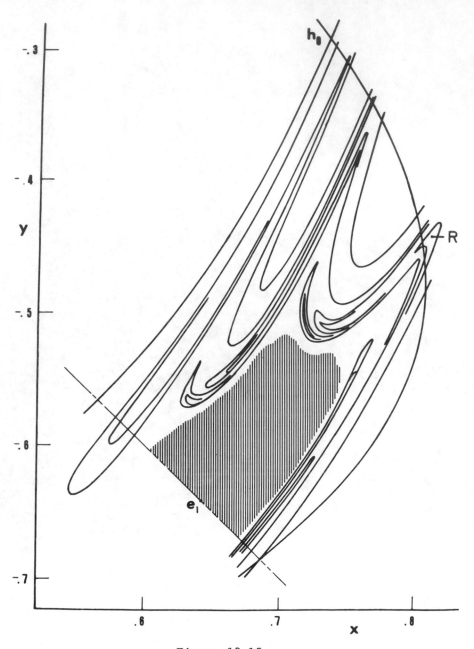

Figure 10.15

Eigencurves for h_0 and h_1, for $a = 1.7$, showing later stages
of oscillation and region of stability about e_1.

shown as a function of a. This angle is measured at the
crossing of the outgoing eigencurve of h_o and the incoming
eigencurve of h_1 at the line of symmetry, $x + y = 0$. (A null
value for this angle would imply the existence of a true
separatrix.) In Figure 10.16a the value of the angle appears
to be zero for $a < 1.5$ but Figure 10.16b, using a logarithmic
scale, shows that this is not the case. It probably becomes
zero at the same time that the sixth order fixed points
collapse into the origin ($a = 1.0$). The increase in the
angle with a is rapid indeed up to about $a = 1.6$.

Stability

As we shall show, the degree to which eigencurves oscillate
may be used to estimate how stable a region is. If the eigencurves from one hyperbolic fixed point go through its
neighbors, the region (cell) they bound is fully stable;
otherwise a certain fraction of the cell area is unstable.
We have examined how the size of the unstable region depends
on the area of an oscillation loop. The loop area is preferable as a parameter to amplitude of oscillation (mentioned
by Mel'nikov) or to the angle of intersection of opposite
eigencurves, because it is an invariant and does not depend
for definition on isolation of a particular loop. Moreover,
its ratio to the area of the cell is probably the signifi-
cant factor, even though shapes and sizes of cells may vary
widely.

The cell boundary will be taken as before: Proceed from
one hyperbolic point along the outgoing eigencurve until the
line of symmetry is reached, then proceed along the eigencurve
going into the other hyperbolic point. Repeat the procedure
along the other side of the cell to close the curve. The area
bounded by the eigencurve segments is then the cell area,

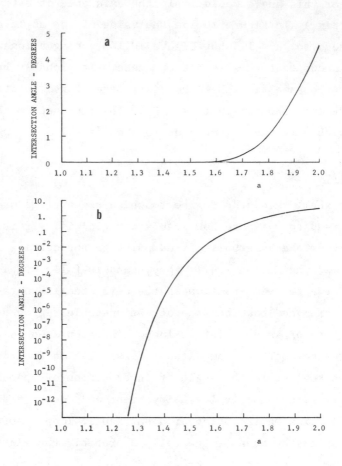

Figure 10.16

Intersection angle of the eigencurve into h_1 with the eigencurve outgoing from h_o, at the line of symmetry $x + y = 0$, as a function of the parameter, a.

denoted by C. The fluctuation or loop area on the outer
eigencurve will be labelled L_o.

The stable area S inside C was determined numerically by a
point-by-point survey under 10^4 mappings. The unstable area
U = C - S was then calculated, and the dependence of U/C on
L_o/C is shown in Figure 10.17. The flagged points are for
a = 1.6, the others for a = 1.7. An examination of Figure
10.17 indicates that for the mapping considered here, if
$L_o/C > 0.2$ almost all the cell is unstable, but for
$L_o/C < 10^{-5}$ almost all the cell is stable. For a = 1.6, this
occurs for hyperbolic points fixed under T^{90}.

The delineation of stability of the plane under repeated
area-preserving mappings onto itself is accounted for by the
study of invariant curves. Boundaries of stable regions are
approached by invariant curves in two limiting processes. In
a stable region such as that surrounding e_1 in Figure 10.15,
i.c.'s are closed and the largest closed invariant curve
which can be drawn in the stable zone is an inner bound of
the margin of stability. An outer bound is formed by the
interior envelope of eigencurves through neighboring hyperbolic points outside the stable region, as may also be seen
from Figures 10.14 and 10.15. Finding either one of these
approximations to the margin of stability requires a very
large number of mappings. The presence of oscillating loops
on the eigencurves through hyperbolic fixed points is an
indicator of instability and the size of these loops is a
measure of their destabilizing tendency.

For the algebraic mapping, Jenkins and Bartlett (1972)
considered (a) the stable region around the origin and (b)
the stable region around the satellite point e_1. Since the
oscillations of the inner eigencurves through h_o, h_1, etc.
are very small, they concluded that the central stable region

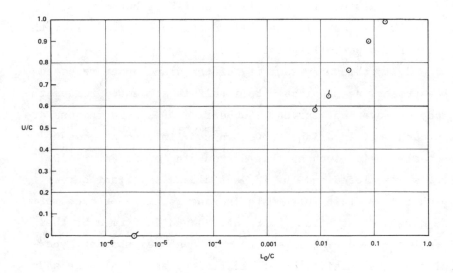

Figure 10.17

Unstable fraction of cell as a function of outer eigencurve oscillation loop area, algebraic mapping.

has them as an approximate boundary. <u>To obtain the stable</u>
<u>region around</u> e_1, <u>a first approximation is obtained by</u>
<u>finding when the satellite hyperbolic points have a stretching</u>
<u>parameter which is approximately unity.</u> <u>The actual region</u>
<u>will be somewhat larger,</u> <u>and this may be found very closely</u>
<u>by determining when the ratio of oscillation loop area to cell</u>
<u>area is very small</u> (<u>in the present case about 10^{-5}</u>). In this
case, the stable region will extend virtually to the hyper-
bolic points and, moreover, practically all the region
enclosed by the inner eigencurves of the connected string of
hyperbolic points will be stable. If the loops are large,
the stable region about the elliptic point contained in the
cell will shrink until for large enough loops (about one-
fifth of cell area in the present case) there is no practical
way of determining whether or not a given point is stable.

When numerical integration of differential equations is
required to produce successive iterations of the mapping, it
becomes extremely difficult to maintain accuracy for a large
number of mappings. However, Hénon (1969) claims that one
can, with less than 10,000 iterations, delimit the region of
stability fairly well, by finding a 'largest' invariant closed
curve which is smooth to the eye. An alternative method is to
find systematically the hyperbolic fixed points and the associated eigenvalues and eigencurves. When the eigenvalue is
close to unity, then the loop area is small and the region
inside the eigencurves is roughly the stable region. The
two methods of determining this region seem to be equally
effective.

No guarantee is given that, for any given system, the
oscillation loop area will tend to zero rapidly as the
central elliptic point is approached, but it is so for the
systems so far studied. When this area is very small, then

successive iterations of the mapping yield even thinner
tongues just inside the eigencurves, and these seem to
approach asymptotically an inner limit which would be the
outer boundary of a stable region. In any case, the number
of mappings required to map an inside point to the outside
appears to increase astronomically, so that the region is, to
all intents and purposes, a really stable one.

Finally, these methods can be applied to find stable regions
for stars moving in a galaxy. Hénon and Heiles (1964) have
displayed invariant curves for typical galactic potentials,
and Contopoulos (1970) shows some invariant curves for our
galaxy. The boundaries of the stable regions have not been
determined, but this can readily be done.

Motion in The Field of a Magnetic Dipole

When an electrically charged particle moves in the field of a magnetic dipole, the trajectories can have fairly complicated shapes. Störmer thought that this could be used to explain phenomena of the polar aurora, but various difficulties arose. Lemaitre and Vallarta assumed cosmic rays to be charged, and they studied many of the properties of the motion of highly energetic charged particles coming from infinity and impinging on the earth, which is approximately a magnetic dipole. Finally, Van Allen interested himself in the oscillatory motion of charged particles in such a magnetic field.

The problem is not as difficult as the restricted 3-body problem, but it is a useful one to study, because some of the techniques for its solution may be used on the more difficult problem. We shall assume that the particle does move in the field of a true magnetic dipole and shall try to find out just what regions of space are accessible to it.

First of all, if the magnetic field is constant in time, the particle will neither lose or gain energy, and its velocity will be constant. The magnetic field just deflects the particle. The equations of motion are

$$\frac{d}{dt} \Gamma m_o \vec{v} = \frac{e}{c} (\vec{v} \times \vec{H}) \qquad (11.1)$$

where $\qquad\qquad \Gamma = [1 - (v^2/c^2)]^{-1/2}$,

and

$$\frac{d\Gamma}{dt} = \frac{\Gamma^3 v}{c^2}\frac{dv}{dt} \qquad\qquad (11.2)$$

The Lagrangian for a charged particle moving in an electro-magnetic field has already been determined (Equation 6.5). It is, with $\Phi = 0$, and charge e,

$$L = \frac{e}{c}(v\cdot A) - m_o c^2 \left(1 - \frac{v^2}{c^2}\right)^{1/2} \qquad\qquad (11.3)$$

The magnetic field does not vary with time, so L does not involve time explicitly and therefore the Hamiltonian $H = mc^2$ is constant and consequently the velocity v is of constant magnitude. This is then a <u>first integral</u> of the motion.

Taking the scalar product of $\vec{v}$ with (11.1), the right-hand side vanishes. The left-hand side, using (11.2), becomes

$$m_o v^2 \frac{\Gamma^3 v}{c^2}\frac{dv}{dt} + \Gamma m_o \vec{v}\cdot\frac{d\vec{v}}{dt}.$$

The first term vanishes because dv/dt = 0, and since the whole expression vanishes, the second term must, also. The acceleration is thus perpendicular to the velocity.

Another first integral may be found because the system has cylindrical symmetry about the axis of the dipole. This integral, or constant of the motion, will be a generalization of the ordinary angular momentum to include a contribution from the magnetic field. For particles of a given energy coming from infinity, we shall then find out, as a function of this angular momentum, what regions of space are accessible.

The vector potential of a magnetic dipole of strength M is

$$\vec{A} = \vec{M} \times \vec{\nabla}(1/r) = -(\vec{M} \times \vec{r})/r^3$$

where $\vec{r}$ is the radius vector. If ρ denotes the distance from the dipole axis and ϕ the longitude, then

$$v \cdot A = \frac{M\rho^2}{r^3} \frac{d\phi}{dt} \qquad (11.4)$$

(Note that longitude on the earth is defined for a left-handed system.)

Let us write $v\, dt = \ell\, ds$ and then adjust ℓ so that the formulae are as simple as possible. A dot will now denote differentiation re s. Then

$$L = \frac{eMv}{c\ell r^3} \rho^2\dot{\phi} - m_o c^2 \left[1 - \frac{v^2}{\ell^2 c^2}(\dot{z}^2 + \dot{\rho}^2 + \rho^2\dot{\phi}^2)\right]^{1/2} \qquad (11.5)$$

The angular momentum p_ϕ conjugate to ϕ is constant, since L is independent of ϕ. It is

$$p_\phi = \frac{\partial L}{\partial \dot{\phi}} = \frac{eMv}{c\ell}\frac{\rho^2}{r^3} + \frac{mv^2}{\ell^2}\rho^2\dot{\phi} \qquad (11.6)$$

The second term on the right-hand side is the angular momentum ordinarily encountered, while the first term is the contribution due to the dipole. The momentum due to the field is $(e/c)A_\phi$, and this contribution is just ρ times the tangential momentum, as is to be expected.

Rewriting (11.6),

$$\frac{\rho\dot{\phi}}{\ell} = \frac{p_\phi \ell}{mv^2\rho} - \frac{eM}{cmv}\frac{\rho}{r^3} \qquad (11.7)$$

If we now adopt as a unit of length the quantity

$$\ell = \sqrt{|e|\,M/mvc}$$

and set $m = 1$ and $v = 1$, Equation (11.7) for electrons ($e < 0$) becomes

$$\rho\dot{\phi} = (p_\phi/\rho) + (\rho/r^3) \qquad (11.8)$$

In these units, conservation of energy is expressed by the equation

$$\dot{z}^2 + \dot{\rho}^2 + \rho^2\dot{\phi}^2 = 1 \qquad (11.9)$$

But we have just calculated the ϕ-component of velocity and found it to be given by (11.8). At large distances from the dipole, it is p_ϕ/ρ as usual, and the square of it is twice the effective potential V_I due to inertial effects (centrifugal potential). Substituting (11.8) into (11.9),

$$\dot{z}^2 + \dot{\rho}^2 + [(p_\phi/\rho) + (\rho/r^3)]^2 = 1 \qquad (11.10)$$

or
$$\dot{z}^2 + \dot{\rho}^2 + 2V_I = 1$$

where
$$2V_I = [(p_\phi/\rho) + \rho/r^3]^2 \qquad (11.11)$$

Equations of Motion

The easiest way to find the equations of motion is to use the Hamiltonian function from (1.28) or (6.8), i.e.,

$$H = c\,[(m_o c)^2 + m^2 v^2]^{1/2}$$

For _cylindrical coordinates_, with $m = 1$, the momenta are $p_z = \dot{z}$ and $p_\rho = \dot{\rho}$. Then

$$H = c\,[(m_o c)^2 + p_z^{\,2} + p_\rho^{\,2} + 2V_I]^{1/2}$$

For spherical coordinates, $p_r = \dot{r}$, $p_\lambda = r^2\dot{\lambda}$, and

$$H = c\,[(m_o c)^2 + p_r^{\,2} + (p_\lambda^{\,2}/r^2) + 2V_I]^{1/2}$$

Störmer found it convenient to use the quantity

$$Q = 1 - 2V_I$$
$$= 1 - [(p_\phi/\rho) + (\rho/r^3)]^2$$

because Q = 0, from (11.10), necessitates $\dot{z} = 0$ and $\dot{\rho} = 0$ and so defines a zero-velocity curve for the (z, ρ) plane.

In terms of Q, the equations of motion are

$$\ddot{\rho} = \dot{p}_{\rho} = -\,\partial H/\partial\rho = \frac{1}{2}\,\partial Q/\partial\rho$$

$$\ddot{z} = \frac{1}{2}\,\partial Q/\partial z$$

(11.12)

for cylindrical coordinates, and

$$\ddot{r} = \dot{p}_r = -\,\partial H/\partial r = r\dot{\lambda}^2 + \frac{1}{2}\,(\partial Q/\partial r)$$

$$\frac{d}{ds}\,(r^2\dot{\lambda}) = \dot{p}_\lambda = -\,\partial H/\partial\lambda = \frac{1}{2}\,(\partial Q/\partial\lambda)$$

(11.13)

for spherical coordinates, with λ = latitude.

The solution of these equations is a matter of numerical integration, and will yield trajectories for whatever p_ϕ may be of interest. Knowing these, then, we may hope to make statements about the intensity of incoming particles of a given energy, as a function of the direction of incidence.

A very simple motion, namely a circular orbit, is possible here. Set $\dot{z} = \dot{\rho} = 0 = \ddot{\rho}$ in (11.9) and (11.12). Then λ = 0, $p_\phi = -2$ and ρ = 1 will give such an orbit. When the radius equals ℓ, the unit of length above adopted, then the particle can move in a circular orbit, with λ = 0. The radius r corresponding to the radius R of the earth (6370 km.) is given by r = R/ℓ, and measures (in Störmers) the energy of the particle. This correspondence is given in Table 1, the energy being expressed in electron volts.

Now, since the velocity is unity, then if θ be the angle between the forward tangent of the trajectory and the meridian plane, the velocity may be resolved into its horizontal components: X = sin θ = $\rho\dot{\phi}$ (11.14a)

TABLE I

(after Lemaitre and Vallarta)

r	Electrons (10^{10} v.)	Protons (10^{10} v.)
0.1	0.0596	0.01722
0.2	0.238	0.1618
0.3	0.536	0.449
0.4	0.954	0.861
0.5	1.490	1.397
0.6	2.145	2.050
0.7	2.920	2.823
0.8	3.821	3.719
0.9	4.830	4.729
1.0	5.96	5.85

perpendicular to the meridian plane, and

$$Y = \cos \theta \sin \eta \qquad (11.14b)$$

parallel to it, where η is the angle made with the zenith by the projection of the velocity vector on the meridian plane, and

$$\tan \eta = r d\lambda/dr. \qquad (11.14c)$$

The direction from which a trajectory strikes the earth is then represented by a point in the (X, Y) plane. If we know p_ϕ, ρ, and r, then $\sin \theta$ can be calculated, and if we have found the slope $d\lambda/dr$, then $\tan \eta$ can be obtained immediately, so that the point may be plotted.

Zero-Velocity Curves

Before discussing the intensity problem, let us find limits
to the motion by locating the zero-velocity curves $Q = 0$ and
other contours $Q = $ const.

Letting $k = \sin \theta$, then from (11.8) and (11.14a)

$$r\rho k = rp_\phi + (\rho^2/r^2) \qquad (11.15)$$

If p_ϕ is positive, then k must be positive also.

Now $\rho = r \cos \lambda$, where λ is the latitude. In these terms,
(11.15) becomes, multiplying by $k \cos \lambda$,

$$k^2 r^2 \cos^2 \lambda - p_\phi \, kr \cos \lambda - k \cos^3 \lambda = 0$$
$$\qquad (11.16)$$

Setting $\gamma = p_\phi/2$, the roots of this are

$$k\rho = kr \cos \lambda = \gamma \pm (\gamma^2 + k \cos^3 \lambda)^{1/2}$$
$$\qquad (11.17)$$

For any value of k, or of $Q = 1 - k^2$, Equation (11.17) gives
us in general two contours in the (z, ρ) plane. All trajectories hitting a given contour will have the same angle θ
with the meridian plane, and no particles can attain the
regions $Q < 0$.

First of all, let p_ϕ (or γ) be positive. If $\cos \lambda$ is
small, i.e., $\rho \ll r$, then from (11.17) $\rho = 2\gamma/k$, a straight
line parallel to the z-axis. The innermost such line is for
$k = 1$, so that particles cannot get closer to the axis than
$\rho = 2\gamma$, if r is large. The intercept of (11.17) with the
equator is

$$\rho = \{\gamma + (\gamma^2 + k)^{1/2}\}/k$$

which becomes infinite as $k \to 0$, and unity if $k = 1$ and $\gamma \to 0$.

For $k = 1$, Figures (11.1) and (11.2) show the forbidden
regions (plots of z vs. ρ) for $\gamma = 0.03$ and $\gamma = 0.2$ respectively. The above-mentioned constancy of ρ for small $\cos \lambda$
is seen to be valid for both values of γ.

Figure 11.1

Allowed area (black) of meridian plane for $\gamma = 0.03$.

Figure 11.2

Allowed area, $\gamma = 0.2$

As $\gamma \to 0$, the band about the z-axis becomes narrower and narrower, until finally, for $\gamma = 0$ and $k = 1$, Equation (11.17) becomes $r^2 = \cos \lambda$, <u>an oval stretching between $\rho = 0$ and $\rho = 1$</u>, with the forbidden region lying inside this oval. The band has vanished completely, and the oval is tangent to the z-axis.

Unless the particles have energy greater than about 6×10^{10} e.v. (see Table I) the radius of the earth will, in our units, be less than unity. Unless γ is extremely small, the earth is then mostly all inside the forbidden region, i.e., where no particles from infinity can get in. Thus, <u>positive values of p_ϕ are of little interest to us.</u>

<u>For negative</u> p_ϕ, let us set $\gamma_1 = -(p_\phi/2) = -\gamma$ and then locate the zero-velocity curves. As matters stand, the potential energy V_I in (11.11) becomes zero on the ρ-axis at $\rho = 1/2\gamma_1$. For convenience of calculation, we can change the scale so that this zero-point always occurs at unit distance. We introduce the new standard coordinates

$$\rho_1 = 2\gamma_1\rho, \quad z_1 = 2\gamma_1 z, \quad r_1 = 2\gamma_1 r \text{ and } \tau = (2\gamma_1)^3 s.$$

Then $\qquad\qquad\qquad Q = 1 - (2\gamma_1)^4\, U$ $\qquad\qquad$ (11.18)

with $\qquad\qquad\qquad U = (\dfrac{\rho_1}{r_1^{\,3}} - \dfrac{1}{\rho_1})^2 = 2V_I/(2\gamma_1)^4$

There has been a new time variable introduced, which makes
the parameter γ_1 disappear from the equations of motion,
which are now

$$\frac{d^2\rho_1}{d\tau^2} = -\frac{1}{2}\frac{\partial U}{\partial \rho_1}$$

$$\frac{d^2 z_1}{d\tau^2} = -\frac{1}{2}\frac{\partial U}{\partial z_1}$$

The energy integral corresponding to (11.10) is now

$$(\frac{d\rho_1}{d\tau})^2 + (\frac{dz_1}{d\tau})^2 = (2\gamma_1)^{-4} - U = Q/(2\gamma_1)^4$$

When working with spherical coordinates, one can use this
standard scale and the Goursat transformation, introducing
as new variables x and σ, defined by the relations

$$e^x = 2\gamma_1 r \quad \text{and} \quad ds = r^2 d\sigma/2\gamma_1.$$

The dipole singularity then is at $x = -\infty$, and the equations
of motion become very simple in terms of the new time variable.
Note that $\dot{r} = r\dot{x}$ and $\tan\eta = r\,d\lambda/dr = d\lambda/dx$.

The original equations of motion in spherical coordinates
are given by (11.13), namely,

$$\ddot{r} = r\dot{\lambda}^2 + (Q_r/2)$$

$$(d/ds)(r^2\dot{\lambda}) = Q_\lambda/2$$

Then

$$(2\gamma_1)^2 \frac{d^2x}{d\sigma^2} = r^2 \frac{d}{ds} (r\dot{r})$$

$$= r^2[Q + (rQ_r/2)]$$

$$= \frac{1}{2} \frac{\partial}{\partial x} (r^2 Q)$$

and

$$(2\gamma_1)^2 \frac{d^2\lambda}{d\sigma^2} = r^2 \frac{d}{ds} (r^2\dot{\lambda}) = r^2 Q_\lambda/2$$

Setting $P = (r/2\gamma_1)^2 Q$, these equations of motion reduce to

$$\frac{d^2x}{d\sigma^2} = \frac{1}{2} \frac{\partial P}{\partial x} \quad \text{and} \quad \frac{d^2\lambda}{d\sigma^2} = \frac{1}{2} \frac{\partial P}{\partial \lambda} \qquad (11.19)$$

which are quite simple, as already announced.

Now we can make one contour map of the surface U and can determine all of the zero-velocity curves from it, because the z.v.c. for γ_1, by (11.18), is just the contour $U = (2\gamma_1)^{-4}$. One single plot thus shows how the zero-velocity curves change with the angular momentum p_ϕ.

First of all, we shall state how the forbidden regions change, using the conventional scale. Then a closer look can be had in terms of the standard potential U.

Figure 11.3

Allowed area, $\gamma = -0.05$

Figure 11.4

Allowed area, $\gamma = -0.5$

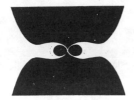   

Figure 11.5   Figure 11.6
Allowed area, $\gamma = -0.97$ Allowed area, $\gamma = -1.016$

As γ becomes negative, the pattern of the forbidden regions
becomes more complicated, because now (11.16) can have two
roots. The (double) oval already mentioned for $\gamma = 0$ becomes
smaller as $|\gamma|$ increases, as is seen in the sequence of
Figures 11.3-11.6. In addition, an hourglass figure about
the z-axis develops, broadens out, and bulges toward the
ρ-axis as shown in Figure 11.5. These bulges meet on the
ρ-axis at $\rho = 1$ when $\gamma = -1$, enclosing an allowed area.
Figure 11.6 shows the forbidden region for $\gamma = -1.016$, with
an outside hourglass boundary, and an inside allowed region
between two ovals.

When the radius of the earth is less than unity, particles
can reach the earth from infinity only if the angular momentum
parameter lies in the range $0 < \gamma_1 < 1$, where $\gamma_1 = -\gamma$. If
the angular momentum is greater than the above value, then
there is just one boundary attainable by the particles from
infinity, and it does not appear that periodic motion will
occur. Consequently, we shall regard this case as of relatively minor interest and shall confine the treatment to the
particles of lower angular momentum, trying to understand
more clearly how to account for Figures 11.3 to 11.6.

The reduced potential energy $U/2$ is proportional to $\sin^2 \theta$
and so has a minimum where $\theta = 0$, at which $U = 0$. Along the

ρ_1 axis, we have the equation

$$U = (\frac{1}{\rho_1{}^2} - \frac{1}{\rho_1})^2,$$

so that U behaves as $1/\rho_1{}^4$ when ρ_1 is small, goes to the minimum U = 0 at ρ_1 = 1, rises to a maximum U = 1/16 at ρ_1 = 2 and then falls off steadily to zero as ρ_1 becomes infinite. The point ρ_1 = 2 is a saddle point, with U rising as we leave the ρ_1-axis and proceed either toward positive or negative values of z_1. The contours of U are drawn in Figure 11.7 for negative values of γ, and this enables us to see how the allowed regions change with γ_1. This can be done with the aid of Equation (11.17).

First, when k = 0, Equation (11.15) gives

$$r = \cos^2 \lambda/2\gamma_1, \text{ or } r_1 = \cos^2 \lambda \qquad (11.20)$$

showing that the line U = 0 is an oval through the origin from r_1 = 0 to r_1 = 1.

In general, $U = k^2/(2\gamma_1)^4$, from (11.18), so that the allowed region for large γ_1 would be a small one near the oval U = 0.

Let now k be negative, and set $k = -k_1$. From (11.17),

$$k_1\rho = \gamma_1 \pm (\gamma_1{}^2 - k_1 \cos^3 \lambda)^{1/2} \qquad (11.21)$$

In order that there be an intercept on the ρ-axis (cos λ=1), we must have $\gamma_1{}^2 > k_1$. This will always be so if $\gamma_1 > 1$, in which case there are two intercepts and in fact two curves. The outer curve will be as close in as possible when k_1 = 1, so that the forbidden region is to the left of

$$\rho = \gamma_1 + (\gamma_1{}^2 - \cos^3 \lambda)^{1/2}.$$

This curve has asymptote (cos λ = 0) $\rho = 2\gamma_1$, and marks the boundary of the hourglass shape in Figure 11.6. We also see

such curves to the right of $\rho_1 = 2$ in Figure 11.7.   For
$\underline{\gamma_1 > 1}$, particles cannot reach the earth from infinity.

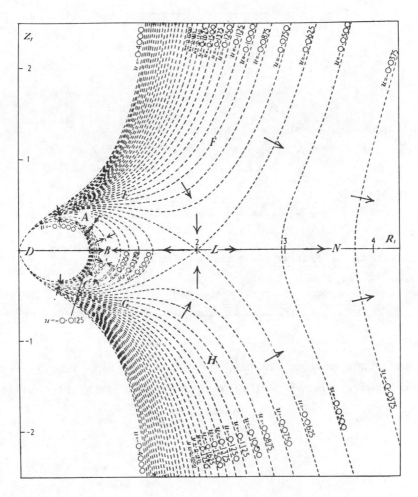

Figure 11.7

Field of force U = constant for all negative values of γ.
(Abscissa R_1 is denoted by ρ_1 in text, u here is U in text.)

The inner curves for $\gamma_1 > 1$ are ovals through the origin and are described by [see (11.15) and (11.17)]

$$r = \frac{\cos^2\lambda}{\gamma_1 + (\gamma_1{}^2 - k_1 \cos^3 \lambda)^{1/2}} \qquad (11.22)$$

These lie inside of the boundary ($k_1 = 1$)

$$r = \frac{\cos^2 \lambda}{\gamma_1 + (\gamma_1{}^2 - \cos^3 \lambda)^{1/2}} \qquad (11.23a)$$

For k positive, (11.17) gives only the one root

$$k\rho = -\gamma_1 + (\gamma_1{}^2 + k \cos^3 \lambda)^{1/2}$$

Multiplying numerator and denominator by

$$\gamma_1 + (\gamma_1{}^2 + k \cos^3 \lambda)^{1/2},$$

this becomes

$$r = \frac{\cos^2 \lambda}{\gamma_1 + (\gamma_1{}^2 + k \cos^3 \lambda)^{1/2}} \qquad (11.24)$$

describing an oval through the origin. The innermost such curve is for k = 1, namely

$$r = \frac{\cos^2 \lambda}{\gamma_1 + (\gamma_1{}^2 + \cos^3 \lambda)^{1/2}} \qquad (11.25)$$

It is now easy to see how the allowed regions develop as γ increases from $-\infty$. At first, there is a small crescent about the contour U = 0, bounded by the curves

$$r = (-\gamma_1 + (\gamma_1{}^2 + \cos^3 \lambda)^{1/2})/\cos \lambda \quad (k = 1)$$
$$(11.25a)$$
$$r = (\gamma_1 - (\gamma_1{}^2 - \cos^3 \lambda)^{1/2}/\cos \lambda \quad (k = -1)$$
$$(11.23b)$$

Setting $\lambda = 0$, the intercepts on the ρ_1-axis are, respectively

$$\rho_1 = 2/[1 + (1 + \gamma_1^{-2})^{1/2}]$$

and

$$\rho_1 = 2/[1 + (1 - \gamma_1^{-2})^{1/2}]$$

The first is just inside the second when γ_1 is large, and both are near $\rho_1 = 1$. However, the interval between them widens as γ_1 decreases. When $\gamma_1 = 1$, the first intercept has become $\rho_1 = 2/(1 + \sqrt{2})$ and the second has reached its maximum value $\rho_1 = 2$, corresponding to the saddle point.

In addition to the crescent bounded by curves (11.25) and (11.23), there is, for $\gamma_1 > 1$, an exterior allowed region with inside boundary

$$r = (\gamma_1 + (\gamma_1^2 - \cos^3 \lambda)^{1/2}/\cos \lambda \quad (k = -1)$$

$$(11.26)$$

Its intercept on the ρ_1-axis is

$$\rho_1 = 2\gamma_1^2 (1 + (1 - \gamma_1^{-2})^{1/2})$$

As γ_1 decreases from ∞, this decreases also from $4\gamma_1^2$ to its minimum value $\rho_1 = 2$, the saddle point. For $\gamma_1 > 1$, curves (11.23) and (11.26) have no point in common, so that particles from infinity cannot reach the crescent. For $\gamma_1 = 1$, these curves just touch at $\rho_1 = 2$, the saddle point, and together comprise the contour $U = 1/16$. As γ_1 decreases below unity, the zero-velocity contour $U = 1/16\gamma_1^4$ rises higher on the surface and has no intercept on the ρ_1-axis. The exterior allowed region and the crescent have merged, and particles can now come from infinity and go close to the origin.

Having treated the case $\gamma_1 > 1$ as well as $\gamma > 0$ and shown that these are not of appreciable significance for the theory of cosmic rays, we are left with the region $-1 < \gamma < 0$, where the allowed space can go from the origin to infinity, so that charged particles can come to the earth from large distances. The essential thing is that curve (11.23) and curve (11.26) have joined to form one continuous and smooth boundary. Let us look at this in more detail.

The values of r determined by (11.23) and (11.26) will be real only for latitudes $\lambda \geq \lambda_c$, where $\cos^3 \lambda_c = \gamma_1^2$. Thus, if $\gamma_1 < 1$, the minimum latitude λ_c is greater than zero, and neither (11.23) nor (11.26) can touch the ρ-axis. The values of r in (11.23b) and (11.26) are the same for λ_c and equal to $\gamma_1/\cos \lambda_c$. Furthermore, $dr/d\lambda$ is infinite at λ_c, showing that the tangent to curve (11.23) or (11.26) at λ_c passes through the origin.

Once we know the contours of the surface of potential energy, we need only draw lines normal to them to indicate the direction of the force. This is shown by arrows in Figure 11.7.

The Allowed Cone

When particles of a certain energy reach the earth from space, they can only come in at certain directions. These directions fill up a complicated cone of many sheets, which is called the allowed cone. It consists of 3 parts, (a) the main cone, within which all directions are allowed, and where the trajectories come directly from infinity, (b) the penumbra, due to shadow effects of the earth, where "patches of directions" are allowed and the rest forbidden, and (c) the circular Störmer cone outside of which all directions are forbidden.

The total intensity of particles of a given energy which
arrive at any point of the earth may be found if we know the
boundaries of the allowed cone, provided that the distribution of particles at infinity is isotropic. From Liouville's
Theorem on the conservation of the element of volume in phase
space, it follows that the number of particles coming in per
unit solid angle per unit time (intensity) is the same as it
is at infinity, and so the intensity is the same in all
allowed directions. Therefore, if we can find the shape of
the allowed cone (for particles of a given energy), we only
need multiply the specific intensity by the solid angle of
the cone to find the total intensity. The detailed study of
the allowed cone can become fairly complicated because of the
role played by the earth's shadow. We shall, therefore, refer
the reader to the original sources for this, and content
ourselves with a discussion of the main cone, which is the
complete allowed cone at the equator. (The shadow effects
are predominant at high latitudes.)

Motion in Equatorial Plane

The general nature of the motion is seen clearly if we set
$\lambda = 0$, i.e. if we confine ourselves to the equatorial plane.
Then Equations (11.8) and (11.10) give

$$\dot{r}^2 = 1 - (\frac{2\gamma_1}{r} - \frac{1}{r^2})^2 \qquad (11.27)$$

and

$$r^2\dot{\phi} = -2\gamma_1 + (1/r)$$

The last equation just gives the change of longitude with
time, which would change sign at $r = (1/2\gamma_1)$. If γ_1 is near
unity, and r also, this will not happen. Let us suppose this
to be so.

Equation (11.27) may be written

$$r^2 \dot{r} = (r^2 + 2\gamma_1 r - 1)^{1/2} (r^2 - 2\gamma_1 r + 1)^{1/2}$$

or, when $r \cong 1$,

$$\dot{r} \cong (2\gamma_1)^{1/2} (r^2 - 2\gamma_1 r + 1)^{1/2}$$

The right-hand side can become zero only if $\gamma_1 > 1$. If $\gamma_1 > 1$, the quantity $r^2 - 2\gamma_1 r + 1$ will be negative near $r = 1$, but positive for $r < r_1$, the root less than unity. Consequently, a particle started off from the earth with energy such that $r < r_1$ will go outward to $r = r_1$ and then return to the earth.

If $\gamma_1 = 1$, then $\dot{r} = - \sqrt{2} (r-1)$ for an outward going particle. This component of velocity remains positive, but diminishes constantly as $r \to 1$. The solution of (11.27) is

$$1 - r = (1 - r_o) e^{-2t} \qquad (11.28)$$

where r_o is the initial value of the radius.

The orbit is asymptotic to the circular orbit $r = 1$, when $\gamma_1 = 1$.

Finally, if $\gamma_1 < 1$, the quantity $r^2 - 2\gamma_1 r + 1$ does not become zero and so always has the same sign as at $r = 1$, when it is positive. Hence, if the particle is started outwards, it will keep going to infinity. (This is provided that the initial radius is such that $r^2 + 2\gamma_1 r - 1$ is positive, which must be so for the radial velocity to be real.)

The asymptotic orbit (11.28) thus separates the particles reentrant toward the earth from those which proceed directly to infinity, and must hence be part of the boundary of the allowed cone.

More generally, when γ_1 is less than, but not close to unity, particles which go directly to infinity and cannot be blocked

by the earth must cross the outer (principal) periodic orbit
associated with this value of γ_1. There will be orbits which
can just get to this periodic orbit (i.e. are asymptotic to
it) and these asymptotic orbits will mark the boundary between
orbits which go directly to infinity and those which return
to the earth. The main cone is bounded by asymptotic orbits.
We need then (1) to locate the periodic orbits and (2) to find
those orbits which are asymptotic to some of the periodic
orbits (P. O.'s). After a few preliminaries, we shall proceed
to do this.

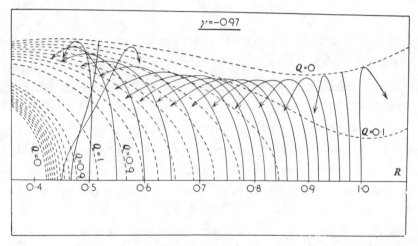

Figure 11.8

(After Störmer, loc. cit., with permission) "Orbits in the
Rz-plane starting at right angles to the R-axis." (Abscissa
R is denoted by ρ in text.)

Motion in the (ρ, z) Plane

Since energy is conserved, the time may be eliminated and
the differential equation for z determined as a function of
ρ. Using subscripts to denote partial derivatives, and letting
$q = \tan \alpha$ and $\tan \psi$ represent the slopes of the trajectory
and of grad Q respectively, we have

$$Q_z = Q_n \sin \psi, \; Q_\rho = Q_n \cos \psi, \; dz/d\rho = \tan \alpha$$

From (11.10) with Q substituted,

$$\dot{\rho} = [Q/(1 + q^2)]^{1/2}, \; \dot{z} = q\dot{\rho},$$

and, from (11.12), we have

$$\frac{d^2z}{d\rho^2} = \frac{dq}{d\rho} = \frac{1}{\dot{\rho}} \frac{d}{dt} \frac{\dot{z}}{\dot{\rho}} = \frac{\dot{\rho}\,\ddot{z} - \dot{z}\,\ddot{\rho}}{\dot{\rho}^3}$$

$$= (1/2Q)\,(1 + q^2)\,(Q_z - qQ_\rho)$$

$$= (Q_n/2Q \cos^3\alpha) \sin\,(\psi - \alpha) \qquad (11.29)$$

If the trajectory is in the direction of the gradient of Q, the curvature will be zero. Otherwise, the curvature is proportional to the sine of the angle between trajectory and gradient of the potential. This is illustrated in Figure 11.8 which shows a family of trajectories ($\gamma = -0.97$) leaving the ρ-axis normally. One member reaches the zero-velocity curve and then retraces its path. Trajectories to the right of this are bent to the right, and trajectories to the left are bent to the left, near Q = 0.

Equation (11.29) is independent of the direction that the particle moves along the trajectory as time increases. The fact that a particle can reach Q = 0 and then retrace its path is understandable when we consider that a particle shot straight up in a gravity field does the same thing. Such an orbit is called self-reversing.

In tracing out where particles go, certain types of trajectories turn out to have special significance. These are (1) orbits through the origin, (2) self-reversing trajectories, associated with Q = 0, (3) periodic orbits, and (4) orbits asymptotic to the periodic orbits. Detailed studies have been made of orbits with $\lambda = 0$ and with λ small, and have made clear what may be expected in general.

Orbits Through the Origin

Let us assume that a particle leaves the origin at time $s = 0$, and that both r and $\cos \lambda$ go to zero when s does.

Tentatively, let us try for an expression for r as an ascending power series in s, i.e.

$$r = a_1 s + a_2 s^2 + \ldots$$

For small r, from (11.15), we have

$$\cos^2 \lambda \cong - r p_\phi \qquad (11.30)$$

Also, since $Q = 1 - k^2$ and $\rho k_\rho + z k_z = -k - (\rho/r^3)$, the equations of motion yield

$$\frac{1}{2} \frac{d^2}{ds^2} (r^2) = \rho \, \ddot{\rho} + z \, \ddot{z} + \dot{\rho}^2 + \dot{z}^2$$

$$= \frac{1}{2} (\rho Q_\rho + z Q_z) + Q$$

$$= 1 + (k\rho/r^3) \qquad (11.31)$$

$$= a_1^2 + 6 \, a_1 a_2 s + \ldots \qquad (11.32)$$

Combining the last 2 equations,

$$k^2 \cos^2\lambda = r^4 (a_1^2 - 1 + 6 \, a_1 a_2 s)^2 \qquad (11.33)$$

which together with (11.30) shows that k^2 is of order r^3.

Differentiating (11.30) to find $\dot{\lambda}$ and substituting in the energy equation $\dot{r}^2 + r^2 \, \dot{\lambda}^2 = 1 - k^2$, we obtain to lowest orders of s

$$a_1^2 + 4 \, a_1 a_2 s - (p_\phi \, a_1^3 s/4) = 1$$

from which $a_1 = 1$ and $a_2 = p_\phi/16$. Consequently

$$r = s + (p_\phi s^2/16) + \ldots$$

and from (11.33)

$$k \cos \lambda = 6 \, a_2 r^2 s = 6 \, a_2 r^3$$

so that (11.15) yields

$$\cos^2 \lambda = -r p_\phi + (3p_\phi/8) r^5.$$

That is, the orbit through the origin has a very high order
of contact with the line of force $r = \cos^2 \lambda/(-p_\phi)$. The
numerical integration can be carried out with the aid of the
above series for r.

Of the resulting trajectories calculated by Störmer,
several are self-reversing. One, at $\gamma_1 = 0.956$, reverses
without any previous rebound from a zero-velocity curve.
Another, near $\gamma_1 = 0.933$, is symmetrical about the ρ-axis and
so returns to the origin. One orbit at $\gamma_1 = 0.9316$ reverses
after one bounce, one near $\gamma_1 = 0.9314$ after 2 bounces, and
another, for γ_1 somewhat less than 0.9314, becomes asymptotic
to a periodic orbit. Systematic integration, varying γ_1,
shows that there are still other self-reversing trajectories,
but these are generally more complicated and less interesting
than those mentioned above.

Periodic Orbits

Among the totality of all possible orbits, the periodic
orbits (to be abbreviated by P.O.'s) occupy a special position,
because they occur only for a discrete set of values of the
initial conditions (for a given γ_1). Furthermore, there are
orbits asymptotic to them (A.O.'s) which may mark the boundary
between orbits coming from infinite and those which do not.

Perhaps the simplest way of locating periodic orbits is to
start with a known one and to vary initial conditions slightly,
then to vary the value of γ_1 until the periodicity condition
is again satisfied.

We have already noted an "equilibrium" solution, namely
$\gamma_1 = 1$ and $\rho = 1$, $\lambda = 0$, where the particle moves along a
circular orbit in the equatorial plane. This orbit proves to
be a member of a family of periodic orbits in the meridian
plane. These orbits are symmetrical about the ρ-axis, cut it
orthogonally, and are also orthogonal to the outer z.v.c.
(P = 0). They are known as the principal P.O.'s, being
defined as those which leave the x-axis normally and go
directly to the outer z.v.c., and are shown in Figure 11.9.
(For figures and text, x is now used for ρ.)

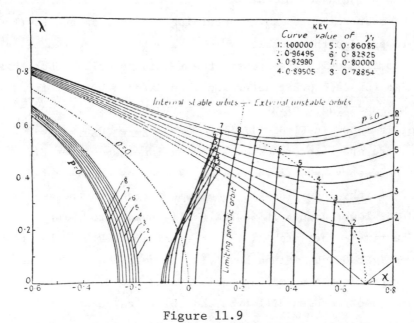

Figure 11.9

(After Vallarta, 1938) Upper part of the family of periodic orbits.

If x_o denote the intercept on the x-axis, then x_o varies
smoothly with γ_1, and a plot of this variation is shown in
Figure 11.10. There is a minimum value of γ_1, namely
$\gamma_1^* = 0.788541$, for which P.O.'s exist. For a value of
γ_1 between γ_1^* and unity, there are 2 P.O.'s, an outer one
and an inner one. The inner branch continues beyond $\gamma_1 = 1$,
but has not been followed.

Störmer was the first to discover this family of principal
P.O.'s, and to show that there is also an infinity of other
periodic orbits. Schremp has since proved that all of these
cross the x-axis between the inner and outer principal P.O.'s.
De Vogelaere has studied, systematically and extensively, the
P.O.'s for the Störmer problem. However, for questions of
cosmic ray intensities, it will suffice to confine ourselves
to the principal P.O. family.

The most accurate way of obtaining a member of the family
is to examine those orbits which leave the z.v.c. (normally)
and to find which one crosses the x-axis normally. For this,
we can for each orbit determine when $dx/d\sigma = 0$, plot its
locus, and find the intersection with the x-axis. Such a
plot is shown in Figure 11.11. For large enough values of
γ_1 there are 2 intersections, giving inner and outer P.O.,
but between $\gamma_1 = 0.78$ and $\gamma_1 = 0.82$ (the exact value is
0.78854) the locus is seen to become tangent to the x-axis
and for lower values of γ_1 the family ceases to exist.

The first-order stability of the principal P.O.'s has been
investigated, one method being (a) to note that an isoenergetic variation δn normal to the periodic orbit satisfies an
equation of Mathieu-Hill type and (b) to calculate the
characteristic exponents (CH.E's). The exponents for the
outer P.O.'s are real, so that these orbits are unstable,

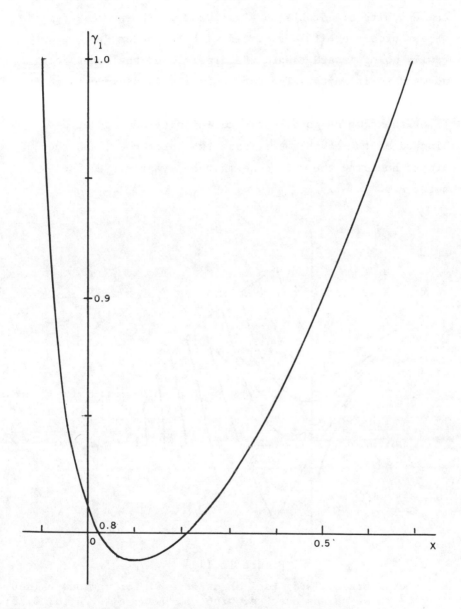

Figure 11.10

Relation between equatorial intercept x of periodic orbit and the value of γ_1

while the exponents for the inner P.O.'s are imaginary and
these orbits are stable to first order. There are 2 families
of asymptotic orbits associated with the outer P.O.'s, one
family going inward toward the dipole, and the other going
outward to infinity. The former family is of importance in
the theory of cosmic rays, because non—reentrant orbits
(i.e. those which go directly to infinity and cannot be
blocked by the earth) must cross the outer P.O., and the
orbits bounding the main cone must be symptotic to the
outer P.O.

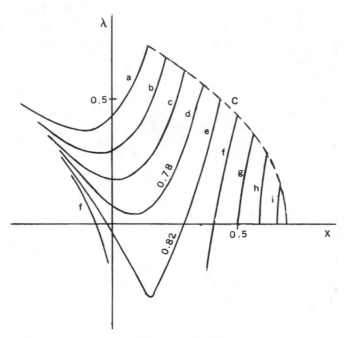

Figure 11.11

(After Schremp, 1938) Loci of $dx/d\sigma = 0$, for various values
of γ_1. These values are: a, 0.66; b, 0.70; c, 0.74; d, 0.78;
e, 0.82; f, 0.86; g, 0.90; h, 0.94; and i, 0.98. The dashed
line is the intersection of the locus of $P = 0$ and
$d^2x/d\sigma^2 = 0$.

Asymptotic Orbits

Now let us consider the case where the orbits are asymptotic to the outer P.O.'s, and see how they may be calculated. The isoenergetic variation δn may be written as

$$\delta n = e^{\Omega\sigma} p(\sigma) = pE \qquad (11.34)$$

where $E = e^{\Omega\sigma}$, Ω is a characteristic exponent, and $p(\sigma)$ is a periodic function of time. Ω will be taken to be real.

When integrating numerically, as with a differential analyzer, one can start a trajectory with arbitrary slope from $\lambda = 0$, in the direction toward the outer P.O. Then adjust the slope until the trajectory neither intersects nor falls short of the P.O., and this will give an asymptotic orbit with a fair degree of accuracy. This method is very fast when many trajectories are required and one doesn't need extremely high precision.

For very accurate results, Lemaitre and Vallarta used a semi-analytical method. They expressed the motion for principal P.O.'s in the form of Fourier series

$$\lambda = \Sigma \mu_q \sin q\omega\sigma \quad (q = 2n + 1)$$

$$x = \Sigma z_p \cos p\omega\sigma \quad (p = 2n)$$

and have then postulated that the asymptotic orbits can be represented by

$$\lambda = \Sigma \ [\mu_q(\sigma) \sin q(\omega\sigma + \phi) + \nu_q(\sigma) \cos q(\omega\sigma + \phi)]$$

$$x = z_0(\sigma) + \Sigma \ [y_p(\sigma) \sin p(\omega\sigma + \phi) + z_p(\sigma) \cos (\omega\sigma + \phi)]$$

$$(11.35)$$

where, as

$\sigma \rightarrow -\infty$, $z_p(\sigma) \rightarrow z_p$, $\mu_q(\sigma) \rightarrow \mu_q$, $y_p(\sigma) \rightarrow 0$ and $\nu_q(\sigma) \rightarrow 0$.
Assignment of a value to ϕ gives a particular member of a

family, and to obtain the whole family it is most expeditious
to calculate the amplitudes $\mu_q(\sigma)$, $\nu_q(\sigma)$, $y_p(\sigma)$, $z_p(\sigma)$ and
then to assign various values to ϕ.

Because of the form of Equation (11.34), it is natural to
write the amplitudes in (11.35) as power series in E, namely

$$y_p(\sigma) = y_p' \, E + y_p'' \, E^2 + \ldots$$

$$z_p(\sigma) = z_p + z_p' \, E + z_p'' \, E^2 + \ldots$$

$$\mu_q(\sigma) = \mu_q + \mu_q' \, E + \mu_q'' \, E^2 + \ldots$$

$$\nu_q(\sigma) = \mu_q' \, E + \nu_q'' \, E^2 + \ldots$$

For a first variation in x, λ of the periodic orbit,

$$\delta x = E \, [z_o' + \Sigma(y_p' \, S_p + z_p' \, C_p)]$$

$$\delta\lambda = E \, (\mu_q' \, \delta_q + \nu_q' \, C_q) \tag{11.36}$$

where $S_p = \sin p(\omega\sigma + \phi)$ and $C_p = \cos p(\omega\sigma + \phi)$, etc.

The variational equations of motion, from (11.19) are

$$\frac{d^2\delta x}{d\sigma^2} = \frac{1}{2} \, (P_{xx}\delta x + P_{x\lambda}\delta\lambda)$$

$$\frac{d^2\delta\lambda}{d\sigma^2} = \frac{1}{2} \, (P_{x\lambda}\delta x + P_{\lambda\lambda}\delta\lambda) \tag{11.37}$$

If one replaces P_{xx}, $P_{x\lambda}$, and $P_{\lambda\lambda}$ by their Fourier series,
combines (11.36) and (11.37), and equates terms independent
of σ and also the coefficients of S_p and C_p, it is possible
to find the characteristic exponents Ω and the first order
coefficients y_p', z_p', etc. A similar calculation will give
the second order coefficients. Thus the asymptotic expansion

is known for small values of E, i.e. in the vicinity of the
periodic orbit, and the further investigation of the trajectories can be made by starting with this expansion.

The two exponents associated with (11.34) are equal in
magnitude and of opposite sign. They are real for the outer
P.O.'s, decreasing in magnitude from $1/\sqrt{2}$ at $\gamma_1 = 1$ to zero
at $\gamma_1 = \gamma_1{}^*$. They are imaginary for the inner P.O.'s, and
increase in absolute value from $\gamma_1 = \gamma_1{}^*$ to $\gamma_1 = 1$. Godart
calculated the exponents for the outer P.O.'s by 2 different
methods, the more accurate one making use of the Mathieu-
Hill equation and Hill's determinant.

Determination of the Main Cone

As mentioned previously, one method of finding an asymptotic
orbit is to adjust the initial slope of a trajectory so that
it just gets to the outer P.O. If the initial slope is on one
side, the trajectory will intersect the outer P.O. and escape
to infinity; if it is on the other side, the trajectory will
fall short of the outer P.O. and will be reentrant toward the
earth. The slope of the asymptotic orbit thus marks the
boundary of the main cone, or of those trajectories which
escape directly to infinity.

Suppose now that the particles have energy such that the
surface of the earth would be at x, that they come in at
latitude λ, and that the angles of incidence are θ (between
the forward tangent and the meridian plane) and the zenith
angle η. Then, as stated previously (Equation 11.14), the
horizontal components of velocity are

$$X = \sin\theta, \quad Y = \cos\theta \sin\eta \qquad (11.14a,b)$$

where $\tan\eta = rd\lambda/dr = d\lambda/dx$. $\qquad\qquad\qquad$ (11.14c)

From (11.8) and (11.14a),

$$\sin \theta = (p_\phi/\rho) + (\rho/r^3) \qquad (11.38)$$

If $p_\phi = -2\gamma_1$ is specified, we can calculate $\sin \theta$ from
(11.38) for any value of r (or x) and λ. That is, if the
angular momentum at infinity is given, then $\sin \theta$ can be
found for any particle energy and observer latitude.

Through any point (λ, x) there are 2 asymptotic orbits.
For each one we may determine the slope $\tan \eta = d\lambda/dx$ and
thus a bounding zenith angle. We thus obtain for each
angular momentum γ_1 two values of η (and one value of θ).
These may be inserted in (11.14a,b) and X, Y may be calcu-
lated as a function of γ_1. The resulting locus gives the
boundary of the main cone for the particular value of x
(particle energy).

The main cones for positive particles and the northern
hemisphere are shown in Figures 11.12, 11.13, and 11.14,
for latitudes $\lambda = 0°$, 20°, and 30°. The allowed regions are
to the west of the boundary, so that there is in general an
east-west asymmetry. For negative particles the main cones
are the mirror images of those shown, reflected in the
north-south plane. For the southern hemisphere the main cones
are the mirror images of those for the northern hemisphere,
reflected in the east-west plane.

We have restricted ourselves to the main cone for simplicity.
The shadow effects due to the earth have been studied, by
calculating with orbits which are tangent to the earth at
some point. The work is straight forward, but for the details
the reader is referred to the lectures of Vallarta. It turns
out that these effects are important for high latitudes.

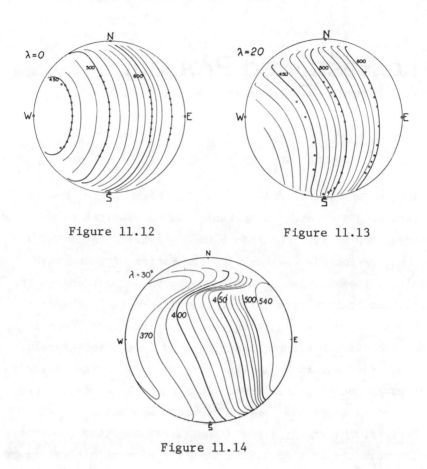

Figure 11.12   Figure 11.13

Figure 11.14

(After Vallarta, 1938)   Main cones, with energy in milli-
störmers.

XII

The Restricted Problem
of Three Bodies

Since the problem of determining the motion of two bodies
interacting according to Newton's Law of Gravitation has been
solved completely, there has been a great temptation to see
to what extent one can describe the motion of three such
bodies. The addition of the extra body does not change the
basic qualitative structure of the differential equations.
There are more singularities, but if we stay away from them,
then the existence theorem applies and we can integrate nine
second order equations to find a trajectory corresponding to
any given set of initial conditions. To this extent, then,
the three-body problem may be regarded as "solved".

However, the really fascinating questions about the problem
enter when one asks about the totality of possible solutions.
For instance, when do periodic solutions occur and when is
the configuration a stable one? Such questions have thus far
received only extremely limited answers, and it is in this
sense that the general three-body problem is "unsolved".

A start toward solution can be made if we assume that one
of the bodies has infinitesimal mass, and hence does not
influence the motion of the other two bodies. The three-
body problem is then said to be restricted. In general, the
massive bodies will move in ellipses, and the problem of
determining the motion of the small body is not very simple.

However, if we specify that the massive bodies move in circles
(about their common center of gravity), the process of finding
the trajectories of the small body is reduced to one which
can be handled readily by modern computational facilities. It
may be noted that the restricted three-body problem is really
the problem of finding how one body moves in two dimensions
under the gravitational influence of two masses which have a
preassigned motion.

   Without giving an exhaustive bibliography of work on the
restricted three-body problem, one can briefly indicate its
various stages of development. HILL in 1878 published his
important work on the motion of the moon, showing how effec-
tively to calculate its orbit around the earth. This was
followed by the comprehensive theoretical work of POINCARÉ
in the early nineties. In 1896, G. H. DARWIN published the
results of his numerical calculations for a ratio of finite
masses of one to ten. A review of work on periodic orbits by
F. R. MOULTON appeared in 1920, and subsequently ELIS
STRÖMGREN wrote a summary of his extensive work up to 1934.
Since at that time it took about a month to calculate one
periodic orbit, the subject was essentially shelved until
1962, when the author used an electronic computer to make a
general survey of periodic orbits and of how they depend upon
the mass-ratio. Over 9000 such orbits were calculated and
initial conditions for more than 2000 selected ones have been
published. HÉNON studied local stability and MULLINS and
BARTLETT global stability. RABE, DEPRIT and others have
calculated many Trojan orbits.

   If the massive bodies m_1 and m_2 are to move in circles, then
the centrifugal force on m_2 (say) must be balanced by the
attractive force. If r_2 is the distance from the center of
mass to m_2, and R that from m_1 to m_2, and ω the angular

velocity, then

$$m_2 r_2 \omega^2 = k^2 m_1 m_2 / R_2 \qquad (12.1)$$

where k^2 is a constant. Letting $m_1 + m_2 = M$, we have $r_2 = m_1 R/M$ and, from (12.1)

$$\omega^2 = k^2 M/R^3 \qquad (12.2)$$

Since the time scale is arbitrary, we shall choose it so that the angular velocity is unity, i.e., $\omega = 1$. Then Equation (12.2) becomes

$$k^2 M = R^3 \qquad (12.3)$$

Suppose, then, we have two bodies S and J with masses m_1 and m_2 respectively, which execute circular motions about the common center of mass. Let us study the motion of a third body P which has vanishingly small mass and which moves in the same plane as S and J do (see Figure 12.1).

Let there be a sidereal coordinate system (x, y) fixed in the plane, with origin 0 at the center of mass. Set $SO = r_1$, $OJ = r_2$, $SP = r$, $PJ = \rho$, and let P, S, and J have as coordinates (x, y), (x_1, y_1) and (x_2, y_2). The motion can also be referred to a rotating (synodic) coordinate system (ξ η), where the ξ-axis lies along SJ. Now

$$r_1/R = m_2/M = \mu_2 \text{ (say)}$$

and

$$r_2/R = m_1/M = \mu_1 \text{ (say)}$$

Introducing the mass-ratio parameter γ defined by

$$m_1 - m_2 = M\gamma,$$

we have

$$m_1/M = (1 + \gamma)/2, \quad m_2/M = (1 - \gamma)/2 \qquad (12.4)$$

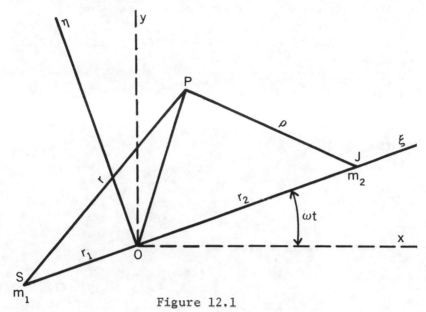

Figure 12.1

Fixed (Sidereal) and Rotating (Synodic) Systems

The potential energy per unit mass for the small particle is

$$V = -k^2 (m_1/r) - k^2 (m_2/\rho)$$

$$= - \frac{R^3}{2} (\frac{1 + \gamma}{r} + \frac{1 - \gamma}{\rho}) \qquad (12.5)$$

where we have used (12.3) and (12.4).

The unit of length has been left open so far. Some authors have chosen R = 1, but we shall follow Strömgren, who set R = 2. Then (12.5) becomes

$$V = -4 (\frac{1 + \gamma}{r} + \frac{1 - \gamma}{\rho}). \qquad (12.5')$$

The kinetic energy per unit mass is $T = (\dot{x}^2 + \dot{y}^2)/2$. The transformation to the rotating system is

$$x = \xi \cos t - \eta \sin t, \quad y = \xi \sin t + \eta \cos t \qquad (12.6)$$

and the kinetic energy per unit mass is given by

$$2T = \dot{\xi}^2 + \dot{\eta}^2 + \xi^2 + \eta^2 + 2(-\dot{\xi}\eta + \dot{\eta}\xi) \qquad (12.7)$$

The corresponding momenta are

$$p_\xi = \dot{\xi} - \eta, \ p_\eta = \dot{\eta} + \xi \qquad\qquad (12.8)$$

and the equations of motion are

$$\ddot{\xi} - 2\dot{\eta} - \xi = -\ \partial V/\partial\xi$$

$$\ddot{\eta} + 2\dot{\xi} - \eta = -\ \partial V/\partial\eta. \qquad (12.9)$$

   The Hamiltonian function is

$$H = \Sigma \ p\dot{q} - L = p_\xi \ \dot{\xi} + p_\eta \ \dot{\eta} - T + V$$

$$= \frac{1}{2} \ [\dot{\xi}^2 + \dot{\eta}^2 - \xi^2 - \eta^2] + V = \bar{h}. \qquad (12.10)$$

Since the time does not enter explicitly, the Hamiltonian
is constant in time and equal to the constant quantity $\bar{h}$.

   It is convenient to regard the Hamiltonian as consisting
of two terms, one being the velocity term $(\dot{\xi}^2 + \dot{\eta}^2)/2$, and
the other an effective potential energy

$$V' = -(\xi^2 + \eta^2)/2 - 4 \ (\frac{1 + \gamma}{r} + \frac{1 - \gamma}{\rho}) = -U \qquad (12.11)$$

Then the energy integral may be written simply as

$$\dot{\xi}^2 + \dot{\eta}^2 = 2U + 2\bar{h} = 2U - K \qquad (12.12)$$

Since one is concerned mainly with orbits which are bound,
i.e., which stay near the masses m_1 and m_2, the energy
constant $\bar{h}$ is usually negative. Strömgren found it convenient to write $2\bar{h} = -K$.

   In order to see in what regions of the plane the motion is
allowed, we note that the velocity must be real, and therefore

$\bar{h} > V'$. The <u>zero-velocity surface</u>, beyond which the motion
cannot go, is given by $\bar{h} = -U$. Contours of the U surface are
shown in Figure 12.2, for $\gamma = 0$.

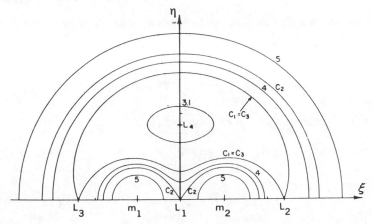

Figure 12.2

Contours of the U surface, $\gamma = 0$

EQUILIBRIUM (LIBRATION) POINTS

The function V' will have a stationary value whenever
$\partial U/\partial \xi = 0$ and $\partial U/\partial \eta = 0$. If a particle be placed at rest at
such a point, then from (12.9) its acceleration will be zero
and so it will stay indefinitely at this point, which is
called a <u>libration point</u>.

To find the libration points, first express U as a function
of r and ρ.

From Figure 12.1, it is apparent that

$$r^2 = \xi^2 + 2\xi r_1 + r_1^{\,2} + \eta^2$$
$$\rho^2 = \xi^2 - 2\xi r_2 + r_2^{\,2} + \eta^2$$

Multiplying the first equation by $r_2 = 1 + \gamma$ and the second
by $r_1 = 1 - \gamma$, and adding, we have

$$2(\xi^2 + \eta^2) = (1 + \gamma)r^2 + (1 - \gamma)\rho^2 - 2\,(1 - \gamma^2)$$

$$\tag{12.13}$$

Then

$$-4V' = 4U = (1 + \gamma)r^2 + (1 - \gamma)\rho^2 - 2(1 - \gamma^2) + 16(\frac{1+\gamma}{r} + \frac{1-\gamma}{\rho})$$

$$(12.14)$$

The conditions for a libration point are

$$U_\xi = U_r \ (\partial r/\partial \xi) + U_\rho \ (\partial \rho/\partial \xi) = 0$$

$$U_\eta = U_r \ (\partial r/\partial \eta) + U_\rho \ (\partial \rho/\partial \eta) = 0$$

To satisfy these, we must have either

$$U_r = 0 \text{ and } U_\rho = 0 \qquad\qquad (12.15)$$

or

$$\partial (r, \rho)/\partial (\xi, \eta) = 0 \qquad\qquad (12.16)$$

Equation (12.15) will hold if $r = \rho = 2$.

There are <u>two such points</u> which satisfy this condition, each one being at the vertex of an equilateral triangle with SJ at its base. That point with $\eta > 0$ will be called L_4, and the point with $\eta < 0$ will be designated L_5. The effective potential energy V' has a maximum here, and, from (12.14)

$$2V' = - 11 - \gamma^2$$

When the energy parameter $-K$ is greater than this, the velocity will always be real, and there is no longer a zero-velocity curve, or "natural" barrier to the motion, in the rotating coordinate system. In the fixed system, the motion can be confined by the potential energy barrier V, in which case the energy constant is less than zero.

For convenience in representing orbits with different values of γ on a common diagram, we can introduce a modified Jacobi constant $\bar{K} = K - \gamma^2$ which equals 11 at the triangular libration points L_4 and L_5. When one sets R = 1 instead of 2 as here, then all energies are divided by 4 (as we see from

(12.5) and a modified Jacobi constant C is introduced which has the value 3 at the points L_4 and L_5. This corresponds to the value K = 12 at these points for $|\gamma| = 1$.

Other libration points are given by (12.16) which demands that $\eta = 0$, i.e., that these points are on the ξ-axis, collinear with m_1 and m_2. Figure 12.3 shows a plot of V' versus ξ for $\gamma = 0$, from which it is apparent that there are 3 libration points. The one to the left of m_1 will be called L_3, that between the masses L_1 and the one to the right of m_2 will be L_2. To find the location of any one, we need only use the condition $U_\xi = 0$. For L_1, (12.14) yields, since

$$\partial r/\partial \xi = 1 \text{ and } \partial \rho/\partial \xi = -1,$$

$$(1 + \gamma)(r - \frac{8}{r^2}) - (1 - \gamma)(\rho - \frac{8}{\rho^2}) = 0.$$

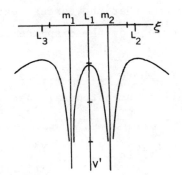

Figure 12.3

Effective Potential Energy

V' vs. ξ, $\gamma = 0$

Setting $|r - 1| = d$ and $r\rho = D$, this equation gives for L_1, with $D = 1 - d^2$,

$$\gamma = d(16 + D^2)/[8(1 + d^2) - D^2] \quad (12.17a)$$

Correspondingly, for L_2, with $D = d^2 - 1$

$$\gamma = [8(1 + d^2) - D^2 d]/(D^2 + 16d) \quad (12.17b)$$

This enables us to calculate directly γ as a function of r for all the libration points. (The inverse process, first fixing γ and then solving a fifth degree equation for r, is unnecessarily laborious for most purposes.) Once γ has been determined, the values of $\bar{K}$ at L_1, L_2 and L_3 can be found from (12.14) as a function of γ. The results have been tabulated by Bartlett and Wagner (1965). For $0 < m_2 < m_1$, the highest energy barrier is at L_3, the next higher at L_2, and the lowest at L_1.

SMALL LIBRATIONS

The libration points may also be regarded as points of equilibrium, so that it might be expected that periodic oscillations about these points can occur. This turns out to be so, as we shall now demonstrate.

As a first approximation, assume that the forces near a libration point are linear functions of the displacements from this point.

Let $\xi = \xi_o + X$ and $\eta = \eta_o + Y$, where ξ_o, η_o denote a libration point and X and Y are small. The potential energy expression is, to this order,

$$U = U_o + (1/2) \, (U_{\xi\xi} \, X^2 + 2U_{\xi\eta} \, XY + U_{\eta\eta} \, Y^2)$$

$$\tag{12.18}$$

The equations of motion are then, from (12.9)

$$\ddot{X} - 2\dot{Y} = U_{\xi\xi} \, X + U_{\xi\eta} \, Y$$

$$\ddot{Y} + 2\dot{X} = U_{\xi\eta} \, X + U_{\eta\eta} \, Y.$$

$$\tag{12.19}$$

The second derivatives of U are as follows:

$$U_{\xi\xi} = 1 - \frac{4(1+\gamma)}{r^5} \, [\eta^2 - 2(\xi+r_1)^2] - \frac{4(1-\gamma)}{\rho^5} \, [\eta^2 - 2(\xi-r_2)^2]$$

$$U_{\xi\xi} = 1 + 2A \text{ for } \eta = 0; \quad U_{\xi\xi} = 3/4 \text{ at } L_4$$

where

$$A = \frac{4(1+\gamma)}{r^3} + \frac{4(1-\gamma)}{\rho^3}, \qquad (12.20)$$

$$U_{\xi\eta} = 12(1+\gamma)(\xi+r_1)(\eta/r^5) + 12(1-\gamma)(\xi-r_2)(\eta/\rho^5)$$

$$= 0 \text{ for } \eta = 0$$

$$= \frac{3\sqrt{3}}{4} \gamma \text{ at } L_4,$$

and

$$U_{\eta\eta} = 1 - \frac{4(1+\gamma)}{r^5}[(\xi+r_1)^2 - 2\eta^2] - \frac{4(1-\gamma)}{\rho^5}[(\xi-r_2)^2 - 2\eta^2]$$

$$= 1 - A \text{ for } \eta = 0$$

$$= 9/4 \text{ at } L_4.$$

The quantity A may be evaluated as a function of d by substituting γ from (12.17a), (12.17b). For L_1, where $d^2 < 1$, we find

$$A = 8(7 + d^2)/(7 + 10d^2 - d^4),$$

so that the values decrease from 8 at $d = 0$ to 4 at $d^2 = 1$. For L_2, $1 \leq d \leq 3$, $D = d^2 - 1$, and

$$A = 32[d + \frac{2}{d + 1}]/(D^2 + 16d).$$

The values of A decrease steadily from 4 at $d = 1$ to 1 at $d = 3$, which corresponds to γ going from +1 to -1. In other words, $A \geq 1$ at the libration points L_1, L_2, and L_3.

From (12.12) and (12.18), the energy integral is

$$\dot{X}^2 + \dot{Y}^2 = 2U_o - K + U_{\xi\xi} X^2 + 2U_{\xi\eta} XY + U_{\eta\eta} Y^2$$

$$= 12 - K + \alpha X^2 + 2\beta XY + \delta Y^2 \text{ at } L_4 \qquad (12.21)$$

where $\alpha = 3/4$, $\beta^2 = (27/16)\gamma^2$, $\delta = 9/4$.

Case I: Libration Point on ξ-Axis

When the libration point is on the ξ-axis, $U_{\xi\eta} = 0$ and Equations (12.19) become

$$\ddot{X} - 2\dot{Y} = (1 + 2A)X$$
$$\ddot{Y} + 2\dot{X} = (1 - A)Y.$$

(12.19a)

Differentiating the first equation twice and eliminating Y, we find

$$(d^4X/dt^4) + (2 - A)\ddot{X} + (1 + 2A)(1 - A)X = 0$$

Setting $X = a \exp mt$, $Y = b \exp mt$,

$$m^4 + (2 - A)m^2 + (1 + 2A)(1 - A) = 0$$

(12.22)

Solving this quadratic equation in m^2, one root ρ^2 is positive, given by

$$2\rho^2 = A - 2 + (9A^2 - 8A)^{1/2}.$$

The other root is negative, $-\sigma^2$ where

$$2\sigma^2 = 2 - A + (9A^2 - 8A)^{1/2}.$$

Since A ranges from 8 to 1, the value of σ^2 will always be positive throughout the allowed region $-1 < \gamma < 1$, for all the collinear libration points L_1, L_2, and L_3, and the corresponding orbit will be periodic. The orbits around L_2 belong to class (a) of Strömgren, and the period $2\pi/\sigma$ equals 2π when $\gamma = -1$ (for which A = 1). There are also librations around L_1 (class (n) of Strömgren) for all values of γ between -1 and +1.

If $m = -\rho$ ($\rho > 0$), the particle leaves the libration point in an asymptotic orbit; if $m = +\rho$, it approaches this point in another asymptotic orbit. The above first equation yields

$$m^2a - 2mb = (1 + 2A)a$$

so that the slope b/a of an asymptotic orbit is determined by the value of γ (through A and m). Since m has two values of opposite aigns, namely ρ and $-\rho$, the two asymptotic orbits will have slopes of opposite signs. For certain very special values of γ, the one asymptotic orbit can bend around and match up with the other asymptotic orbit, to form a symmetric doubly asymptotic orbit. Such orbits have been calculated by Deprit and Henrard (1965).

Librations about L_1 and L_2 (m_2 small)

When the mass m_2 is small (γ near 1) the collinear libration points L_1 and L_2 lie close to the mass m_2, L_1 being on the inside and L_2 on the outside. The condition $U_\xi = 0$ is, by differentiating (12.14),

For L_1:

$$\frac{1}{2}(1 + \gamma)r - \frac{1}{2}(1 - \gamma)\rho - 4(1 + \gamma)/r^2 + 4(1 - \gamma)/\rho^2 = 0$$

For L_2:

$$\frac{1}{2}(1 + \gamma)r + \frac{1}{2}(1 - \gamma)\rho - 4(1 + \gamma)/r^2 - 4(1 - \gamma)/\rho^2 = 0$$

The second terms, for ρ small, may be neglected. Since $\rho = 2 - r$ for L_1 and $\rho = r - 2$ for L_2, the first and third terms combine to give $\mp \frac{3\rho}{2}(1 + \gamma)$, which is the net force due to the mass m_1 and to inertia. Equating this to the gravitational force due to m_2, we have

$$(3/8)\rho_o{}^3 = (1 - \gamma)/(1 + \gamma)$$

which locates L_1 and L_2 when m_2 is small. They are on opposite sides of, and equidistant from, the small mass, as shown in Figure 12.4.

Setting this relation for ρ_o into (12.20), with r = 2 and

$\gamma = 1$, the result is $A = 4$ and the libration equations are

$$\ddot{X} - 2\dot{Y} = 9X$$

$$\ddot{Y} + 2\dot{X} = -3Y$$

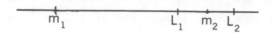

Figure 12.4

Relation of L_1 and L_2 to m_2 when m_2 is small

These equations are of course only valid for very small values of X, negligible with respect to ρ_o for the libration point, and this gets smaller as γ approaches unity. The motion in the fixed system is that of a particle which is riding around with the planet m_1, executing small oscillations about a circle of radius $2 \pm \rho_o$, the synodic period of these being $2 \pi / \sigma$, where

$$\sigma^2 = -1 + \sqrt{28} .$$

The limiting orbit is the circle, and there is no commensurability relation. The particle is staying near the mass m_2, the influence of which cannot be neglected, and the period is the result of the combined forces due to m_1 and m_2, so that the motion cannot be regarded as a two-body motion even in the limit. If, on the other hand, the particle were allowed to go around m_2, then we would have a two-body problem in the limit, but that is not the case here.

Librations about L_2 (m_1 small)

When the mass m_1 is small (γ near -1) the libration point L_2 is near $\xi = 2$, i.e., on the opposite side of m_2 from m_1.

The equilibrium condition here is the same as before, except that now $r \simeq 4$, $\rho = r - 2 \cong 2$.

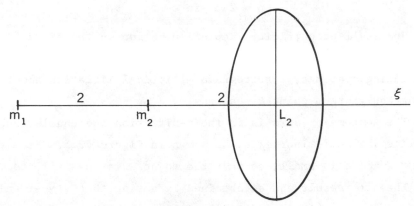

Figure 12.5

Libration around L_2 (m_1 small)

The particle oscillates about a point which describes a circle in the fixed system, and this point is far removed from the mass m_1, so that as $m_1 \to 0$ the motion becomes governed solely by m_2, so must be along an ellipse in the fixed system. If the angular frequency is n, then

$$nt = \varepsilon - e \sin \varepsilon \qquad (4.13)$$

where ε is the eccentric anomaly.

If the fixed system coordinates are (x,y) referred to m_2, then, with $b = a (1-e^2)^{1/2}$, $x = a (\cos \varepsilon - e)$, $y = b \sin \varepsilon$.

Let us now take n = 1 and suppose the eccentricity e is small. Then

$$\cos t \simeq \cos \varepsilon + e \sin^2 \varepsilon$$
$$\sin t \simeq \sin \varepsilon - e \sin \varepsilon \cos \varepsilon.$$

The coordinates in the rotating system are given by

$$\xi = x \cos t + y \sin t, \quad \eta = -x \sin t + y \cos t.$$

If terms in e^2 are neglected, then substitution yields

$$\xi = a - ae \cos \varepsilon$$

$$\eta = 2ae \sin \varepsilon.$$

For small eccentricity, Keplerian motion in the fixed system with unit angular velocity is represented in the rotating system by a retrograde elliptical libration about the equilibrium point $\xi = a$, $\eta = 0$.

The semi-major axis is in the η-direction and equals 2ae, twice the semi-minor axis, as shown in Figure 12.5. Here $a \cong 2$ and e is small, so that the major axis is small, too.

The same result can be obtained by noting that $\gamma = -1$ and $\rho = 2$ substituted into (12.20) yield $A = 1$ and the libration equations

$$\ddot{X} - 2\dot{Y} = 3X$$
$$\ddot{Y} + 2\dot{X} = 0$$

$$(12.19b)$$

The angular frequency for libration about $\xi = 2$ is then $\sigma = 1$, and $X = -ae \cos t$, $Y = 2ae \sin t$ describes the motion approximately, in accordance with our previous result.

Case II: Motion near L_4 (or L_5)

At L_4, Equation (12.19) becomes

$$\ddot{X} - 2\dot{Y} = \alpha X + \beta Y$$
$$\ddot{Y} + 2\dot{X} = \beta X + \delta Y$$

$$(12.23)$$

where

$$\alpha = \frac{3}{4}, \quad \delta = \frac{9}{4}, \quad \beta^2 = \frac{27}{16} \gamma^2 .$$

Differentiating the first equation twice re the time, and eliminating Y, we have

$$(d^4X/dt^4) + (4 - \delta - \alpha)\ddot{X} + (\delta\alpha - \beta^2)X = 0$$

so that, if $X = a \exp mt$, $Y = b \exp mt$,

$$m^4 + m^2 + \frac{27}{16} (1 - \gamma^2) = 0,$$

the roots of which are

$$m^2 = -\frac{1}{2} \pm \frac{1}{2} [1 - \frac{27}{4} (1 - \gamma^2)]^{1/2} \qquad (12.24)$$

These will be real for $|\gamma| \geq (23/27)^{1/2} = 0.922958$, but complex otherwise.

When m^2 is real, it is negative and m is a pure imaginary, $m = \pm iq$ $(q > 0)$. The larger value of $|m^2|$ gives large angular velocity, and these orbits are called short period orbits. The smaller value of $|m^2|$ gives the long period orbits. The amplitudes a and b of horizontal and vertical oscillations are related to each other through Equation (12.23) and will in general be of different magnitude, so that the orbit will be an ellipse.

In the limit of small m_1 or m_2, $\gamma^2 = 1$ and the roots of (12.24) yield $q = 1$ and $q = 0$. The latter results in $X = 0$, $Y = 0$, from (12.23). The short-period orbits are physically indistinguishable in this limit from the librations about L_2 for $\gamma = -1$, since the mass m_1 has no appreciable influence. So, to obtain these orbits about L_4, when the mass is small, just rotate the set of orbits about L_2 by $2\pi/3$. The major axis will be normal to the radius vector from m_2 to L_4 and will be twice the minor axis. The motion in the synodic system will be retrograde, just as it is for L_2.

When $\mu \neq 0$, if the amplitude of libration about L_4 is increased, the orbits, which are asymmetrical about the ξ and η axes, swell up until an orbit is reached which is symmetrical about the ξ-axis and is a member of the class (a) of librations about L_3. After this, the evolution is essentially shrinkage onto L_5.

When $|\gamma|$ is below the critical value, the quantity m will be complex, i.e., $m = p + iq$. If $p > 0$, the amplitude of the oscillations grows as $\exp(pt)$, so that the trajectory is a spiral outwards leaving the libration point at $t = -\infty$. In order to carry out the integration, one can start at $X = 0$ and $Y = Y_o$, where Y_o is sufficiently small. If the trajectory is for $K = 12$, the energy integral from (12.21) is

$$\dot{X}^2 + \dot{Y}^2 = \alpha X^2 + 2\beta XY + \delta Y^2 \qquad (12.25)$$

$$= \delta Y^2 \text{ if } X = 0.$$

Equation (12.23) gives $\dot{Y}$ and $\dot{X}$ in terms of X and Y, if we use $\ddot{X} = m^2 X$, $\ddot{Y} = m^2 Y$. The ratio $\dot{Y}/\dot{X}$ is the slope. This, used in conjunction with (12.25), will give the initial values of $\dot{X}$ and $\dot{Y}$ separately, and one can then proceed to integrate numerically. Explicitly, let

$$X = (A - iB)\, e^{mt} + c.c.$$

$$Y = (C - iD)\, e^{mt} \pm c.c.$$

where the values at $t = 0$ yield $A = 0$ and $C = Y_o/2$. Then

$$\dot{X} = (+p + iq)(A - iB)\, e^{mt} + c.c.$$

$$\dot{Y} = (+p + iq)(C - iD)\, e^{mt} + c.c.$$

from which the initial slope at $t = 0$ is

$$\frac{dY}{dX} = \frac{+pC + qD}{+pA + qB} \qquad (12.26)$$

The first equation (12.23) gives

$$m^2(A - iB) - 2m(C - iD) = \alpha(A - iB) + \beta(C - iD).$$

Equating real and imaginary parts on both sides, and setting $\theta = \beta + 2p$, we have $(p^2 - q^2 - \alpha)B + 2qC = \theta D$

$$+2pqB - \theta C = 2qD.$$

The quantity D can be eliminated to give C/B and then D/B, after which substitution into (12.26) results in the initial slope for an outgoing orbit, namely

$$\frac{dY}{dX} = +p - \beta \; \frac{p\theta + \frac{5}{4}}{\theta^2 + 4q^2} \qquad (12.27)$$

(Here we have used A = 0, α = 3/4, $p^2 + q^2$ = -1/2.) For an incoming orbit, replace +p by -p in θ and in Equation (12.27).

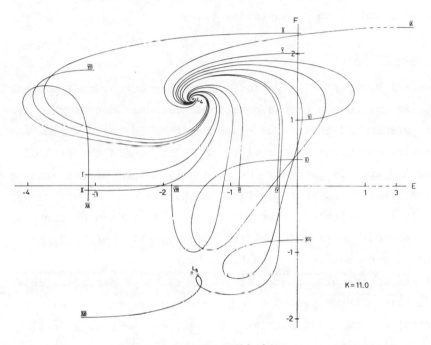

Figure 12.6

Periodic Limiting Orbits (K = 11 .0), Symmetric with respect to the E- or F-axis and Asymptotic to L_4 and L_5. (For complete half-orbits, take the proper mirror images to couple L_4 and L_5.)

Some of the resulting orbits, which evolve from spirals issuing from L_4, may be symmetric about the ξ-axis or the η-axis or return to L_4. In these cases they will be curves

onto each of which a whole class of orbits may converge, and
so are called asymptotic periodic orbits (A.P.O.). For each
value of γ, there will be a whole set of A.P.O.'s, and the
relative locations of the members of a set will change continuously with γ. Two A.P.O.'s may move toward each other to
form a common one, after which they have both disappeared.
Naturally, none exist for $|\gamma| > (23/27)^{1/2}$. Several are shown
in Figure 12.6 for $\gamma = 0$ (equal masses) in regularized coordinates E and F, to be explained below.

PERIODIC SOLUTIONS

General Remarks: The periodic solutions occupy an important
place in the theory of the equations, because some of them are
quite stable and the system stays together for a long time.
Therefore our first task will be to locate where the periodic
solutions are, in general. We need only determine the simpler
periodic solutions, because BIRKHOFF has proved the existence
of solutions which have periods that are multiples of the
basic period. A revision of this proof of Birkhoff's Fixed
Point Theorem has been given by SIEGEL.

If the motion of a dynamical system is to be periodic, this
means that after a period τ the variables return to their
original values. Alternatively stated, one must in general
solve a system of non-linear ordinary differential equations,
subject to the boundary conditions that the final positions
and velocities must have the same values as the initial ones.

Let us first consider motion in one dimension subject to a
force which is not explicitly dependent on the time. The
equation of motion $\ddot{u} + f(u) = 0$ has a first integral
$\frac{1}{2}\dot{u}^2 = h - V(u)$, where h is constant and V is the potential
energy. If V has a minimum, then librations will occur in

the valley of V, and the period may be determined by a duad-
rature. Given h and u, one can determine the velocity u
except for sign. The periodic motions may be easily visualized by drawing the trajectories in the phase plane (u, u̇).

It is somewhat more difficult but still feasible to characterize periodic motion in two dimensions, such as is the case
for the restricted 3-body problem. Let the Jacobi constant
$\bar{K}$ have a definite value, and consider the totality of periodic
motions belonging to this value. They will be closed curves
in the (E, F) plane, which we shall call eigencurves. The
curves can be symmetric with respect to (1) the ξ-axis,
(2) the η-axis, (3) both the ξ-axis and the η-axis, or
(4) neither the ξ-axis nor the η-axis. Strömgren confined
his attention mainly to orbits which were symmetrical either
with respect to the E-axis, or to the F-axis, or perhaps both.
This makes the location of a periodic orbit rather easy for a
given $\bar{K}$, because one knows that the initial inclination is
perpendicular to one of these axes. Then it is only necessary
to vary the distance along the axis until the final boundary
conditions are fulfilled, provided of course that a periodic
orbit of the desired type does exist for the value of $\bar{K}$ in
question.

[The symmetry properties of the equations do not by any
means exclude asymmetric solutions, but such solutions for
γ = 0 have hitherto been largely ignored because they are
somewhat more complicated and also less easy to locate.
However, Strömgren (see Tableau V, Figure 8) gives an example.]
When $\bar{K}$ is varied continuously, the eigencurves also change
continuously, and generate eigensurfaces in (E, F, $\bar{K}$) space.
The totality of these surfaces is thus a representation of
periodic motion for our problem. More generally, one can

let γ, the mass-ratio parameter, vary and see how the eigen-surfaces change. Each distinct surface is said to represent a <u>class</u> of periodic solutions.

Periodic Motions for $m_1 = 0$ (2-Body Problem)

Poincaré considered the case where the mass-ratio m_1/M is small, and reasoned that the motion of the infinitesimal body ought not to differ greatly from what its motion would be if m_1 were actually zero. In general, <u>solutions of the equations</u> <u>of motion should vary continuously</u> when any parameter, whether γ or K or one of the initial conditions, <u>is varied by a small</u> <u>amount</u>. Queer things, such as singular points or the occur-rence of double roots, may act to limit the application of this principle at times, but one can usually chart his course so as to detour around such obstacles rather than to meet them head-on. In the present case, one will be on safe ground if the small body does not come near m_1.

When $\gamma = -1$, or $m_1 = 0$, the motion in the fixed (sidereal) system will be an ellipse with the origin (m_2) at one focus. Since the motions for $\mu \neq 0$ are referred to the rotating (synodic) frame of reference, we shall see what the motion for $\mu = 0$ looks like in the rotating system, so that compari-son can be made. If the eccentricity $e = 0$, then the motion will be circular in the rotating frame also. But if $e \neq 0$, the motion in the rotating system will be closed and periodic only if the periods in the two systems are commensurable.

K vs. ρ Profile ($m_1 = 0$)

If ρ be the distance from the origin (at m_2) to one end of the ellipse, we can find a relation between the Jacobi constant K and the distance ρ, exact for $m_1 = 0$ and approxi-mately true for small $m_1 \neq 0$.

Let the motion start at $\xi = \rho$, $\eta = 0$ and in the η direction, $\dot{\xi} = 0$, $\dot{\eta} = \rho$ $(\dot{\theta} - 1) = (J/\rho) - \rho$, where $\dot{\theta}$ is the angular velocity, and J the angular momentum, in the fixed system. The potential energy/unit mass for $\gamma = -1$ is, by (12.5'), equal to $- 8/\rho$. Consequently, from (4.11), the semi-major axis is given by (with $\kappa = -8$)

$$a = - 4/h \qquad (12.28)$$

and the eccentricity, from (4.10) by

$$1 - e^2 = (-2h) (J^2/\kappa^2) = J^2/8a \qquad (12.29)$$

The energy integral, from (12.11) and (12.12), is

$$K = - \dot{\eta}^2 + (16/\rho) + \xi^2$$

$$= (-J^2/\rho^2) + 2J + (16/\rho)$$

$$= -2h + 2J \qquad (12.30)$$

where we have used the above initial conditions and (4.3) with $\dot{r} = 0$, $r = \rho$ substituted. That is, with (12.28),

$$J^2 = 2h\rho^2 + 16\rho$$

$$= (8/a) \ [(2a-\rho)\rho] \qquad (12.31)$$

so that

$$K - \frac{8}{a} = \pm (32/a)^{1/2}[(2a - \rho)\rho]^{1/2} \qquad (12.32)$$

There are two types of orbit, one being circular in both fixed and rotating systems, and the other an ellipse in the fixed system, but a complicated shape in the rotating system.

Circular Orbits ($m_1 = 0$)

All the circular orbits about m_2 are described by the relation (12.32) if ρ is set equal to a, namely

$$K = (8/a) \pm (32a)^{1/2} \qquad (12.33)$$

where the second term is $2J = \pm (32a)^{1/2}$. The initial value of $\dot{\eta}$ is $(J/a) - a = \pm (8/a)^{1/2} - a$, and for positive J is positive when $a < 2$.

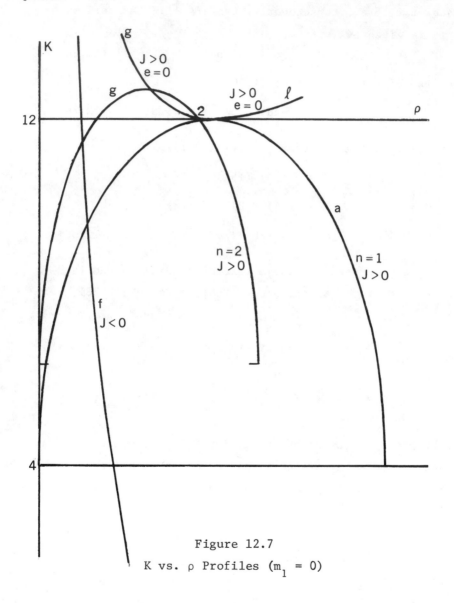

Figure 12.7

K vs. ρ Profiles ($m_1 = 0$)

For $J > 0$, K has a minimum at $a = 2$. The orbits for $a < 2$
are direct about m_2, since $\dot{\eta}$ (initial) is greater than
zero, the value of K decreases from ∞ to 12, and they
correspond to the (g) class of Strömgren. The orbits for
$a > 2$ are the retrograde (ℓ) class about both masses, and
the value of K increases from 12 ($a = 2$) to ∞ ($a = \infty$). For
$J < 0$, the orbits for $a < 2$ are the retrograde (f) class of
motions about m_2, with K decreasing from ∞ to -4 (at $a = 2$).
The orbits for $a > 2$ are the retrograde (m) class of motions
about both masses, with K decreasing from -4 to $-\infty$ ($a = \infty$).
Figure 12.7 shows the overall behavior of K vs. a.

Eccentric Orbits ($m_1 = 0$) $e \neq 0$

When $e \neq 0$, the commensurability condition is to be applied.
The mean motion n will be commensurable with unity, the rate
of rotation of the synodic frame, if it equals the ratio of
two integers. Then, from (4.14), with $|\kappa| = 8$,

$$-(h/2) = 2/a = n^{2/3} \qquad (12.34)$$

For the motion to be periodic in the rotating system, the
major axis of the ellipse can only have the values given by
(12.34) for those values of n which are ratios of two integers.
If the motion is to be simple, this restricts the value of
"a" very greatly.

For $n = 1$, the semimajor axis $a = 2$, and from (12.29), for
$J > 0$,

$$K = 4 + 8(1 - e^2)^{1/2}$$

$$= 12 - 4e^2 \text{ for e small.}$$

When the eccentricity e is small, we have already seen that

there is a retrograde elliptical libration about the equilibrium point (L_2 or L_4) with semimajor axis 4e and semiminor
axis 2e and period 2π. The Jacobi constant is a maximum
for e = 0, and decreases as e increases. These orbits, with
n = 1, represent the (a) class of Strömgren for γ = -1. The
circular orbits for a = 2 have K = 12 and K = -4, which are
at the top and bottom, respectively of the oval described by
(12.32) in the (K,ρ) plane.

The equation for the oval can be recast as

$$K - \frac{8}{a} = \pm \ (32a)^{1/2} \ [(2 - \frac{\rho}{a}) \ \frac{\rho}{a}]^{1/2} \qquad (12.35)$$

from which we see that the form remains the same for all
values of a (or n), but the height, width, and Jacobi
constant for J = 0, i.e., K = 8/a, will depend on the particular value of the invariant index n. The upper halves of
the ovals for n = 1 and n = 2 are shown in Figure 12.7.

Direct and Retrograde Motion at Aphelion

Examination of Figure 12.7 shows that the oval for n = 2
intersects the e = 0 curve for a value of ρ less than 2.
This means that an elliptical orbit can have the end of its
major axis touching a possible circular orbit, so that K and
ρ have a common value. This is puzzling at first sight, and
warrants a closer look.

Reference to (12.30) shows that

$$\dot{\eta}^2 = -K + (16/\rho) + \rho^2 \qquad (12.36)$$

so that K and ρ only determine $\dot{\eta}$ up to its sign. Consequently, it is possible that there would exist two different
periodic orbits (in the synodic system), one with $\dot{\eta} > 0$ and
the other with $\dot{\eta} < 0$, for the same K and ρ.

For $e \neq 0$, from (12.31) and (12.28)

$$\frac{J}{\rho} = (\frac{16}{\rho} - \frac{8}{a})^{1/2} \qquad\qquad (12.37)$$

and for $e = 0$

$$\frac{J}{\rho} = (\frac{8}{\rho})^{1/2} \qquad\qquad (12.38)$$

The value $\rho = 1.825$, when substituted into (12.37) and (12.38), gives a pair of $\dot{\eta}$'s with opposite signs and thus the same value of K.

For $e \neq 0$, $J/\rho = (8.77 - 6.35)^{1/2} = 1.556$, and

$$\dot{\eta} = (J/\rho) - \rho = -.269.$$

For $e = 0$, $J/\rho = (4.385)^{1/2} = 2.094$ and $\dot{\eta} = +.269$.

The circular orbit is direct in the synodic system, while the elliptical orbit is initially retrograde, and the angular momenta in the fixed system are different. The values of J in (12.37) and (12.38) are only equal when $\rho = a = 1.26$, in which case there is a common orbit which is direct.

Synodic (g) class orbits ($m_1 = 0$)

In order to understand the shapes of the (g) class orbits for a general mass ratio, it is valuable to learn what they look like when $m_1 = 0$. (In fact, there is great similarity between the general behavior at $m_1 = 0$ and at $m_1 = m_2$, if one compares the early stages of evolution of the class and neglects the complications associated with motion near L_4 or L_5.)

We have just seen that, for $n = 2$, there is a direct orbit at $\rho = 1.26$ (K = 12.70) and a retrograde ($\dot{\eta} < 0$) orbit at $\rho = 1.825$. Since there is general continuity, one would expect there to be a transition point $\dot{\eta} = 0$ for an intermediate value of ρ. This is actually so, and occurs when $\dot{\eta} = (J/\rho) - \rho = 0$, or from (12.37) when $\rho = \rho_o$ satisfies

$$\rho^3 + (8/a)\rho - 16 = 0 \qquad (12.39)$$

or when $a = 8\rho/(16 - \rho^3) > \rho/2$. That is, the motion will change from direct ($\dot{\eta} > 0$) to retrograde ($\dot{\eta} < 0$) for a value of ρ less than 2a. When this happens, there is a zero-velocity point on the ξ-axis, since $\dot{\xi} = 0$, $\dot{\eta} = 0$, and the trajectory must have a cusp there, as will be shown presently. When the root $\rho = \rho_o$ of (12.39) is plotted as a function of a, the curve is as shown in Figure 12.8.

   It starts out with tangent $\rho = 2a$, and intersects the line $\rho = a$ (e = 0) at a = 2. Thus, for n > 1 and therefore a < 2, the zero-velocity point $\rho = \rho_o$ occurs between $\rho = a$ and $\rho = 2a$.

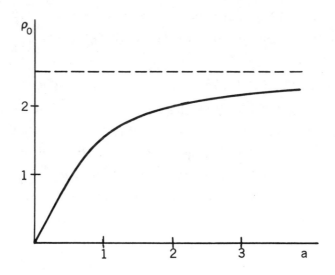

Figure 12.8

Transition point, $m_1 = 0$, g class, direct to retrograde, ρ_o vs. a.

If there is a zero-velocity point on the ξ-axis, the trajectory must have a cusp there. If $\xi = \xi_o + X$, $\eta = \eta = \eta_o + Y$, where (ξ_o, η_o) now denotes the point of zero velocity, then the equations of motion are, from (12.9) and (12.11),

$$\ddot{X} - 2\dot{Y} = U_\xi$$
$$\ddot{Y} + 2\dot{X} = U_\eta.$$

On the axis, $U_\eta = 0$, so that $\ddot{Y} = 0$ at $\xi = \xi_o$. The acceleration is in the direction of negative ξ, so that X will be proportional to t^2 and $\dot{X}$ to t. Since U_η is proportional to Y, which contains no terms lower than t^3, then $\ddot{Y} = -2\dot{X}$ and Y is of order t^3 or $|X|^{3/2}$. Thus we have a cusp at the zero-velocity point, and have found its approximate shape.

Figure 12.9 shows how the synodic orbits for n = 2 (g-class) change from the circular shape as the aphelion distance ρ increases from a = 1.26. They first are ovals, with motion direct, but at $\rho = 1.7194$, the root of (12.39), a cusp appears. For $\rho > 1.7194$, the lower left part of the curve continues to move in the same direction as before, so that it intersects the ξ-axis and then turns around to form a retrograde loop between the intersection and maximum $\xi(=\rho)$.

The equations of motion indicate the shape of the loop, when $\dot{\eta}$ is small. Then

$$\ddot{X} = U_\xi, \quad \dot{X} = U_\xi t, \text{ and } X = U_\xi t^2/2.$$

where the motion is to start at maximum ξ. Now

$$\ddot{Y} \simeq -2\dot{X} = -2U_\xi t$$

and

$$Y = \dot{\eta}_o t - U_\xi t^3/3.$$

The trajectory again crosses the ξ-axis when $t_1^2 = 3\dot{\eta}_o/U_\xi$

and at $X_1 = 3\dot{\eta}_o/2$. Since $\dot{\eta}_o < 0$ and $U_\xi < 0$, t_1 will be real
and X_1 negative. For this crossing, $\dot{Y} = -2\dot{\eta}_o > 0$ and
$\dot{X} = U_\xi t_1 < 0$. Furthermore, changing t to $-t$ just changes
the sign of Y, so that the trajectory is symmetric about the
ξ-axis.

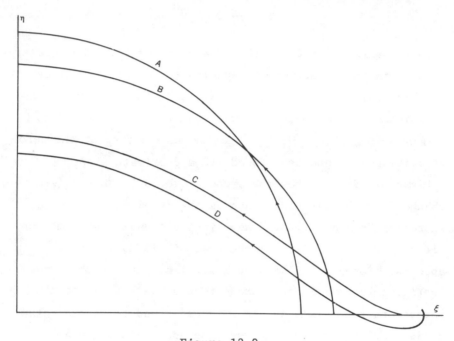

Figure 12.9

Synodic Trajectories $m_1 = 0$, $n = 2$, $\rho > a$. Each curve from
the ξ-axis to the η-axis represents a quarter orbit, the full
orbit being obtained by reflections about both the ξ-axis and
the η-axis. Curve A: $\rho = 1.26$; Curve B: $1.26 < \rho < 1.7194$;
Curve C: $\rho = 1.7194$; Curve D: $\rho > 1.7194$.

REGULARIZED EQUATIONS

Before launching on an extended discussion of how to locate
the periodic solutions for any mass ratio, it is appropriate
to introduce a system of regularized equations, with the aid
of which the integration can be performed systematically and
expeditiously for all except a few cases.

In the early days of work on the restricted 3-body problem, the equations of motion (12.9) were integrated numerically as is, and a certain amount of trouble was encountered because these equations have singularities at r = 0 and ρ = 0. When, however, an appropriate transformation was introduced to remove these singularities, these troubles vanished and one could treat orbits involving collisions with m_1 or m_2 on a par with other orbits. Whereas it had been thought that a collision orbit was the end of a family (class) of periodic orbits, now it was realized that it is merely one member and that the family can be extended smoothly beyond it, with the Jacobi constant varying continuously with the orbit parameter.

Removal of the singularities is called regularization and is accomplished readily by a change of the time variable. Suppose for instance that the motion is in one dimension, and that the energy integral is

$$(\frac{dr}{dt})^2 = \frac{A}{r} + h \qquad (12.40)$$

At r = 0, the right-hand side becomes infinite and so cannot be handled directly by a computer. However, if a new time variable τ is introduced such that $d\tau/dt = 1/r$, (12.40) is replaced by the new energy integral

$$(\frac{dr}{d\tau})^2 = Ar + hr^2 \qquad (12.41)$$

Now the right-hand side is finite when r is, and the singularity has been removed. The regularized velocity $dr/d\tau$ is finite for finite r.

A solution of (12.41) is

$$r = -(A/2h)(1 + \cos \omega\tau) \qquad (12.42)$$

where $\omega^2 = -h$. Using $t = \int r d\tau$, this integrates to

$$t = -(A/2h)[\tau - (1/\omega) \sin \omega\tau] \qquad (12.43)$$

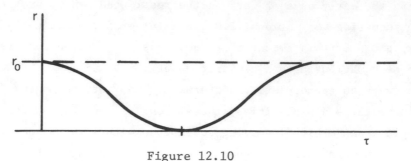

Figure 12.10

Regularized Solution of (12.40), r vs. τ

Equation (12.42) is plotted in Figure 12.10, and we see that r oscillates between r_o and 0, with τ-period equal to $2\pi/\omega$. The velocity $dr/d\tau$ is negative from the initial point $r = r_o$ until the collision at $r = 0$, $\omega\tau = \pi$, after which it becomes positive while the particle returns to its original position, to repeat the cycle. From (12.43), $\omega\tau$ can be interpreted as the eccentric anomaly, our motion being that for an ellipse (degenerate) with e = 1.

The generalization to the restricted problem is easy, because here there are two singularities instead of just one, and the regularization can be accomplished by setting

$$dt = r\rho\,d\psi \qquad\qquad (12.44)$$

the variable ψ replacing the variable τ above. (The reader can verify that the singularities are thus removed from the equations of motion (12.9).)

Since the distances r and ρ now enter on an equal footing, it would appear advantageous to introduce some sort of elliptical coordinates for ease of computation. Various schemes have been proposed, each of which has some merit. The author has found it very convenient to use the transformation of THIELE-BURRAU, which introduces variables E and F which are constant along confocal hyperbolas and ellipses, respectively.

Naturally, for physical interpretation, it may be necessary
to go back to the synodic variables (ξ, η) or to the sidereal
variables (x, y). This merely entails looking up either the
natural or the hyperbolic sines and cosines.

To introduce the transformation in question, we note that
the point midway between the two masses has coordinates
$\xi = \gamma$, $\eta = 0$. Accordingly, consider the complex variable
$\zeta = \xi - \gamma + i\eta$, which has the above midpoint as origin. Then
set

$$\zeta = \cos \omega, \text{ with } \omega = E + iF.$$

This gives

$$\zeta = \xi - \gamma + i\eta = \cos E \text{ ch } F - i \sin E \text{ sh } F$$
$$(12.45)$$

which separates into

$$\xi - \gamma = \cos E \text{ ch } F$$
$$(12.46)$$
$$\eta = - \sin E \text{ sh } F$$

If $E = 0$, then $\xi - \gamma = \text{ch } F$ gives that segment of the ξ-axis
from m_2 outward to the right.

If $E = \pi$, then $\xi - \gamma = -\text{ch } F$ represents the segment of the
ξ-axis from m_1 to the left. The η-axis through the midpoint,
on which lie L_4 and L_5, is given by $E = \pi/2$.

If $F = 0$ then $\eta = 0$ and $\xi - \gamma = \cos E$ represent the segment
of the ξ-axis between m_1 and m_2.

Confocal ellipses, with $F = \text{const}$, are given by

$$\frac{(\xi - \gamma)^2}{\text{ch}^2 F} + \frac{\eta^2}{\text{sh}^2 F} = 1.$$

Confocal hyperbolae, with $E = \text{const}$, are represented by

$$\frac{(\xi - \gamma)^2}{\cos^2 E} - \frac{\eta^2}{\sin^2 E} = 1.$$

All these relations are shown in Figure 12.11.

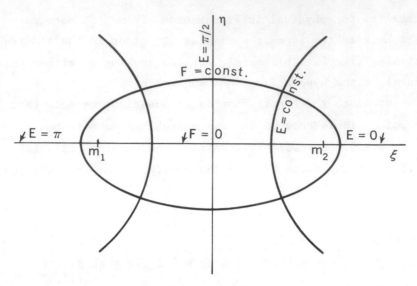

Figure 12.11

Thiele Coordinates

The distances $r = |1 + \zeta|$ and $\rho = |1 - \zeta|$ can be computed
from (12.45) and are

$$r = \text{ch } F + \cos E$$

$$\text{(12.47)}$$

$$\rho = \text{ch } F - \cos E$$

whence their product

$$D \equiv r\rho = (1/2)(\text{ch } 2F - \cos 2E) \qquad \text{(12.48)}$$

The kinetic energy in (12.7) is the sum of a quadratic term
T_2, a linear term T_1, and a zero-order term T_0. Substituting
(12.44) and (12.46) into (12.7) and (12.5), we find

$$T_0 = \frac{1}{4} (\text{ch } 2F + \cos 2E) + \gamma \text{ ch } F \cos E + (\gamma^2/2)$$

$$DV = -8[\text{ch } F - \gamma \cos E] \qquad \text{(12.49)}$$

$$T_1 = a(dE/dt) + b(dF/dt)$$

with $-a = \operatorname{sh} F\,(\operatorname{ch} F + \gamma \cos E)$

$\qquad\qquad -b = \sin E\,(\cos E + \gamma \operatorname{ch} F)$

Note that

$$-a_F + b_E = \operatorname{ch} 2F - \cos 2E = 2D \qquad (12.50)$$

The quadratic term is

$$T_2 = \frac{D}{2} \left[\left(\frac{dE}{dt}\right)^2 + \left(\frac{dF}{dt}\right)^2 \right] = \frac{1}{2D}(\dot{E}^2 + \dot{F}^2)$$

where the dots now shall signify differentiation re ψ rather than re t. The Lagrange equation for E is

$$\frac{d}{dt}\left[D\,\frac{dE}{dt} + a \right] = \frac{D_E}{D} T_2 + a_E \frac{dE}{dt} + b_E \frac{dF}{dt} + \frac{\partial(T_o - V)}{\partial E}$$

$$(12.51)$$

The energy integral is

$$T_2 - T_o + V = h = -K/2 \qquad (12.52)$$

Substituting this in (12.51) and using (12.44), (12.46) and (12.50), the equation of motion for E is

$$\ddot{E} = 2D\dot{F} + \partial\bar{H}/\partial E \qquad (12.53)$$

where

$$\bar{H} = D(h + T_o - V) = (\dot{E}^2 + \dot{F}^2)/2 \qquad (12.54)$$

The corresponding equation of motion for F is

$$\ddot{F} = -2D\dot{E} + \partial\bar{H}/\partial F \qquad (12.55)$$

Now none of the terms in $\bar{H}$ are singular, as is seen by inspection of (12.49), nor are the derivatives.

Therefore Equations (12.53) and (12.55) can be integrated with no trouble, whether or not the trajectory is a collision orbit, i.e., one going to r = 0 or to $\rho = 0$.

The essential step in the regularization is the introduction of a new time variable ψ such that $d\psi = dt/rp$. As long as this or its equivalent is done, other details are relatively unimportant. In some cases, people have preferred to avoid computations which necessitate looking up trigonometric and hyperbolic functions, but this ordinarily occasions no difficulty.

The equations of motion are, explicitly

$$\ddot{E} = 2D\dot{F} + (1/4) \sin 4E - (\bar{K}/2) \sin 2E$$

$$- (\gamma/4)(\sin E \operatorname{ch} 3F - 3 \sin 3E \operatorname{ch} F - 32 \sin E)$$

$$(12.56)$$

and

$$\ddot{F} = -2D\dot{E} + (1/4) \operatorname{sh} 4F - (\bar{K}/2) \operatorname{sh} 2F + 8 \operatorname{sh} F$$

$$-(\gamma/4)(-3 \cos E \operatorname{sh} 3F + \cos 3E \operatorname{sh} F).$$

$$(12.57)$$

For $\gamma = 0$, the equations are still the same if E is replaced by $E + \pi$, which amounts, if $\gamma = 0$, to replacing ξ by $-\xi$ and η by $-\eta$. In other words, for equal masses the physical system remains invariant under a rotation of 180°.

The reversed motion is obtained either by replacing ψ with $-\psi$ or by changing F to $-F$. The latter is equivalent to the transformation $\xi' = \xi$, $\eta' = -\eta$, or reflection about the ξ-axis, and also, from the preceding, to reflection about the η-axis, i.e., $\xi' = -\xi$, $\eta' = \eta$. The equations are thus invariant under $E' = \pi - E$, $F' = -F$.

For $\gamma \neq 0$, the equations of motion are invariant under the transformation

$$E' = E + \pi, \quad F' = F, \quad \gamma' = -\gamma$$

Accordingly, we need obtain $(\bar{K}, E)$ profiles only from $\gamma = -1$ to $\gamma = 0$, if $0 \leq E \leq 2\pi$. This transformation amounts to interchanging the masses and replacing ξ, η by $-\xi$, $-\eta$, from which it is apparent that the equations of motion will remain the same.

LOCATION OF PERIODIC SOLUTIONS

The simplest type of periodic solution is that where the particle is at rest in the rotating system, such as when it is at a libration point. Also, one might expect relatively simple periodic orbits near one of the masses, since the influence of the other mass would be small there.

We have already found some periodic solutions which may be grouped together and from which, by continuity considerations, we may find others. Firstly, when one mass m_1 is zero, then for any fraction n = p/q, where p and q are integers, we obtain a whole class of periodic solutions by varying the distance from Sun to aphelion (perihelion). Secondly, about each libration point there will, for small amplitude, be a set of ellipses, the major axis being the variable parameter. Finally, there is a set of circular orbits about the Sun, whose energy depends on the radius. From all these, with appropriate techniques, we can find periodic orbits when the one mass is not negligible.

Alternatively, direct numerical exploration can reveal where periodic orbits lie for any mass ratio, in particular when the two heavy masses are equal. For example, the Jacobi constant K can be held fixed for $\gamma = 0$, and all those trajectories found which are normal to either the E- or F-axis. Inspection of the resulting graphs quickly reveals when a particular trajectory will again strike one of these axes normally. Since the equations of motion themselves are symmetric about these axes, the result is a symmetric periodic

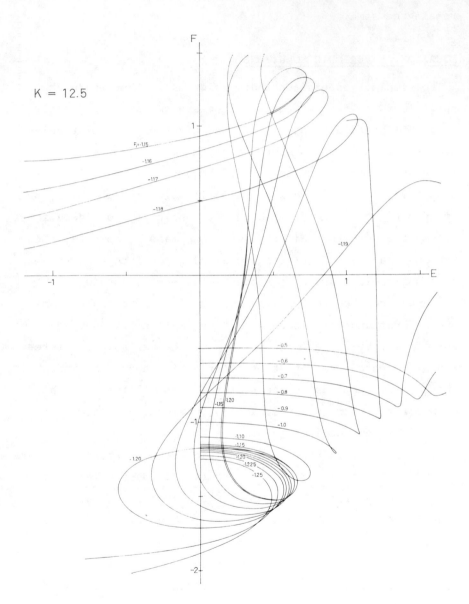

Figure 12.12

Trajectories Normal to the F-Axis, K = 12.5, γ = 0

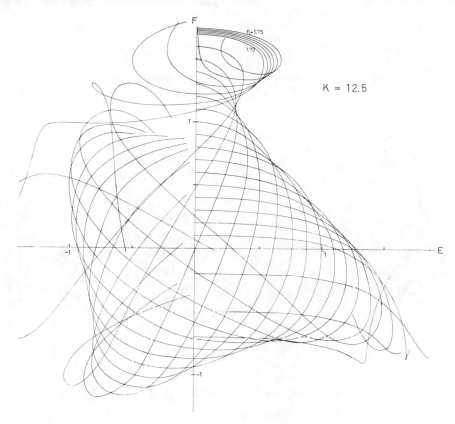

Figure 12.13

Trajectories Normal to the F-Axis, K - 12.5, γ = 0

solution. Typical trajectories for K = 12.5, γ = 0, are
given in Figures 12.12-12.15 and will be discussed in some
detail shortly.

Another type of periodic solution, when $|\gamma| < |\gamma_c|$, is
associated with the spiral trajectories issuing from L_4 or
L_5. Numerical integration shows that several of these are
normal to either the E- or F-axis. Take any one of them and
we can replace the inner part of the spiral by a short curve

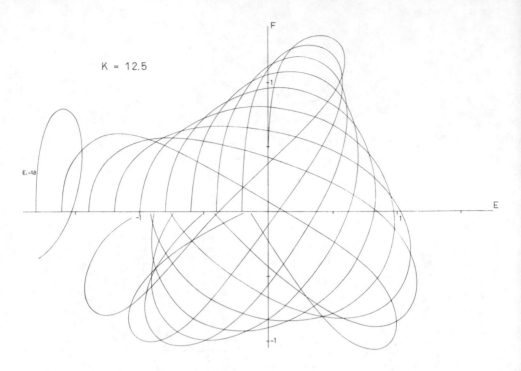

Figure 12.14

Trajectories Normal to the E-Axis, K = 12.5, $\gamma = 0$

normal to the η-axis, and at the same time change the Jacobi constant somewhat, to yield a symmetric periodic solution.

As a starting point, we thus have sufficient material to use for an extensive exploration of the various eigensurfaces. First, hold γ = constant and set E = 0. Vary the initial value F_i by a small amount from that for the known periodic solution and find out by trial how much $\bar{K}$ must be changed for a new periodic solution. This gives a short arc of the $(\bar{K}, F)$ profile, which is the section of the eigensurface with E = 0 and this value of γ. Then, if appropriate, set F = 0 and determine a $(\bar{K}, E)$ profile in similar manner. Finally,

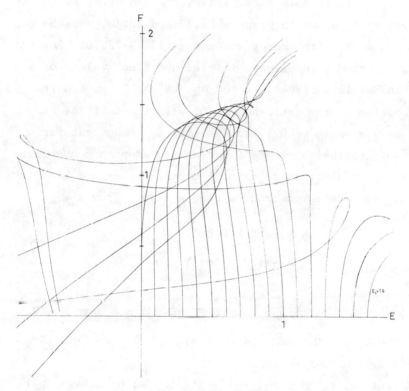

Figure 12.15
Trajectories Normal to the E-Axis, K = 12.5, γ = 0

take E and F constant and determine a $(\bar{K},\gamma)$ profile. Provided that we have one member of the class, it is thus
possible to trace out the whole class in $(E,F,\bar{K},\gamma)$ space.

Periodic Solutions, K = 12.5, γ = 0

An appreciation of how large a variety of periodic solutions are present may be gained by examining the trajectories
for a typical energy (K = 12.5) and a typical mass ratio,
$m_1 = m_2$.

They are shown in Figures 12.12–12.15, and Figure 12.16
gives the periodic solutions. The variation of E_i for these
with energy is shown in Figures 12.17 and 12.18, and the
variation of F_i with energy in Figures 12.19–12.20. For the
K = 12.5 periodic solutions, initial and final values of E or
F are in Bartlett (1964). Later on, we shall compare the
trajectories of the g-class, γ = 0, with those of the same
class at γ = -1 (m_1 = 0), and shall find common features.

To avoid confusion, it is important to adopt a convention
concerning the initial signs of $\dot{E}$ on the F-axis and $\dot{F}$ on the
E-axis. We have taken both to be positive. From the
relations

$$\dot{\eta} = -shF \; \dot{E} \text{ for } E = 0$$

$$\dot{\eta} = -sinE \; \dot{F} \text{ for } F = 0$$

we have:

For E = 0, $\dot{E} > 0$, then $\dot{\eta} < 0$ if F > 0 and $\dot{\eta} > 0$ if F < 0.

For F = 0, $\dot{F} > 0$, then $\dot{\eta} < 0$ if sin E > 0 and $\dot{\eta} > 0$ if
sin E < 0.

A direct orbit about m_2, belonging to the (g) class, will
have $\dot{\eta}_i > 0$ initially (E = 0) and so will start out with
F < 0. Figure 12.16 shows such an orbit, with a start at
F_i = - 0.793292 and E_f = 1.22329 for F = 0 and E > 0. This
orbit is symmetric about both E and F axes.

A retrograde orbit about m_2, belonging to the (f) class,
will have $\dot{\eta} < 0$ initially (E = 0) and will start out with

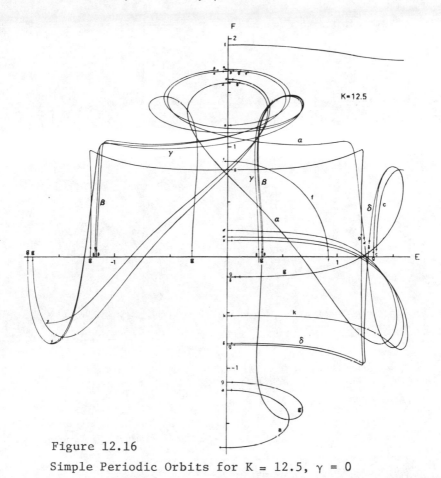

Figure 12.16

Simple Periodic Orbits for K = 12.5, γ = 0

F > 0. Such an orbit (Figure 12.16), symmetric about both
axes, has F intercept at 0.870369 and E intercept at 0.922300.

Trajectories normal to the F-axis (ξ-axis outside m_2) are
shown in Figures 12.12 and 12.13. (These diagrams are symmetric on reflection in the origin, i.e., for E → -E, F → -F).
The <u>libration about</u> L_2 (class a orbit) has intercepts
F_i = -1.200282, F_f = -1.71821 or F_i = 1.71821, F_f = 1.20082.
As F_i becomes less negative (12.12) the curvature increases,
and the orbit temporarily loses its loop, via a cusp.

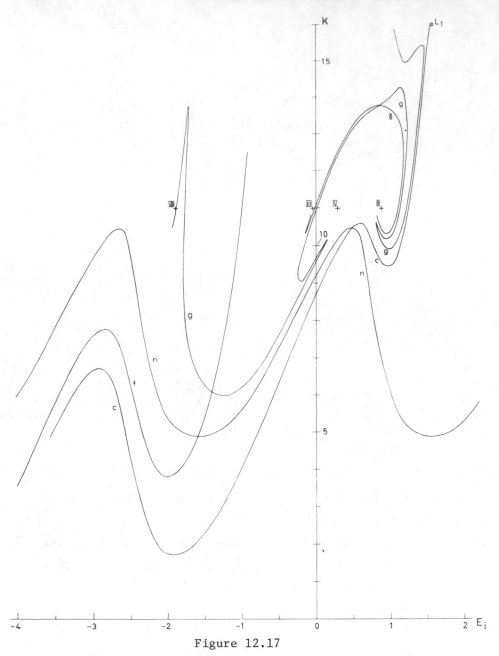

Figure 12.17

General Profile of Eigensurfaces, K vs. E_i, $\gamma = 0$. A
cross (+) with a Roman numeral, at K = 11.0, represents
a limiting orbit, shown in detail in Figure 12.6

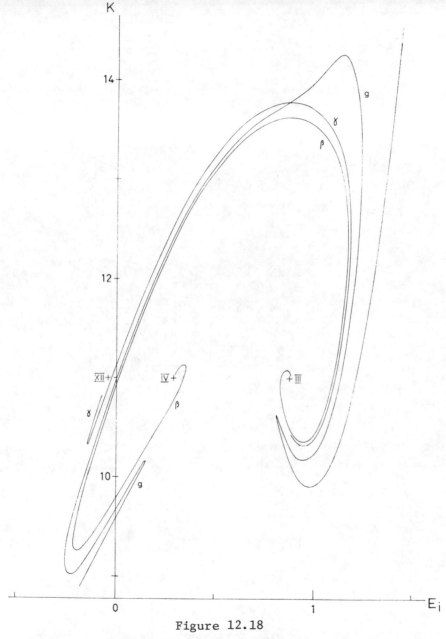

Figure 12.18

For γ = 0, Detailed Profile, K vs. E_i, showing the (β), (γ) and (g) classes. The crosses (+) at $K^1 = 11.0$ represent limiting orbits.

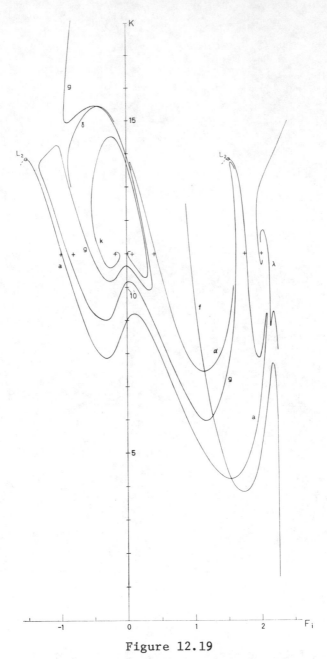

Figure 12.19

For $\gamma = 0$, General Profile of Eigensurfaces, K vs. F_i.
The crosses (+) at K = 11.0 represent limiting orbits.

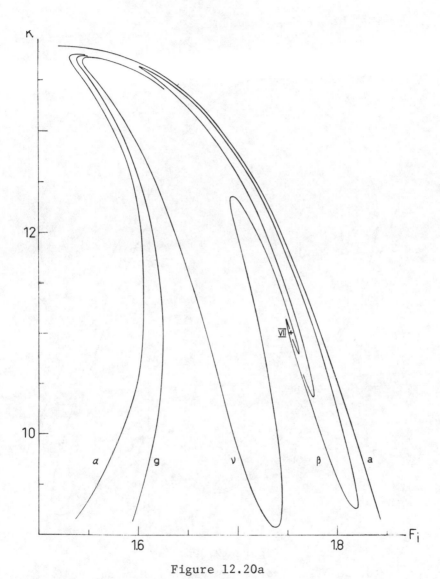

Figure 12.20a

For $\gamma = 0$, Detailed Profile, K vs. F_i, showing the (α), (g), (ν), (β) and (a) classes.

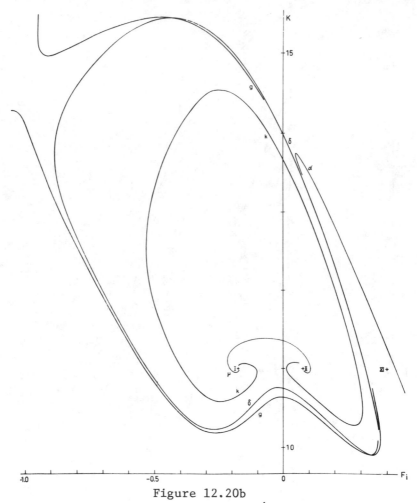

Figure 12.20b

For $\gamma = 0$, Detailed Profile, K vs. F^i, showing the (g),
(δ), (k), (μ), and (α) classes. The crosses (+) at
K = 11.0 represent limiting orbits.

The periodic orbits of Figure 12.16 are many and varied, e.g., $F_i = -1.131696$, $E_f = 0.310989$, (g); $F_i = -0.793292$, $E_f = 1.22329$, (g); orbit with large loop, $F_i = -0.777582$, $F_f = -0.181433$, (δ); $E_i = 1.328125$, $F_f = -0.170220$, (g); libration about L_1, $E_i = 1.350854$, (c); $F_i = -0.531857$, symmetric about $F = -0.830982$, $E = \pi/2$, (k), direct about both masses.

This survey of the development of various types of trajectories indicates (1) that the motion can be a general reverberation, where the particle can come close to, and sometimes touch, a zero-velocity curve, and (2) that a trajectory with a loop in it may be quite unstable, since a small change in direction on entering may result in a large change of direction on leaving (the loop). For the velocity to be zero, $\dot{\xi} = 0$, or $\dot{\eta} = 0$, which amounts to $\dot{E} = 0$, $\dot{F} = 0$, since

$$\dot{\xi} = -\,\text{ch}F\,\sin E\,\dot{E} + \text{sh}F\,\cos E\,\dot{F}$$

$$\dot{\eta} = -\,\text{sh}F\,\cos E\,\dot{E} - \text{ch}F\,\sin E\,\dot{F}$$

and the determinant $= \text{ch}^2F\,\sin^2 E + \text{sh}^2F\,\cos^2 E \neq 0$. A cusp in the (E,F) trajectory is therefore indication that the velocity is zero there. One cusp occurs (Figure 12.12) when $F_i \cong -1.0$ and another (Figure 12.16) at about $F_i = -.78$ (or $E_i > 1.223$). Still another is found (Figure 12.15) when $E_i \cong 0.5$. So contact with the zero-velocity surface can occur often, and hence the idea of comparing the motion to reverberation.

The influence of a loop on the stability of a trajectory may be considerable. Starting with the libration orbit at $F_i = -1.200282$, a change to $F_i = -1.19$ is sufficient to send the particle off to other parts of the plane, so that it no

longer librates about L_2. Similarly, the loops near the
(c) orbit about L_1 are quite sensitive to initial conditions,
as witness the change from (g) orbit, $F_i = -.793292$, to
(δ) orbit, $F_i = -.777582$, to (g) orbit, $E_i = 1.328125$, to
(c). orbit. Finally, there is extreme sensitivity associated
with the loops with center at approximately $E = 0.5$, $F = 1.35$.
One orbit, called γ, leaves the E-axis at $E = 0.282896$,
crosses the F-axis downwards after making the loop, and hits
the E-axis normally again at $E = -1.178148$. The second orbit,
class g, leaves from $E_i = 0.310989$, makes the loop and hits
the F-axis normally at $F_f = 1.131696$. A third orbit, class
β, leaves from $E_i = 0.320434$, makes the loop, crosses the
F-axis upwards and then turns back to hit the F-axis normally
at $F_i = 1.714170$. The actual positions of the loops is not
something that one would expect to predict beforehand, but
they are rather to be found by calculation. Once we know
their location, then the trajectories in the neighborhood
should in general be unstable.

GENERAL PROPERTIES OF EIGENSURFACES

   To find all members of a class of periodic solutions, we
can start with one such solution, which is a relation between
E and F, satisfying certain boundary conditions, for given
$\bar{K}$ and γ. For simplicity, let us consider only solutions
which are symmetric to the E- or F-axes. If $F = 0$ and $\dot{E}_i = 0$,
then let ΔE_i be a small change in the initial value of E and
$\Delta\bar{K}$ a small change in the (modified) Jacobi constant. Increments can be found such that $\bar{K} + \overline{\Delta K}$, $E_i + \Delta E_i$ give a new
periodic solution, and so one can trace out a $(\bar{K}, E_i)$ profile
for each class. Similarly, for $E = 0$ and $\dot{F}_i = 0$, a $(\bar{K}, F_i)$
profile can be constructed. The periodic solutions for a

given class and given γ lie on one or more surfaces in
$(E, F, \bar{K})$ space, which we shall call collectively an eigensurface.

Next, we may hold E_i or F_i constant and trace out a $(\bar{K}, \gamma)$
profile for each class. Typical profiles are shown in
Figure 12.21, and it is seen that a curve, for instance Curve
D, may reverse itself. For some values of γ, there may be
more than one value of $\bar{K}$, which means that the $(\bar{K}, E_i)$
profile can consist of more than one curve, which might not
be suspected if one should trace out only one of these for
a fixed value of γ. Thus, the (n) class (Curve D) for $\gamma = 0$
has three branches, but for $\gamma = 0.2$ there is only one. Consider now how two branches $[(K, E_i)$ profiles, say] change as
γ is changed. The two curves may move vertically toward each
other and then touch. When γ is changed still more, there
will be a separation laterally from the point of tangency, so
that now there is just one hourglass curve. The two modes of
separation can be designated as the upper-lower mode and the
right-left mode, respectively. In general, as γ changes
continuously in one direction, two branches, or two parts of
one branch, of a class may move toward each other, touch, and
separate in the other mode.

Figure 12.22 shows how a $(\bar{K}, F_i)$ profile may change as γ
does. Part of the profile for class (a) is sketched in
Figure 12.22-a for a γ near 0.63. There is an upper curve C_1
and a lower curve C_2, convex to the first. As γ becomes
more positive, C_2 swells up while C_1 does not move very much.
The two curves become tangent for some value γ_a of γ, as
shown in Figure 12.22-b. As γ increases still further, there
is a sideways separation at the point of tangency, to give
the curves C'_1 and C'_2 of Figure 12.22-c, with sharp hairpins.

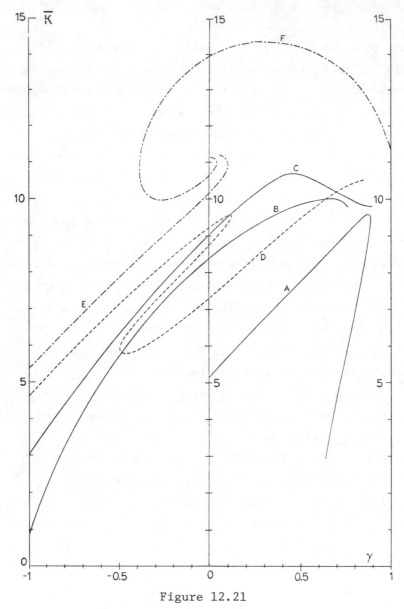

Figure 12.21

$(\bar{K},\gamma)$ Profiles of Eigensurfaces
(a) class: F = 1.1 (A); F = 0.4 (B); F = 0.0 (C); (n) class:
E = 0.0 (D); (g) class: F = 0.0 (E), (F).

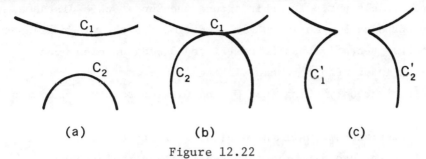

Figure 12.22

Rearrangement of Profile with change of γ

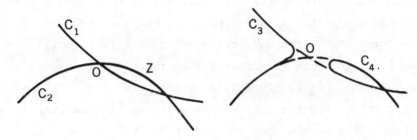

Figure 12.23

g-class change from $\gamma = -1$ to $\gamma > -1$

The same sort of behavior occurs at $\gamma = -1$. (See Figure 12.23). C_1 represents the curve for direct circular orbits while C_2 up to Z shows the profile for direct elliptical orbits with mean motion $n = 2$. The curves intersect at 0, where the Jacobi constant for C_2 is a maximum. When γ is increased to -0.998, the Sun-Jupiter ratio, the curves pull away from 0, resulting in C_3 and C_4 and leaving a gap in the vicinity of 0. In other words, there exists a certain distance ξ from the Sun such that, if a particle starts out normal to the ξ-axis, then, no matter what the energy in the neighbourhood of that for a circular orbit, it will not hit the ξ-axis normally on its next crossing. The presence of Jupiter has thus introduced sufficient disturbance to make

impossible a periodic orbit around the Sun at this distance.
Such gaps have been known for some time for the minor planets,
but only recently has the restricted 3-body problem been
solved in an effort to explain them.

Since the only thing that C_1 and C_2 seem to have in common
is the circular orbit through 0, we need to specify just what
characterizes C_3. All the orbits of C_3 start out normal to
the ξ-axis, are direct and approximately circular above 0,
and strike the ξ-axis normally at the next crossing. The
orbits below 0 are approximate ellipses in the fixed system,
with the same major axis (or period), but they too are direct
near 0 and start from and arrive at the ξ-axis normally. The
transition from one set of orbits to the other is a smooth
one. On one side far from 0 the orbits are nearly circles,
on the other side they are nearly ellipses, in the fixed
system, while near 0 they are something in between.

Actually, it would be ridiculous to search for some hidden
invariant along C_3. This curve has been constructed as the
locus of those points which represent the solution of a
special eigenvalue problem. Each point is the trace of a
trajectory which leaves the ξ-axis normally and strikes it
normally at the next crossing. This is a common feature
built in to define the class, and we should not expect any
further invariants.

For successful numerical exploration, there are a few
pitfalls which must be avoided. If a periodic orbit exists,
it is not isolated, for it is a section of a smooth eigensurface. The profiles may have some very sharp hairpin turns,
so that a very small increments are needed if a profile is
to be followed. However, variation of the mass ratio can
result in profiles without such hairpin turns, so we can
often make a detour around the turns by changing γ.

THE (g) CLASS

Rather than enter on an extensive discussion of a number of classes of periodic solutions, we shall concentrate mainly on the (g) class, which starts out as composed of direct orbits around one mass m_2. This will be done because this class presents most of the typical problems and because some of its orbits are reasonably stable.

Behavior at $m_1 = 0$ ($\gamma = -1$)

At $\gamma \neq -1$, there are circular direct orbits and elliptical direct orbits (n = 2), both periodic in the rotating system. A half orbit will start normal to the + ξ-axis and end normal to the -ξ-axis, or start normal to the -F-axis ($\dot{E}_1 > 0$) and end normal to the E-axis.

The elliptical orbits (in the fixed system) for n = 2 are shown in (E,F) coordinates in Figure 12.24. Orbit 1 is for ρ = 1.7194 (see Figure 12.9) where the initial motion changes from direct to retrograde. Cusps appear at both ends of the half orbit [at + ξ (-F) and at - ξ (+E)]. Going around the n = 2 oval counterclockwise, the next orbit of interest is the circular one, orbit 2, at ρ = a = 1.26, with intercepts F = ch^{-1} 2 = 1.3169 and E = π/2. A loop appears and becomes larger (orbits 3,4 at seq.) so that the motion is in part retrograde. When F → 0, the trajectory approaches an ejection orbit, which starts out tangent to the +E-axis (- ξ-axis) and concludes the half-orbit by being tangent to the -F-axis (+ ξ axis) and hitting m_2 head on [thus making a double collision]. This orbit is one for which J = 0, e = 1, $\bar{K}$ = 5.3496. Proceeding still further, orbits 6,7,8 and 9 are wholly retrograde, with J < 0. Note that these orbits go through six quadrants in the (E,F) plane, or three times around m_2, to return to the starting point, and are thus triply-periodic.

Between 8 and 9, at minimum $\bar{K}$, there will be a circular orbit,
coinciding with orbit 2, but retrograde, and also belonging
to the (f) class of retrograde circular orbits, (Figure 12.7).
Orbit 10 is for J = 0, $\bar{K}$ = 5.3496, ρ = 2a = 2.52, e = 1 and
corresponds to straight line motion in the fixed system, from
ρ = 2a in to m_2 and back. The half orbit starts at
ξ = - 2.52 (E = 0, F = 1.931) and ends at (E=π/2, F=ch^{-1} 1.52).
It is a skew collision orbit through E = 0, F = 0.
Orbits 11 and 12 have J > 0, but are initially retrograde
in the rotating system. This changes back to direct at orbit
13, which is the same as orbit 1. With this, we have completed
the (E,F) description of the orbits with invariant index n = 2,
γ = -1. Some are simply-periodic and some are triply-periodic,
with the transition at the collision orbits. The motion is
direct at times and retrograde at other times.

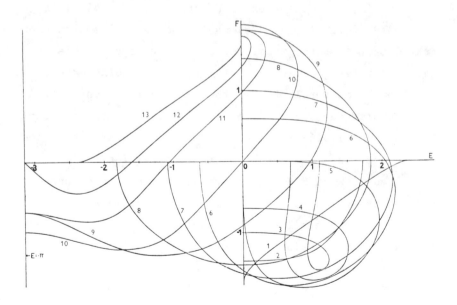

Figure 12.24

Elliptical Orbits, Fixed System, n = 2

The whole class at $\gamma = -1$ is determined by $n = 2$, for this
gives the semimajor axis a by (12.34), as well as the value
of h. The angular momentum J can range from zero to $\pm\,(8a)^{1/2}$,
by (12.29), the maximum being for a circular orbit. For any
value of ρ between 0 and 2a, we can calculate J^2 from (12.31)
and then the initial velocity (at end of ellipse) $\dot{\eta} = (J/\rho) - \rho$.

The orbits can be calculated as ellipses in the fixed
system, and then transformed to the (E,F) system via the
(ξ, η) coordinates. The result is Figure 12.24, the subject
of the above discussion.

Behavior of g-class, small m_1

When m_1 is small, then the direct circular orbits link up
with those for $n = 2$ in the manner shown in Figure 12.23. As
γ increases from -1, the double collision orbit is replaced
by 2 simple collision orbits, one occuring shortly after the
other. At $\gamma = -9/11$ the g-class has repeated contacts with
the zero-velocity surface, and the eigensurface itself may be
endless, consisting of sheets close to each other and not
too far away from the $n = 2$ surface for $\gamma = -1$. Apart from
this, however, there do not seem to be any other complications
of note. The trajectories are probably qualitatively about
the same as for $\gamma = -1$.

Behavior of g-class, $m_1 = m_2$ $(\gamma = 0)$

The development of the g-class orbits for $\gamma = 0$ is shown in
Figure 12.25, a-c, which is to be compared with Figure 12.24
for $\gamma = -1$ (circular orbits not shown). A view of the relation between orbits of different classes is given by Figure
12.16. Vertical sections (profiles) of the eigensurfaces,
including that for the g-class, are shown in Figures 12.17-
12.21.

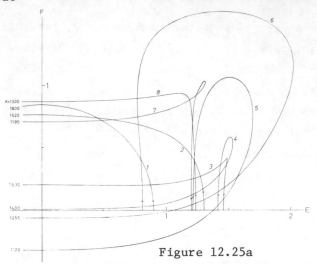

Figure 12.25a

For γ = 0, Development of the (g) Class, near mass m₂. The curves are numbered in order along the profile (E or F) starting with 1 near the mass m₂ and increasing until the limiting orbit in Figure 12.25c is reached.

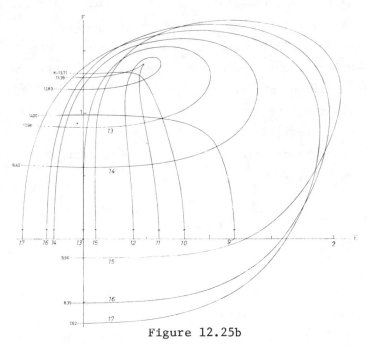

Figure 12.25b

For γ = 0, Development of the (g) Class, Intermediate Part.

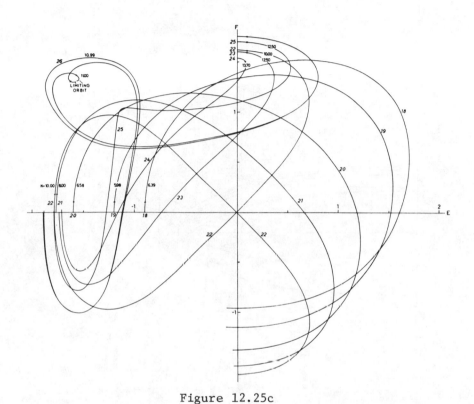

Figure 12.25c

For $\gamma = 0$, Development of the (g) Class, Termination.
The curves from 22 onwards are started on the F-axis, instead
of on the E-axis, for clarity of representation and for comparison with Figure 12.25b.

Typical g-class orbits, starting with K = 18 (close to m_2)
and going along the profile, are numbered from 1 to 26.   The
initial value E_i and the final (half-orbit) value F_f are
given in Table 12.2, for these numbered orbits and for other
orbits of interest.

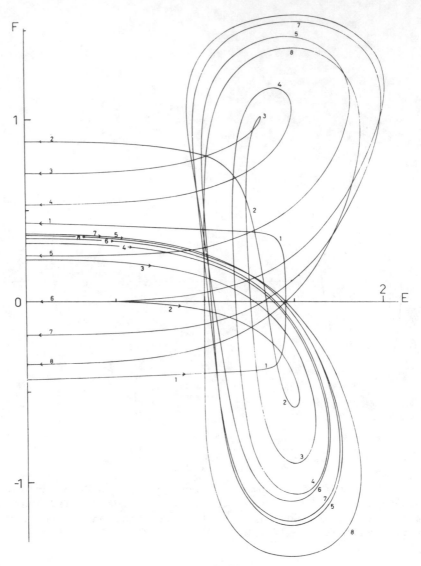

Figure 12.26

For $\gamma = 0$, Development of the (δ) Class (a closed group),
from maximum K to minimum K. (For the reverse development,
take the mirror images about the E-axis.) The curves are
numbered consecutively from 1 to 8, and have K-values of
15.428, 13.988, 12.0, 11.0, 10.23, 10.75, 10.30 and 9.915
respectively.

Table 12.1   δ-class Data   γ = 0

ORBIT	K	F_f	F_f	REMARKS
1	15.42751	-0.4311	0.4311	g-class
2	13.98826	0	0.878648	collision orbit
3	12.5	0.181433	0.777582	see Figure 12.26
	12.0	0.231782	0.703796	
4	11.0	0.319742	0.530173	
5	10.23082	0.371252	0.25	
6	10.74980	0.346634	0	collision orbit
7	10.3	0.371880	-0.184561	
8	9.9152	0.3439	-0.3439	g-class

Table 12.2   g-class Data   γ = 0

ORBIT	K	E_i	F_f	REMARKS
2	15.153	1.300	0.766	
3	15.00000	1.456591	+0.203998	
4	14.000	1.4104	0.009	almost collision
	12.500000	1.328125	-0.178220	see Figure 12.26
	12.000	1.2949	-0.230	
5	11.000	1.2109	-0.317	
	9.915	1.000	-0.344	δ minimum
6	10.55390	0.810757	-0.062000	
	10.62969	0.804456	-0.000002	Collision
7	11.9487	1.200	0.707	
	12.50000	1.223290	+0.793292	see Figure 12.26
8	13.00000	1.233426	+0.866735	
9	14.00000	1.200304	+0.987776	
10	13.708	0.800	1.319	γ maximum
11	13.392	0.600	1.285	
12	12.8324	0.400	1.186	cusp
	12.50000	0.310989	+1.131698	see Figure 12.26
13	11.00000	0.003349	+0.887609	near collision
14	9.4	-0.2392	0.570	
	9.65813	0.000	0.125	collision
	10.15088	0.146658	0	collision
15	9.958141	0.100000	-0.150372	
	9.575164	0	-0.255922	collision
16	8.390	-0.300	-0.511	
17	7.463546	-0.543351	-0.700000	
18	6.391237	-0.900000	-0.960491	
19	5.979706	-1.200000	-1.152400	
20	6.575256	-1.600000	-1.375607	

Table 12.2 continued

ORBIT	K	E_i	F_f	REMARKS
21	8.000000	−1.767032	−1.533985	
22	10.00000	−1.784111	−1.617194	skew collision
23	12.500000	−1.737042	−1.601685	
24	13.70000	−1.712301	−1.538452	
25	12.50000	−1.786034	−1.701971	
26	10.99000	−1.881730	−1.752751	

The δ-class consists of direct doubly-periodic orbits,
symmetric about the F-axis. Its development is shown in
Figure 12.26, and the corresponding orbits 1-8, plus that for
K = 12.5, are specified by F_i and F_f in Table 12.1. Orbit 1,
K = 15.42751 and orbit 8, K = 9.9152 are symmetric about the
E-axis and so are common orbits of both δ-class and g-class.
As we see from Figure 12.20b, the g-class profile is close to
the δ-class profile for a considerable interval, and, where
this is the case, the δ-class orbits each roughly consist of
one half-orbit of g-type connected to another half-orbit of
g-type by a line skew to the E-axis.

For instance, at K = 12.5 (see Figure 12.16) the δ-orbit is
specified by F_i = .777582, F_f = −0.181433, which is a hybrid
of two g-orbits, one with F_i = −.793292, E_f = 1.223290 and
the other with E_i = 1.328125, F_f = −0.178220. At K = 12.0,
the similarity is quite striking between the g-orbit 7 (12.25a)
and the upper portion of the δ-orbit (12.26).

From the maximum of the δ-class (K = 15.428, δ-orbit 1), the
g-class keeps close company with it until after g-orbit 7,
when a transition occurs with E_i getting much smaller and
F_f much larger. Orbit 10 (Figure 12.25b) is at the maximum
K of another direct class, the γ-class, which is symmetric
with respect to the E-axis and doubly-periodic. Another
cusp is gone through, and a loop is formed, very prominent

in Figure 12.16. Now the g-orbits are initially sandwiched
in between the γ-orbits and the orbits of another class, the
β-class. The loop becomes larger and larger, while F_f
constantly decreases. Orbits 15 to 23 are very similar to
orbits 5-11, Figure 12.24 for γ = -1, with orbit 22 the same
sort of collision orbit that orbit 10, γ = -1, is.

The β-class was already mentioned under "Periodic Solutions,
K = 12.5, γ = 0." Its orbits differ from g-orbits in that
they cross the F-axis with negative slope but then return to
hit it normally. The K vs. E_i profile is shown in Figure
12.18, from which it is seen that the class goes between the
two asymptotic orbits III and IV. These coalesce at γ ≅ -0.24
and no longer exist when γ becomes still more negative.

Behavior of g-class as function of mass ratio

The g-class at γ = 0 displays some similarity to that at
γ = -1, but there are also large differences. It starts out
with circular orbits around m_2, but for γ = 0 develops a
collision orbit early, right after Orbit 4. Classes (γ) and
(δ), both doubly-periodic and direct, are present at γ = 0
but absent at γ = -1, and this allows the g-orbits to have
more latitude and to show more diversity. After this dis-
traction, the class takes on a more regular behavior, with
large retrograde orbits until the skew collision orbit
similar to that at γ = -1, n = 2, J = 0, ρ = 2a. Then the
orbits go in the limit to an asymptotic periodic orbit
associated with the libration points L_4 and L_5. This is
apparently just a temporary stopping point, and further
calculations have not been made. However, it is clear that
the class, in these later stages, is nowhere near as simple
for γ = 0 as it is at γ = -1.

Since the g-class is in part close to the (β) and (δ)
classes, it is instructive to see how they behave as γ
changes. The (β) and (δ) classes probably behave similarly
as γ becomes more negative. For γ = 0, the (β) class has an
open profile, ending in spirals about points representing
asymptotic orbits III and IV of Strömgren. As γ goes more
negative, these two points come closer together until they
finally coincide for a value of γ about -0.24. [For γ still
more negative, these asymptotic orbits (normal to the ξ-axis)
do not exist.] As the points come closer, so do the associated spirals until they finally touch. The (β) class then
becomes closed (via the upper-lower splitting mode), but
surrounds an open spiral branch between III and IV. This in
turn closes, and generates another pair of closed and open
curves, so that an infinite nested set of closed curves
evolves as a result of the above coincidence. As γ becomes
still more negative, the outer branch shrinks down (and with
it the inner branches), finally to disappear at about
γ = -0.8. The (δ) class is already closed at γ = 0, but
surrounds its open (μ) class offspring between orbits I and
II, which are close together, and coalesce at $\gamma \cong - 0.059$.
The (δ) class itself shrinks to zero at about γ = -0.48.

In general, once a class has become closed by a change of
γ in some direction, further change of γ in the same direction will bring about its shrinkage to zero. The particular
value of γ at which the class disappears does not seem to
have any special significance.

Figure 12.27 shows how the (β) class disappears as γ
becomes negative and what happens to the nearby portion of
the (g) class at the same time. Curve β_o, representing the
(β) class at γ = 0, is open at the bottom and stretches

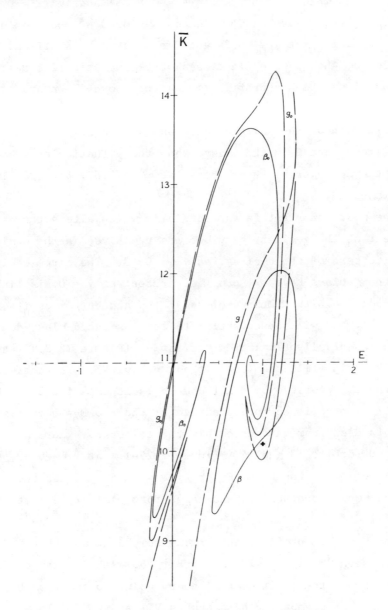

Figure 12.27

Detailed ($\bar{K}$,E) Profiles for the (β) and (g) classes,
β,g (γ = $-$ 0.59); β_o, g_o (γ = 0).

between asymptotic orbits III ($E = 0.8706$) and IV ($E = 0.2957$).
Curve g_o [(g) class at $\gamma = 0$] detours around β_o and does not
intersect it.   At $\gamma \cong -0.24$ and $E \cong 0.75$, orbits III and IV
coalesce, the (β) profile closes off (generating its nested
set of inner closed profiles), and then moves downward and
to the right.   The (g) profile also moves downward, with
elimination of the hairpin turns and upward bulge.   The curves
labelled g and β in the figure show the situation at $\gamma = -0.59$,
and the heavy cross shows where, at $\gamma \approx -0.83$, the (β) class
vanishes.

The most important factor initially which alters the development of the g-class, as γ becomes negative, is the shrinkage to zero of the δ-class, as $\gamma \rightarrow -0.48$.   The large cavity
in the g-class $\bar{K}$ vs. F_i profile, present at $\gamma = 0$, is linked
by common orbits to the δ-class at top and bottom of its
profile, and will hence shrink to zero as the δ-class does.
This essentially eliminates orbits 3-8 (Figure 12.25a) and
makes the profile simpler (without a cavity).   The coalescence
of the asymptotic orbits III and IV is accompanied by a shift
of the β-profile to more positive E, and two $E = 0$ collision
orbits disappear, and with this the temporary excursion to
$E_i < 0$ (orbits 13 and 14, Figure 12.25b).   At $\gamma = -0.59$, then,
the section of the K vs. E_i profile with $E_i > 0$ has been
rather well smoothed out and has a form similar to that at
$\gamma = -1$.

The $\bar{K}$ vs. E_i profile has various complications when $E_i < 0$,
as is seen in Figure 12.25c.   Orbit 26, with its large loop,
is close to the limiting trajectory going to and from L_4,
composed of asymptotic half-orbits VII and VIII.   To each of
these there is a conjugate asymptotic orbit, namely VII* and
VIII*, respectively.   The asymptotic orbits disappear in

pairs, VIII and VIII* at $\gamma \cong -0.71$, $E_i \cong -1.791$, and VII and
VII* at values as yet unknown. The apparent termination of
the g-class, $\gamma = 0$, on the above asymptotic orbit is only
illusory, since there will be other conjugate(s). Eventually,
as γ becomes more negative, the asymptotic orbits will have
vanished, and with them the associated complicated orbit
forms. The details have not been studied, but the process
has been completed by the time $\gamma = -9/11$ has been reached.

ORBITAL STABILITY--FIRST ORDER

Even though the survey of simple periodic solutions of the
restricted 3-body problem is reasonably complete for all
values of the mass-ratio, the study of their stability has
not been as systematic. Hénon (1965) has found characteristic exponents for $m_1 = m_2$, as a function of the Jacobi
constant K, and these determine the first-order isoenergetic
stability (see Chapter X, B). Since this case may be
regarded as the maximum deviation from the two-body problem,
an answer to stability questions here may to some extent
eliminate the need to answer then for all mass ratios.
Accordingly, we shall proceed to summarize the results of
Hénon.

First, one uses as a phase plane a "surface of section."
For a given energy K, each time the particle crosses the
ξ-axis ($\eta = 0$), its motion will be specified by the coordi-
nates ξ, $\dot{\xi}$, $\eta = 0$, and $\dot{\eta}$. But $\dot{\eta}$ is determined up to its sign
by the energy equation with a given K, so we can just examine
how points in the (ξ, $\dot{\xi}$) plane map as the ξ-axis is traversed
in the same direction successively. Using the analysis of
Chapter X, B, one can, after locating the fixed points,
classify them as elliptic or hyperbolic and thus stable or

TABLE 12.3

K Intervals of First Order Stability

$$\gamma = 0$$

A line between the entries for $\lambda = 1$ and $\lambda = -1$ indicates that all values of a between -1 and $+1$ are taken on. A broken line beginning and ending with $\lambda = 1$ means that the values of a do not extend to $\lambda = -1$. These energy values are of $K = 4C$, where C has been tabulated by Hénon.

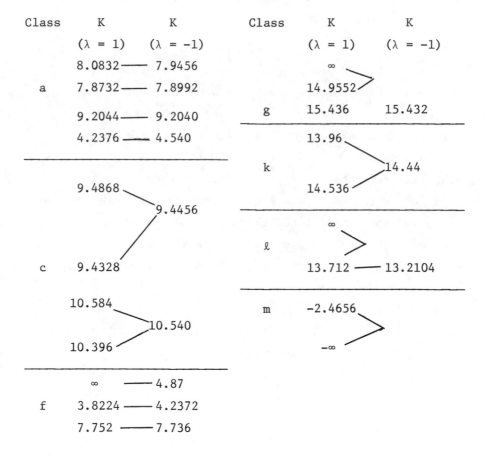

Class   K ($\lambda = 1$)   K ($\lambda = -1$)

a   8.0832 —— 7.9456
   7.8732 —— 7.8992
   9.2044 —— 9.2040
   4.2376 —— 4.540

c   9.4868 \ 9.4456
   9.4328
   10.584 \ 10.540
   10.396

f   ∞ —— 4.87
   3.8224 —— 4.2372
   7.752 —— 7.736

Class   K ($\lambda = 1$)   K ($\lambda = -1$)

   ∞
g   14.9552
   15.436   15.432

k   13.96 \ 14.44
   14.536

ℓ   ∞
   13.712 —— 13.2104

m   -2.4656
   -∞

unstable to first order, or locally stable or unstable. The
point will be elliptic if $-1 < a < 1$, where a is the
coefficient in (10.23), and hyperbolic if $|a| > 1$.

Table 12.3 shows the intervals of K, $\gamma = 0$, for which the
periodic orbits are stable to first order, i.e., have
elliptic fixed points (a lies between -1 and $+1$). Class (a)
orbits are unstable from the libration point L_2 (K = 13,8272,
F = 1.5206) down to K = 8.0832, after which there are very
narrow bands of stability near the extrema of K. Class (c)
orbits are unstable from L_1 (K = 16.0, E = 1.5708) down to
K = 9.4868, after which similar narrow bands of stability
near extrema of K also occur. Class (k) has a small stable
band below its maximum K. These classes are to be considered
as mostly unstable. Class (g), of direct orbits about one
mass (m_2), is stable down to the first minimum of K, after
which it is generally unstable, again except for very narrow
bands. The same holds for class (f), whose orbits are retrograde about mass m_2. Class (ℓ) orbits, direct about both
masses in the fixed system, are stable from K = ∞ to
K = 13.2104, after which they are unstable. Class (m) orbits,
retrograde about both masses in the fixed system, are not
stable at small distances (say $F_i < 0.65$), but are stable
from K = -2.4656 to $-\infty$ along the outer branch of the profile.

Figure 12.28 shows where the motion about mass m_1 is stable
to first order. The retrograde orbits (corresponding to class
(f) around m_2) are stable in the unshaded regions. The direct
orbits, corresponding to class (g) around m_2, are stable
inside the innermost curve.

Figure 12.29 shows where the motion about both masses is
stable to first order. The retrograde orbits, class (m), are
stable outside the shaded region. The class (ℓ) orbits,

direct in the fixed system, are stable outside the outmost
curve.   (These figures are due to Hénon.)

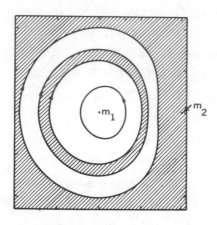

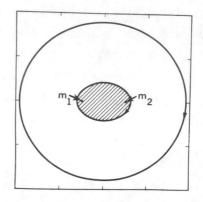

Figure 12.28	Figure 12.29
Regions of Local Stability for Motion around m_1, $\gamma = 0$	Regions of Local Stability for Motion around both Masses, $\gamma = 0$

General Conclusions--Local Stability

Most of the periodic orbits for the restricted 3-body
problem are unstable, essentially because the small body
encounters regions where a small change in initial direction
can become greatly amplified while in such a region. A good
example of this is shown in Figure 12.16, K = 12.5, $\gamma = 0$,
where the only known stable orbit is that of the f-class.
There are loops not far from the zero-velocity curve, as well
as the a-class libration, all of which contribute to instability.   Also, for the various classes, and especially for the
g-class, collision orbits occur when K is varied, and these
are all unstable.   (The g-class has not been followed much
beyond the β-class, and it may well be that the triply-
periodic retrograde orbits, to the extent that they are similar
to the $\gamma = -1$, n = 2 orbits, will show a certain degree of

stability because of their resemblance to the elliptical
orbits in the fixed system.)

The most stable classes at $\gamma = 0$ are (f), (g), (ℓ), and (m).
The region of stability for the direct orbits around one mass
is small indeed (Figure 12.28), not even going out to
L_1 ($\xi = 1$). The retrograde orbits, class f, fill out a region
extending somewhat beyond L_1, with the exception of the small
shaded region. The m-class orbits, retrograde, are around
both masses and stable if they are not too close to the masses,
because the orbit may be approximated by a Keplerian one about
the total mass concentrated at the center. The ℓ-class
orbits, direct in the fixed system, must be farther out
(Figure 12.29) to be stable.

ORBITAL STABILITY--GLOBAL

The "local stability" analysis can only reveal, for a given
energy K, whether the fixed point in the surface of section
phase plane $(\xi, \dot{\xi})$ is elliptic or hyperbolic, and it does not
give much in the way of clues as to how large the region of
global stability is. The basic question to be settled is:
How far from the fixed point may we displace the particle,
keeping K constant, and still be sure that the motion will
not leave the domain (of stability)? General methods for
studying this expeditiously have been given in Chapter X, but
their application to the restricted 3-body problem has been
limited.

Mullins and Bartlett (1972) have determined rather closely
the extent of the stable region for the g-class, as a function of the Jacobi constant K. Figure 12.30 shows, for
$K = 16.0$, a simple elliptic fixed point E at about $\xi = -0.5$
(distance between primaries is 2 units) and the elliptic (e_i)
and hyperbolic (h_1) fixed points of order 6. The oscillations

of the outer eigencurves through h_i are small (and are not shown), while those of the inner eigencurves are not detectable. Therefore, the latter delineate very closely the boundary of the stable region around the elliptic fixed point E. At $\dot{\xi} = 0$, it goes from $\xi \equiv -0.23$ to $\xi \cong -0.72$.

As the Jacobi constant decreases from K = 16.0, oscillations of the inner eigencurves make their appearance. The dependence of L/C on K is shown in Figure 12.31, and we see that there is a rapid rise from about 10^{-6} at K = 15.7 to about 10^{-3} at K = 15.6, which means that the stable area around E has decreased very rapidly. At K = 15.5, we can expect that almost all of the central cell will be unstable.

Whereas the first order calculations showed that the fixed point is elliptic from K = ∞ to K = 14.9552, the present analysis shows that the stable region has already become rather small at K = 15.6, where a is appreciably less than unity.

From the calculations, the fixed point will be elliptic for certain ranges of K such that the diagonal matrix element a (=d) of (10.23) has absolute value $|a| < 1$. As K decreases, $|a| \rightarrow 1$ and the stable area decreases sharply. For the stable area to have a non-negligible value, it can well be that $|a|$ will differ appreciably from unity, which is apparently so for the g-class.

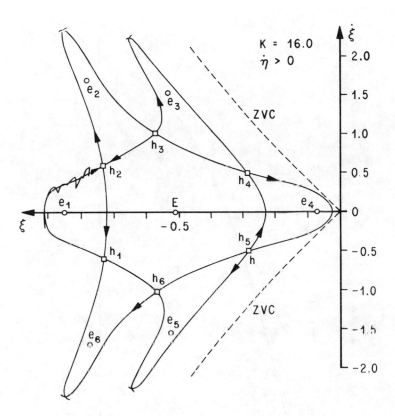

Figure 12.30

Global Stability Region, K = 16.0, γ = 0

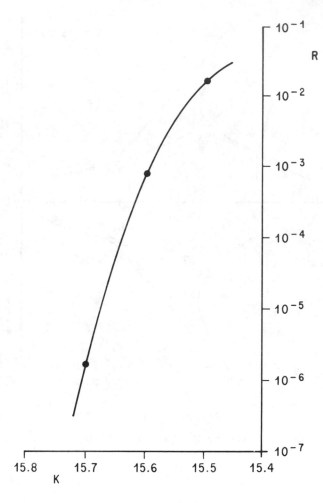

Figure 12.31

Relative Oscillation Area vs. K, $\gamma = 0$

Epilogue

The preceding treatment has been concerned with the development and generalization of Newton's Equations of Motion, with
the ultimate aim of finding how a particle, a system of particles, or a rigid body moves as a function of time.

As a general rule, the more elementary and basic ideas have
been presented in the earlier chapters, while more advanced
and specialized notions have been reserved until later. A
familiarity with the contents of Chapters I and II is a
necessary preliminary to the reading of the rest of the book.
In Chapter I, besides a discussion of the various conservation laws and the nature of rectilinear motion in one dimension, velocity-dependent forces are mentioned, followed by a
treatment of the motion of fast particles in one dimension,
where the mass varies with the velocity. The chapter concludes
with a short discussion of irrotational (conservative) forces
fields in two dimensions.

In Chapter II, we have dealt with general properties of
systems of particles and have introduced the powerful
Lagrange form of equations (which of course has no more
physical content than the original Newton's Equations). This
enables one to exploit the symmetry that the potential energy
term may have, and also to eliminate from explicit consideration the internal constraints of the system. When a potential

energy function exists, the equations of motion may be derived
from a Hamiltonian function. When this, in turn, either does
not involve the time explicitly or is periodic in the time,
the time behavior of the system is "more or less orderly."
One is able, as is seen in later chapters (especially
Chapter X), to make statements about orbital stability for
simple systems. Finally, in this chapter, the powerful
method of moments is introduced and applied briefly to figures
of equilibrium of rotating masses.

The simpler problems of dynamics are perhaps those where
the motion is slow (non-relativistic) and where the forces
depend neither on time or on the velocity. Despite the com-
plexity due to the non-linearity of the equations of motion,
some of these problems can be solved completely. (1) The
spinning symmetrical top, of which the plane pendulum, conical
pendulum, and spherical pendulum are special examples (zero
spin), may be described by the equation for the Jacobian
elliptic functions, which are to be found in standard tables.
The various cases are presented in Chapter III, together with
a short description of these functions and their properties.
(2) If the system consists of two masses interacting according
to an inverse-square law of force then each mass will describe
a conic section about the center of mass. This is shown in
Chapter IV, and some properties of conic sections are detailed
in Appendix B. (3) The restricted 3-body problem, where a
particle of infinitesimal mass is acted on gravitationally by
two large masses rotating in circles about their common
center of mass, has been essentially solved. The equilibrium
points and the simple classes of periodic orbits have been
found numerically, and the stability of some orbits has been
determined. A description of this work is given in Chapter
XII. If some further special questions need to be answered,

then the appropriate computations can be readily carried
out.

The motion of a charged particle in an electromagnetic field
can be handled readily, since the system can be described by
a Hamiltonian function. After a resumé of some properties of
electromagnetic fields in Chapter V, the Hamiltonian formulation is given in Chapter VI, and applied to the betatron and
to the linear accelerator. A further application to the
alternating gradient synchrotron is made in Chapter X, which
deals with situations where the Hamiltonian is periodic in
time. A rather special case, important for the theory of
cosmic rays, concerns the motion of a particle in the field
of a magnetic dipole (see Chapter XI). Here the Hamiltonian
does not involve the time explicitly, so there is an energy
integral. Periodic orbits exist, as well as orbits asymptotic to them, and their stability is discussed. This problem,
which has been thoroughly studied, is a simpler one than the
circular restricted 3-body problem, for which one now knows
the general morphology of the solutions. Finally, a whole
discipline concerning the behavior of a collection of charged
particles, known as a plasma, has sprung up. The main
problem here is whether or not a plasma can be contained for
an appreciable time in a finite region, so that nuclear
fusion processes can be controlled. This is outside the
scope of the present treatment, but the interested reader
will find a readable account in the book by Schmidt "Physics
of High Temperature Plasmas."

For many applications, and especially for the design of
high-energy accelerators, it is vital to know what happens to
a beam of particles when it encounters a force field. For
maximum usefulness, the beam must stay together (particles
not being lost to the walls of the apparatus) during the

course of the experiment. This problem is entirely analogous
to that of geometrical optics, where a light beam encounters
a region of varying index of refraction, and one is interested
in the focusing properties of lenses. The analogy is demonstrated in Chapter VIII, which deals with the propagation of
wave fronts, contact transformations, and a method of transforming Hamilton's Equations. Especially important is the
notion of area-preservation in the phase plane for Hamiltonian
systems, this being generalized to phase space (Liouville's
Theorem) and having application in statistical mechanics. The
area preservation property is used in the treatment of stability in Chapter X.

For particles which are fast or moving in a strong gravitational field, one needs to modify the dynamical description
by taking into account special and/or general relativity. We
have shown in Chapter VII, p. 175, that the principles of
special relativity requires that Newton's Equations of Motion
be modified to the forms (7.37) and (7.58), where the mass
varies with velocity. Further development of these equations
is carried out in Chapter I, pp. 22-24, and in Chapter VI,
pp. 131-133. For general relativity, we have (a) shown how
inertial effects in a rotating system appear in an invariant
kinetic part of a Lagrangian (essentially the "metric" ds^2),
(b) shown how scalar and vector potentials appear in the
equations, (7.70), (c) assumed the equivalence of gravitational forces and inertial forces, (d) adopted Einstein's
generalization of Poisson's equation and the Schwarzschild
metric (7.74) outside a spherically symmetric mass, and
(e) deduced the motion of a planet, as well as the trajectory
of a light ray, in the field of the Sun.

Motion of a particle under the influence of velocity-
dependent forces has its peculiarities when the system does

not admit a Hamiltonian function and there is as a result no
conservation of energy. Such systems are treated in Chapter
IX, for the case of motion in one dimension. The motion in
the phase plane can be understood qualitatively once one has
located the singular (critical) points, at which the motion
may be quite peculiar, of types called elliptic, hyperbolic,
radial, nodal or facal. Periodic orbits can occur and, when
they are approached asymptotically by other orbits, are called
limit cycles. If the damping term is small, the periodic
oscillation in time is closely sinusoidal, while if the damping
term is large, the wave form is quite different, and may be
more like a square wave, where the amplitude seems to undergo
very rapid changes at certain times, giving rise to "relaxation oscillations." A very detailed treatment of this general
area is to be found in Minorsky's "Non-Linear Oscillations."
If the system is controlled by a servomotor or other feedback
device, this just adds further complication. This is discussed
by Minorsky, and in still more detail by other writers (see
bibliography).

It is of prime interest in accelerator design, in the
restricted 3-body problem, in the motion of stars in a
galaxy, and in statistical mechanics in general to study the
behavior of a particle in a field of force which varies periodically with time. When the force is linear, the equation
of motion in one dimension is the Mathieu-Hill equation; where
it is non-linear, the equation is more complicated and has
been investigated more fully for the cubic case. Chapter X
treats the location of periodic orbits and their stability.
The examination of stability reduces to determining the long-
term behavior of points which represent the position and
momentum of the particle at intervals of one period. These
points may be regarded as the successive maps of an initial

point in the phase plane. Since the systems studied here are
Hamiltonian in character, the mapping is an area-preserving
one. It is shown in the text how one may find the approximate
boundary of a region of global stability by looking at how
various eigencurves from hyperbolic points intersect. When
hyperbolic points have been located such that the intersection (loop) area is very small, then the eigencurves are
approximately the stability boundary. On the other hand, when
the intersection area is larger, there may be a good deal of
complicated interlacing of the eigencurves, so that the plane
becomes divided like a checkerboard, with black squares stable
and white squares unstable. If one has not determined this
pattern in advance, then the points appear to behave in an
unpredictable fashion, as mentioned by Heisenberg in "Topics
in Nonlinear Physics." The accelerator designer can avoid
this pitfall by finding those regions where there is a small
intersection area of (higher order) eigencurves, and consequently no appreciable checkerboard pattern inside.

   As we have seen, the tractable dynamical problems, i.e.,
those for which a comprehensive picture can be readily
constructed, are mainly those which can be treated with the
aid of the phase plane. For instance, the circular restricted
3-body problem is really a one-body problem, involving the
motion of a particle in two dimensions under a given force
field. Periodic solutions can be located by appropriate
scanning techniques, and their variation with initial conditions and mass ratio is easily worked out. With the aid of
the energy integral and a surface of section, the motion is
reduced to mapping of points on this surface of section
(phase plane) and so the regions of global stability can be
located. The same holds for motion in a given galactic potential. However, complications already set in when we consider

motion in an alternating-gradient synchrotron, if radial
and vertical deviations from a standard orbit are regarded
as coupled (Sturrock, 1955). Energy is not conserved, so
that it appears that one must look at four variables $(x, y, \dot{x}, \dot{y},)$
simultaneously, which is difficult to do. When energy was
conserved, we could set $y = 0$ and then know that specification
of x and $\dot{x}$ determined the motion. However, when energy is
not conserved, the motion at $y = 0$ will be specified by a
point x, $\dot{x}$, $\dot{y}$ in space rather than in a plane. Mappings in
such a space may not be simple, and have not been studied.

When n bodies interact with one another gravitationally,
one can assume one or more sets of initial conditions,
integrate numerically, and hope for representative samples.
Saslaw et al. (1973) have studied 13000 3-body collisions
where one body collides with a binary and there results one
body (not necessarily the same) and a companion binary. Also,
computer runs for the n-body problem have been made by
various investigators, including v. Hoerner (1963), Aarseth
(1966), Miller et al. (1970), Hohl (1970), and Quirk (1971).
In the latter investigation, the evolution of a cold rotating
disk of 50000 stars with initial random surface density was
found to produce a spiral pattern after a few rotations.

Due to the computer, the circular restricted 3-body
problem is now essentially solved, the problem of global
stability in the phase plane is well-understood even though
rigorous mathematical proofs are still desirable, and the
motions of collections of stars are beginning to be studied.
In short, tremendous progress has been made and one has
reached a horizon, only, of course, to be confronted with a
new horizon, as in inevitable. A judicious combination of
such results with statistical mechanics techniques may enable
one to make still further advances with the dynamical theory.

Appendices

APPENDIX A--EXISTENCE OF SOLUTIONS

In the introduction, we have assumed that one can find a
unique solution of Equation (1), but we did not concern
ourselves with showing precisely when this is possible. In
this appendix, we shall assemble for reference the basic
existence theorems for solutions of equations such as (1),
and shall give some indication of how these theorems are
derived.

First, consider the first-order nonlinear equation

$$dy/dx = f(x, y) \qquad\qquad (A.1)$$

THEOREM I. If $f(x,y)$ be single-valued and continuous in a
region D about the point $(x_o\ y_o)$, and if $\left| \Delta_y f/\Delta y \right| < K$ in D
(Lipschitz condition), then there exists a unique continuous
function $y(x)$, in a region E about x_o, which satisfies (A.1)
and such that $y = y_o$ for $x = x_o$.

Notes to theorem. The region D is taken to be the rectan-
gular one $\left| x-x_o \right| \leq a$, $\left| y-y_o \right| \leq b$. The function f is bounded,
i.e., $\left| f(x,y) \right| < M$. If h is the smaller of a and b/M, the
region E is $\left| x-x_o \right| < h$. The difference quotient $\Delta_y f/\Delta y$ is
taken between any two points in D which have the same value
of x.

Proof of theorem. If there is a solution of A.1, it will satisfy the integral equation

$$y(x) = y_0 + \int_{x_0}^{x} f(t, y(t))\, dt \qquad (A.2)$$

Accordingly, let us try to solve A.2 by successive approximation, constructing a sequence of functions

$$y_1(x),\ y_2(x)\ \ldots\ y_n(x)\ \ldots$$

where

$$y_1(x) = y_0 + \int_{x_0}^{x} f(t,\ y_0)\, dt \qquad (A.3)$$

$$y_n(\dot{x}) = y_0 + \int_{x_0}^{x} f(t, y_{n-1})\, dt$$

and x is the interval $(x_0,\ x_0 + h)$.

Our real goals will be to show (1) that as $n \to \infty$, the sequence $y_n(x)$ tends toward a limit $y(x)$, which function is continuous in x and satisfies the differential equation, and (2) that this solution is the only continuous solution which equals y_0 when $x = x_0$.

The first part is achieved by noting the relation

$$y_n = y_0 + \sum_{r=1}^{n} (y_r - y_{r-1}) \qquad (A.4)$$

and calculating bounds on the terms $y_r - y_{r-1}$. We obtain

$$|y_1 - y_0| < \int_{x_0}^{x} |f(t, y_0)|\, dt \le M\, |x - x_0| \le b$$

$$|y_2 - y_1| < \int |f(t, y_1) - f(t, y_0)|\, dt < K \int |y_1 - y_0|\, dt$$

$$< KM \int |t - x_0|\, dt < KM\, |x - x_0|^2/2$$

$$|y_r - y_{r-1}| < M\, (K|x - x_0|)^r/Kr! \qquad (A.5)$$

As $n \to \infty$, the terms of A.4 are bounded by the terms (right-hand side of A.5) of a series representing an exponential function, so that $y_n(x)$ does tend to a limit function which is continuous.

If Y is some solution of A.2, not necessarily equal to y, then we may calculate bounds on $Y - y_r$ and arrive at

$$|Y - y_n(x)| < K^n b |x-x_o|^n / n!$$

Since the r.h. side goes to zero as $n \to \infty$, the solution Y must be the same as y, and uniqueness is demonstrated.

THEOREM 2. Given the system of equations

$$dy_1/dx = f_1(x \; y_1 \; y_2); \; dy_2/dx = f_2(x \; y_1 \; y_2) \quad (A.6)$$

where f_1 and f_2 are single-valued and continuous in the region $|x - x_o| \le a$, $|y_1 - y_1^o| \le b_1$, $|y_2 - y_2^o| \le b_2$. Let M be the greatest of the upper bounds of f_1 and f_2 and let h be the least of a, b_1/M, b_2/M.

If the Lipschitz condition

$$|f_r (x, Y_1, Y_2) - f_r(x, y_1, y_2)| < K_1|Y_1-y_1| + K_2|Y_2-y_2|$$
$$(A.7)$$

is satisfied, then there exists a unique set of continuous solutions y_1, y_2 of the system A.6, and these solutions take on the values y_1^o, y_2^o when $x = x_o$.

(The proof follows along the same lines as that for a single first-order equation, and will therefore be omitted).

THEOREM 3. Given the linear system of equations

$$dy_1/dx = p_{11}y_1 + p_{12}y_2 + r_1 = f_1(x, y_1, y_2)$$
$$dy_2/dx = p_{21}y_1 + p_{22}y_2 + r_2 = f_2(x, y_1, y_2) \quad (A.8)$$

where the coefficients p_{ij} and r_i are continuous functions of x in the interval $a \le x \le b$.

There exists a set of continuous solutions (y_1, y_2) which is unique in the interval (a,b).

Proof. We have

$$|f_1 (x, y_1, y_2) - f_1 (x, y_1, y_2)| \quad |p_{i1} (y_1 - y_2) + p_{i2} (y_2 - y_2)|$$

$$\le |p_{i1}||y_1 - y_1| + |p_{i2}||y_2 - y_2|$$

But the coefficients p_{ij} are bounded in the interval (a,b), so that the Lipschitz condition A.7 is valid and our theorem is essentially a corollary of Theorem 2.

THEOREM 4. Given the linear second-order differential equation

$$p_0 (d^2y/dx^2) + p_1 (dy/dx) + p_2 y = r \quad (A.9)$$

where p_0, p_1, p_2 and r are continuous functions of x in the interval $a \le x \le b$ and where $p_0 \ne 0$ for any point in (a,b).

There exists a unique solution $y(x)$ of Equation A.9, which, together with $y'(x)$, is continuous in (a,b) and such that $y(x_0) = y_0$, $y'(x_0) = y_0'$ where x_0 is in (a,b) and y_0 and y_0' are arbitrary constants.

Proof. Equation A.9 is equivalent to the system

$$dy/dx = y_1 \qquad\qquad (A.10)$$

$$dy_1/dx = (r/p_0) - (p_1/p_0)y_1 - (p_2/p_0) y$$

Since $p_0 \ne 0$, the coefficients (r/p_0), (p_1/p_0), and (p_2/p_0) will be continuous and so Theorem 3 may be applied. There exists a set of continuous solutions y, y_1 such that $y = y_0$, $y_1 = (y_1)_0$ at $x = x_0$. This is, with $y_1 = y'$, just another statement of Theorem 4.

General Comments

For the simple equation $dy/dx = f(x,y)$, we can be assured
of a unique solution if (a) $f(x,y)$ is continuous and single-
valued, and (b) $\partial f/\partial y$ is bounded (Lipschitz condition). If a
point of the (x,y) plane is such that the conditions for
existence of a unique solution are met, then this point is
called an ordinary point. All other points are called singular points, and it is of vital interest to us to find out
what may happen at such points. Consider the equation

$$dy \,|\, dx = y/x. \qquad\qquad (A.11)$$

For this, $\partial f/\partial y = 1/x$, and so at $x = 0$ the conditions for
existence are not met. Likewise, for $dx/dy = x/y$, the conditions are not met for $y = 0$. Therefore, the point $x = 0$,
$y = o$ is a singular point of the equation.

Equation A.11 has $y = cx$ for solutions, where c is a parameter. These are straight lines from the origin, so that the
singular character of the origin is revealed by the fact that
there is an infinite number of solutions passing through it.
For any other finite point (x_o, y_o), however, there is just
one solution through it, namely $y = y_o \, x/x_o$. If $y_o = 0$,
$x_o \neq 0$ then $y = 0$ is the solution, while if $x_o = 0$, $y_o \neq 0$,
then $x = 0$ is the solution.

A more detailed discussion of singular points is given in
Chapter 9. Suffice it to say that at such a point there may
be no solution, a discontinuity in a solution, or an infinite
number of solutions through the point.

APPENDIX B--CONIC SECTIONS

A conic is the locus of a point P whose distance FP from a
fixed point F (the locus) is e times its distance PK from a

fixed line (directrix) HX. We shall suppose first that the
directrix is to the right of F.

That is, referring to the figures e.g., (B1) and (B4),

$$\frac{FP}{PK} = \frac{FL}{LH} = e = \text{eccentricity}$$

If we have polar coordinates (r,θ) with the initial line
$\theta = 0$ (FX) perpendicular to the directrix, then, if
$\ell = FL$ = latus rectum,

$$r = FP = ePK = e(LH - r \cos \theta) = \ell - er \cos \theta$$

$$\text{(B.1)}$$

or
$$\frac{1}{r} = \frac{1}{\ell} (1 + e \cos \theta) \qquad \text{(B.2)}$$

This is the basic equation of a conic section with eccentricity e. The conic is called an _ellipse_ if e < 1.

For an ellipse, from (B.2), r is always finite, so the
ellipse is a closed curve. For a parabola, e = 1, 1/r will
vanish when $\theta = \pi$, so that the parabola is a curve extending
to infinity in one direction. For an hyperbola, e > 1, 1/r
will vanish if $\cos\theta = -1/e$ and will be finite whenever
$1 + e \cos \theta > 0$, i.e., $\cos \theta > -1/e$.

Properties of an Ellipse

For the ellipse, there are two intercepts with the initial
line. When $\theta = 0$, $r = \ell/(1 + e)$; and when $\theta = \pi$, $r = \ell/(1-e)$.
The sum of these distances represents the major axis 2a of
the ellipse, and $a = \ell/(1 - e^2)$.

If we now introduce Cartesian coordinates (x,y) with the
origin O halfway between the two intercepts A and A', then
(see Figure B.1) OF = OA-FA = a - a(1-e) = ea and
$$r \cos \theta = x - ea; \quad r \sin \theta = y \qquad \text{(B.3)}$$

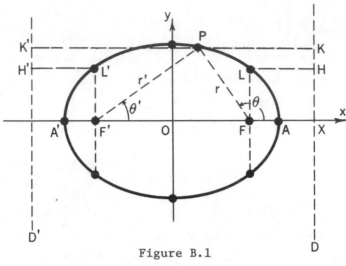

Figure B.1

Ellipse (e < 1)

Squaring (B.1) and (B.3) and inserting $\ell = a(1-e^2)$, we have

$$r^2 = (1-e^2)^2 \, a^2 - 2ea \, (1-e^2)(x-ea) + e^2 \, (x - ea)^2$$

$$= (x-ea)^2 + y^2$$

from which, when terms are combined, there results

$$y^2 = (1-e^2)(a^2 - x^2)$$

or, in standard form,

$$\frac{x^2}{a^2} + \frac{y^2}{a^2(1-e^2)} = 1 \qquad\qquad (B.4)$$

This is the equation of an ellipse with center at the origin and a focus at $x = ea$ ($r = 0$). The semi-minor axis b is the value of y on the ellipse at $x = 0$, and is $b = a \, (1 - e^2)^{1/2}$.

The ellipse is symmetrical with respect to reflection in the
minor axis and in the major axis. Therefore, there is another
focus at x = - ea. Let now P be any point on the ellipse, F
the original focus and F' the conjugate focus to the left.
Let D be the original directrix to the right of F, and let
D' be the conjugate directrix to the left of F'. Then we have
another equation which is a companion to (B.1) namely

$$r' = \ell + er'\cos \theta' \qquad (B.1)'$$

Also

$$r = \ell - er \cos \theta \qquad (B.1)$$

Adding these two equations together,

$$r + r' = 2\ell + e (r'\cos \theta' - r \cos \theta)$$

$$= 2\ell + e.2ea$$

$$= 2a (1 - e^2) + 2e^2 a = 2a \qquad (B.5)$$

The sum of the distances from any point on the ellipse to
the two conjugate foci is constant and equal to the length
of the major axis of the ellipse.

Eccentric Anomaly

The equation for the ellipse is then, from (B.4) with
$b^2 = a^2 (1 - e^2)$,

$$\frac{x^2}{a^2} + \frac{y^2}{b^2} = 1 \qquad (B.6)$$

It will be convenient to define an angle ε, called the
eccentric anomaly, by

$$\cos \varepsilon = x/a, \sin \varepsilon = y/b, \qquad (B.7)$$

and to express r in terms of ε.

Now

$$r'^2 - r^2 = (ea + x)^2 - (ea - x)^2 \qquad (B.8)$$

$$= 4ea\,x$$

Dividing (B.8) by (B.5), and substituting (B.7)

$$-r' + r = -2ex = -2ea\cos\varepsilon$$

which, when added to (B.5), yields

$$r = a\,(1 - e\cos\varepsilon) \qquad (B.9)$$

This is the desired expression of r in terms of ε. The extreme values are

$$r_1 = a(1 - e) \text{ and } r_2 = a(1 + e). \qquad (B.10)$$

Properties of Hyperbolas

For the case e > 1, that of an hyperbola, the argument needs only slight modification. We now set $a = \ell/(e^2 - 1)$ and proceed as before.

Referring to Figure (B.2), the intercept A is at a distance from the focus equal to

$$FA = \ell/(e+1) = a\,(e - 1)$$

If we now take the origin of coordinates 0 at a distance ea to the right of F, then

$$AO = FO - FA = a$$

For any point on the hyperbola AL with focus F,

$$r\cos\theta = ea + x$$
$$\qquad\qquad (B.11)$$
$$r\sin\theta = y$$

Squaring (B.1) and (B.11), and reducing, the result is

$$y^2 = (e^2 - 1)\,(x^2 - a^2)$$

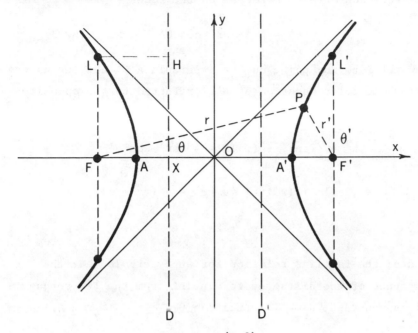

Figure (B.2)

Hyperbolas (e > 1)

or, in standard form, with $b = a \, (e^2 - 1)^{1/2}$,

$$\frac{x^2}{a^2} - \frac{y^2}{b^2} = 1 \qquad (B.12)$$

The hyperbola which we have been considering is such that
all the values of x are negative, and the focus is at x = -ea,
y = 0. Since (B.12) will still be satisfied if x is replaced
by -x, there will be a companion hyperbola with x-intercept
A' and x = a and with focus F' at x = ea.

The polar equation of AL referred to the focus F is just
(B.1):

$$r = \ell - er \cos \theta \qquad (B.1)$$

The equation of A'L' referred to the focus F' is

$$r' = \ell + er' \cos \theta' \qquad (B.13)$$

Now let P be any point on the hyperbola A'L' and let us try to find the polar equation of A'L' referred to the opposite focus F. By the law of cosines, and (B.13),

$$r^2 = (2ea)^2 + (r')^2 + 4 \, ear' \cos \theta'$$

$$= (2ea)^2 + (r')^2 + 4a \, [r' - a \, (e^2 - 1)]$$

$$= (r' + 2a)^2$$

Hence, the familiar relation for an hyperbola that the difference of the distances to a point from the two conjugate foci is constant, namely, since $r > 0$,

$$r = r' + 2a \qquad (B.14)$$

The sum of the projections on the x-axis equals FF', i.e.

$$r \cos \theta - r' \cos \theta' = 2ea \qquad (B.15)$$

Combining (B.13), (B.14), and (B.15), we find

$$r = \ell/(e \cos \theta - 1) \qquad (B.16)$$

This is the equation of an hyperbola referred to the opposite focus. It is useful when we describe the path of a particle which is repelled by the scattering center ($\kappa > 0$). The original equation (B.1) is appropriate for an attractive force ($\kappa < 0$).

Asymptotes

When $|x|$ becomes infinite, the hyperbolas described by (B.12) approach the lines with slopes b/a and $-b/a$ asymptotically.

Proof: If P be any point on the hyperbola A'L', then the perpendicular distances d and h from P (x,y) onto the above lines are given by (see Figure B.3):

$$d = OP \sin (\theta_o - \theta)$$

$$h = OP \sin (\theta_o + \theta) = (bx + ay)/(a^2 + b^2)^{1/2} \qquad (B.17)$$

where

$$\tan \theta_o = b/a, \quad \tan \theta = y/x. \qquad (B.18)$$

The product is

$$dh = (OP)^2 [\sin^2\theta_o \cos^2\theta - \cos^2\theta_o \sin^2\theta]$$

$$= (b^2x^2 - a^2y^2)/(a^2 + b^2)$$

$$= a^2b^2/ (a^2 + b^2) = const. \text{ from (B.12)}$$

Now $y = (b/a) (x^2 - a^2)^{1/2}$, from (B.12) also. As $x \to \infty$, $y \to \infty$, and, from (B.17), $h \to \infty$. Hence, as $x \to \infty$, $d \to 0$, since dh is constant. Also,

$$\frac{dy}{dx} = \frac{b}{a} \frac{x}{(x^2 - a^2)^{1/2}} \to \frac{b}{a} \text{ as } x \to \infty.$$

The line $bx - ay = 0$ is therefore approached asymptotically by the hyperbola. By symmetry, $bx + ay = 0$ is the other asymptote.

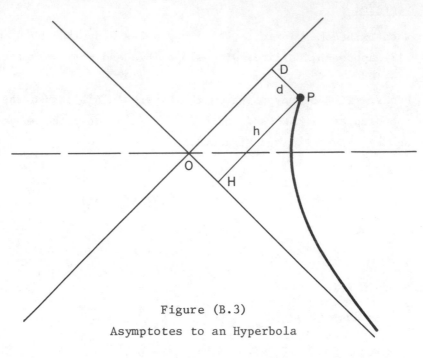

Figure (B.3)

Asymptotes to an Hyperbola

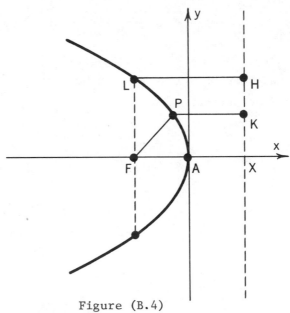

Figure (B.4)

Parabola (e = 1)

Properties of a Parabola

In Figure (B.4) we show a parabola which is tangent to the y-axis at A and whose focus is at F. Its polar equation is that of a conic section with e = 1, namely

$$r = \ell/(1 + \cos \theta) \qquad \text{(B.19)}$$

Its Cartesian equation, referred to A as origin, is $r\cos\theta - (\ell/2) = x = -y^2/2\ell$, consistent with (B.19).

APPENDIX C--PROOF THAT (7.7) IS LINEAR

Suppose we follow the path of a particle moving uniformly in a straight line in an inertial frame S. Then from our assumption, to an observer in another inertial frame S' the same particle is also moving in a straight line uniformly.

There is no loss in generality, if we assume the motion is in the x-direction, and that S' moves with velocity v, in the x-direction, with respect to S.

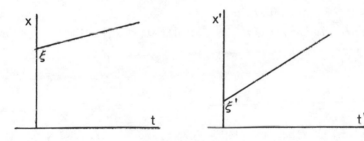

Figure C.1

Uniform Motion in Two Inertial Systems

Let the motion in the S frame be described by

$$x = \xi + ut \qquad \text{(C.1)}$$

and that in the S' frame by

$$x' = \xi' + u't' \qquad \text{(C.2)}$$

where u and u' are constant velocities.

We also require that the following relationships hold
simultaneously:

$$u^2 = c^2 \qquad (C.3)$$

$$(u')^2 = c^2 \qquad (C.4)$$

The problem is then to find the nature of the connection
between the coordinates in the system S and those in the
system S'.   Let us assume a general functional dependence

$$t' = f(t,x) \qquad (C.5)$$

$$x' = g(t,x) \qquad (C.6)$$

and find what f and g must be if (C.1 and (C.2), and also
(C.3) and (C.4), are to be compatible with each other.

Differentiating (C.5) and (C.6), we have

$$dt' = f_t dt + f_x\, dx \qquad (C.7)$$

$$dx' = g_t dt + g_x\, dx \qquad (C.8)$$

where $f_t = \partial f/\partial t$, etc.

For these equations to be soluble for dt and dx, the Jacobian must not vanish:

$$J \equiv \begin{vmatrix} f_t & f_x \\ g_t & g_x \end{vmatrix} \neq 0 \qquad (C.9)$$

Along any trajectory, from (C.1) and (C.2), the velocities
u and u' are constant, so that

$$u' = \frac{g_t + g_x\, u}{f_t + f_x\, u} = \text{const} \qquad (C.10)$$

Now d ln u'/dt = 0, whence

$$\frac{g_{tt} + 2u\, g_{xt} + u^2\, g_{xx}}{g_t + u g_x} (f_t + u f_x) = f_{tt} + 2u f_{xt} + u^2 f_{xx}$$
$$(C.11)$$

Now, if the denominator of the left-hand side of (C.11) is zero, then $u = -g_t/g_x$. Then

$$f_t + uf_x = (1/g_x)(f_t g_x - g_t f_x) \neq 0,$$

by (C.9). But the right-hand side of (C.11) is finite, so that

$$g_{tt} + 2ug_{xt} + u^2 g_{xx} = 0 \qquad (C.12)$$

and

$$g_t + u g_x = 0 \qquad (C.13)$$

must hold simultaneously. The quadratic in (C.12) may be resolved into its two factors, of which $g_t + ug_x$ must be one. Setting $g_{tt} + 2u g_{xt} + u^2 g_{xx} = 2(g_t + ug_x)(h_t + uh_x)$

$$\qquad (C.14)$$

where h_t and h_x are some functions of x and t, we have from (C.11)

$$f_{tt} + 2uf_{xt} + u^2 f_{xx} = 2(f_t + uf_x)(h_t + uh_x)$$

$$\qquad (C.15)$$

Comparing coefficients of like powers of u in (C.14),

$$g_{tt} = 2 h_t g_t$$

$$g_{xt} = g_t h_x + g_x h_t \qquad (C.16)$$

$$g_{xx} = 2 h_x g_x$$

and similarly for the function f.

Now from (C.3) and (C.4), we require from (C.10) that

$$(g_t + ug_x)^2 - c^2(f_t + uf_x)^2 = 0 \qquad (C.17)$$

when u = c and when u = -c. This gives

$$g_t g_x - c^2 f_t f_x = 0 \qquad (C.18)$$

Differentiating (C.18) re t, and substituting (C.16),

$$2h_t \, g_t \, g_x + g_t \, (g_t \, h_x + g_x \, h_t)$$

$$- c^2 \, [2h_t \, f_t \, f_x + f_t \, (f_t \, h_x + f_x \, h_t)] = 0$$

or, using (C.18)

$$(g_t^2 - c^2 \, f_t^2) \, h_x = 0 \qquad\qquad (C.19)$$

Similarly, the derivative of (C.18) re x yields

$$(g_x^2 - c^2 \, f_x^2) \, h_t = 0 \qquad\qquad (C.20)$$

Since the coefficients of h_x and h_t in (C.19) and (C.20) are
not in general zero, then h_x and h_t must be, and the second
derivatives of f and g must all be zero, from (C.16) and its
companion. Therefore the functions f and g are linear in x
and t, which was to have been shown.

APPENDIX D--COVARIANCE AND CONTRAVARIANCE

When dealing with curvilinear coordinates or with crystal
physics, it is useful to introduce basic vectors, not necessarily orthogonal to each other, and to resolve some arbitrary
vector $\vec{A}$ as a linear combination of them.

For instance, if we consider spherical coordinates r, θ,
and φ, then an infinitesimal vector $\vec{ds}$ may be written as

$$\vec{ds} = \vec{e}_r \, dr + \vec{e}_\theta \, d\theta + \vec{e}_\phi \, d\phi$$

where

$$\vec{e}_r, \ \vec{e}_\theta \ \text{and} \ \vec{e}_\phi$$

point in the r, θ, and φ directions respectively, are ortho-
gonal, and have magnitudes $|e_r| = 1$, $|e_\theta| = r$,
$|e_\phi| = r \sin \theta$. The basic vectors are in this case not unit
vectors.

An arbitrary vector $\vec{A}$ in n-dimensional space will be written

$$\vec{A} = A^1 \, \vec{e}_1 + A^2 \, \vec{e}_2 + 2 \ldots A^n \, \vec{e}_n$$

$$= A^i \, \vec{e}_i$$

where we use the <u>summation convention</u> that terms with an index occuring twice are to be summed over the possible values of the index. The quantities A^i are called the <u>contravariant</u> components of $\vec{A}$ with respect to the basic vectors $\vec{e}_i$.

We can construct a set of reciprocal vectors $\vec{e}^1$ such that

$$(\vec{e}^i \cdot \vec{e}_i) = \delta_{ij} = \begin{cases} 1 \text{ if } i = j. \\ 0 \text{ if } i \neq j. \end{cases}$$

In 3-dimensions, for instance, the vector $\vec{e}^1$ will be normal to the plane of $\vec{e}_2$ and $\vec{e}_3$, not necessarily in the direction $\vec{e}_1$, and will have magnitude equal to the reciprocal of that of $\vec{e}_1$.

The vector $\vec{A}$ can be resolved alternatively as

$$\vec{A} = A_1 \, \vec{e}^1 + A_2 \, \vec{e}^2 + \ldots A_n \, \vec{e}^n$$

$$= A_i \, \vec{e}^i$$

and the quantities A_i are called the <u>covariant</u> components of $\vec{A}$.

The components obey the relations

$$A^i = (\vec{e}^i \cdot \vec{A}) = (\vec{e}^i \cdot \vec{e}_j) \, A^j$$

$$A_i = (\vec{e}_i \cdot \vec{A}) = (\vec{e}_i \cdot \vec{e}^j) \, A_j$$

The square of a vector is

$$A^2 = A^i \, A_j \, (\vec{e}_i \cdot \vec{e}^j) = A^i \, A_j \, \delta_{ij} = A^i \, A_i$$

and the scalar product of 2 vectors is

$$\vec{A} \cdot \vec{B} = A^i \, B_i = A_i \, B^i$$

An infinitesimal vector may be written as

$$\vec{ds} = \vec{e}_1 \ dx^1 + \vec{e}_2 \ dx^2 + \ \dots \ \vec{e}_n \ dx^n$$

$$= \vec{e}_i \ dx^i$$

Alternatively

$$\vec{ds} = \vec{e}^i \ dx_i$$

The square of the length is called the line element and is

$$ds^2 = (\vec{e}_i \cdot \vec{e}_j) \ dx^i \ dx^j = g_{ij} \ dx^i \ dx^j$$

where

$$g_{ij} = (\vec{e}_i \cdot \vec{e}_j).$$

The covariant component of a vector is

$$A_i = (\vec{e}_i \cdot \vec{A}) = (\vec{e}_i \cdot \vec{e}_j) \ A^j = g_{ij} \ A^j$$

and may thus be calculated from the contravariant components. If $x_o = ct$, $x_1 = x$, $x_2 = y$, $x_3 = z$, then for

$$ds^2 = c^2 \ t^2 - x^2 - y^2 - z^2,$$

we have $g_{oo} = 1$, $g_{11} = g_{22} = g_{33} = -1$, $g_{ij} = 0$ if $i \neq j$.

If the vector ct, x, y, x, is contravariant, then ct, -x, -y, -z is covariant, for the Lorentz transformation.

Transformation Properties

The quantities dx^i are the coordinate differentials, and also the contravariant components of $\vec{ds}$. If another coordinate system x_i' be introduced, then

$$dx_i' = \frac{\partial x_i'}{\partial x_j} \ dx_j \quad \text{(contravariant)}$$

From now on. coordinate differentials will be regarded as automatically contravariant components, and we shall use

subscripts to label them. This will be an exception to the
usual rule that a superscript indicates contravariance and a
subscript convariance.

Now

$$d\vec{\sigma} = \vec{e}_j \, dx_j = \vec{e}_j \, \frac{\partial x_j}{\partial x_i'} \, dx_i' = \vec{e}_i' \, dx_i'$$

so that

$$\vec{e}_i' = \frac{\partial x_j}{\partial x_i'} \, \vec{e}_j$$

Taking the scalar product with any vector $\vec{A}$

$$(\vec{e}_i' \cdot \vec{A}) = A_i' = \frac{\partial x_j}{\partial x_i'} \, A_j \quad \text{(covariant)}$$

which shows how <u>covariant</u> components transform. Since the
<u>partial derivatives</u> behave in the same way, i.e.

$$\frac{\partial}{\partial x_i'} = \frac{\partial x_j}{\partial x_i'} \, \frac{\partial}{\partial x_j}$$

we see that they <u>can be regarded as covariant components</u> of
a vector.·

The quantity $g_{ij} = (\vec{e}_i \cdot \vec{e}_j)$ transforms according to the
scheme

$$g_{ij}' = \frac{\partial x_k}{\partial x_i'} \frac{\partial x_\ell}{\partial x_j'} \, g_{k\ell}$$

and is therefore called a <u>covariant tensor</u> of the <u>second rank</u>.

APPENDIX E--EXPONENTS FOR A HAMILTONIAN SYSTEM:
MOTION IN A PLANE

Let us suppose that the motion takes place in a plane, and
that the Hamiltonian does not depend explicitly on the time.
Also, for convenience of notation, let $y_1 = \xi$, $y_2 = \eta$,
$y_3 = p_\xi$, $y_4 = p_\eta$, where ξ, η denote the rectangular coordinates, and p_ξ, p_η the corresponding momenta.

Let the system have a periodic solution y^o, and consider
two adjacent solutions $y^o + \Delta y$ and $y^o + \delta y$, where Δy and δy
are small variations. Now

$$\frac{d}{dt} \Delta y_1 = \Delta \frac{\partial H}{\partial y_3} = \Sigma \frac{\partial^2 H}{\partial y_3 \partial y_i} \Delta y_i$$

and accordingly the values of Δy_k when t is increased by the
period τ will be linear combinations of the original variations.

That is, if $\Delta y_k (t + \tau) = \Delta \bar{y}_k$, then

$$\Delta \bar{y}_k = \Sigma a_{km} \Delta y_m \qquad (E.1)$$

and

$$\delta \bar{y}_\ell = \Sigma a_{\ell n} \delta y_n \qquad (E.2)$$

where the coefficients a_{km} are determined by integration over
the period τ.

But we also know that the equation

$$\overline{\Delta y_1} \; \overline{\delta y_3} + \overline{\Delta y_2} \; \overline{\delta y_4} - \overline{\Delta y_3} \; \overline{\delta y_1} - \overline{\Delta y_4} \; \overline{\delta y_2}$$

$$= \Delta y_1 \delta y_3 + \Delta y_2 \delta y_4 - \Delta y_3 \delta y_1 - \Delta y_4 \delta y_2 \qquad (E.3)$$

holds, being the expression of Lagrange invariance.

In matrix notation, Equations (E.1) and (E.2) are

$$\overline{\Delta y} = \Delta y \; A' \qquad (E.4)$$

$$\overline{\delta y} = A \; \delta y \qquad (E.5)$$

where A' is the transpose of A, $\overline{\Delta y}$, Δy are one-row matrices,
and $\overline{\delta y}$, δy are one-column matrices.

Introducing the matrix

$$S = \begin{pmatrix} 0 & 0 & 1 & 0 \\ 0 & 0 & 0 & 1 \\ -1 & 0 & 0 & 0 \\ 0 & -1 & 0 & 0 \end{pmatrix},$$

Equation (E.3) becomes $\overline{\Delta y}\, S\, \overline{\delta y} = \Delta y\, S\, \delta y$, and from (E.4) and (E.5) $\Delta y\, A'SA\, \delta y = \Delta y\, S\, \delta y$, or

$$A'SA = S \qquad\qquad (E.6)$$

(Any matrix A satisfying Equation (E.6) is called <u>symplectic</u>.) Solving this equation, we have $(A')^{-1} = SAS^{-1}$, so that $(A')^{-1}$ has the same eigenvalues as A itself, and so does A^{-1}. In other words, <u>the reciprocal of every eigenvalue λ of A is itself an eigenvalue of A</u>.

If we now set $\lambda_k = e^{-s_k \tau}$ as before, then this last result means that the exponents occur in pairs with opposite signs, i.e., s_k and $-s_k$.

When the <u>time is not contained explicitly in the Hamiltonian</u>, if $y = \phi(t)$ is a solution, then $\phi(t + \varepsilon)$ is also a solution, and so $\frac{\partial}{\partial \varepsilon}\, \phi(t + \varepsilon)$ is a solution of the variational equations. But if $\phi(t)$ is periodic, $\frac{\partial}{\partial \varepsilon}\, \phi(t + \varepsilon)$ is also periodic and so one eigenvalue is unity and the exponent is zero. Since the exponents can be arranged in pairs, the result is that for <u>motion of a particle in a plane, where a Hamiltonian exists and does not involve the time explicitly, the exponents of any periodic solution are $(0, 0, \alpha, -\alpha)$ where α is some number</u>.

For the circular restricted 3-body problem, this theorem is directly applicable, since the Hamiltonian in the rotating system is independent of time.

References

Chapter I

Librations

Pars, L. A., 1965, "Analytical Dynamics," Wiley.

Forced Oscillation

Heitler, W., 1936, "Quantum Theory of Radiation," Oxford.

Chapter II

Problems in Dynamics

Spiegel, M. R., 1967, "Theoretical Mechanics," Schaum
Publishing Co.

Wells, D. A., 1967, "Lagrangian Dynamics," McGraw-Hill.

Fowles, G. R., 1962, "Analytical Mechanics," Holt, Rinehart
and Winston.

Existence of a Lagrangian

Mayer, A., 1896, Leipziger Berichte 48, 519.

Equilibrium of Rotating Fluids

Jeans, J. H., 1929, "Astronomy and Cosmogony," Cambridge
University Press.

Chandrasekhar, S., 1969, "Ellipsoidal Figures of Equilibrium," Yale University Press.

Physics of Plasmas

Allis, W. P., 1960, "Nuclear Fusion, Stellarators, etc.,"
Van Nostrand.

Lehnert, B., 1964, "Dynamics of Charged Particles (Confine-
ment)," Interscience.

Schmidt, G., 1966, "Physics of High Temperature Plasmas,"
    Academic Press.

Shercliff, J. A., 1965, "Textbook of Magnetohydrodynamics,"
    Pergamon.

Spitzer, L., 1962, "Physics of Fully Ionized Gases," Wiley.

Huang, K., 1966, "Statistical Mechanics," Ch. 5 Wiley.

Chapter III

Elliptic Functions

Byrd, P. F. and Friedman, M. D., 1971, Handbook of Ellip-
    tic Integrals, Springer.

Top and Pendulum

Pars, L. A., 1965, "Analytical Dynamics," Wiley.

Chapter IV

Corben, H. C. and Stehle, P., 1960, "Classical Mechanics,"
    Wiley.

Chapter V

Mason, M. and Weaver, W., 1929, "The Electromagnetic Field,"
    University of Chicago Press.

Landau, L. and Lifshitz, E., 1951, "The Classical Theory of
    Fields," Addison-Wesley.

Chapter VI

Sloan, D. H. and Lawrence, E. O., 1931, Phys. Rev. 38, 2021;
    Proc. Nat. Acad. Sci. 17, 64.

Kerst, D. W., and Serber, R., 1941, "Electronic Orbits in the
    Induction Accelerator," (Betatron), Phys. Rev. 60, 53.

Slater, J. C., 1948, Rev. Mod. Phys. 20, 473.

Livingston, M. S., 1954, "High Energy Accelerators,"
    Interscience.

Handbuch d. Physik v. 44, 1959, "Nuclear Instrumentation I,"
    Springer.

Wilson, R. R. and Littauer, R., 1960, "Accelerators,"
    Doubleday-Anchor.

Lichtenberg, A. J., 1968, "Phase Space Synamics of Parti-
    cles," Wiley.

Persico, E. et al., 1968, "Principles of Particle Accelera-
    tors," W. A. Banjamin.

                        Chapter VII

Møller, C., 1952, "The Theory of Relativity," Oxford.

Fock, V. A., 1963, "The Theory of Space, Time & Gravita-
    tion," Pergamon.

Oppenheimer, J. R. and Snyder, H., 1939, "On Continued
    Gravitational Contraction," Phys. Rev. 56, 455.

Thorne, K. S., 1967, "General Relativistic Theory of
    Stellar Structure and Dynamics" in "High Energy
    Astrophysics," Vol. 3, DeWitt, C., et al., Eds.,
    Gordon & Breach.

Tausner, M. J., 1966, "General Relativity, etc.," Technical
    Report No. 425, Lincoln Laboratory, M.I.T.

Weinberg, S., 1972, "Gravitation and Cosmology," Wiley.

Shapiro, I. I., 1968, "Radar Observations of the Planets,"
    Sci. Amer. 219,28.

                        Chapter VIII

Born, M. and Wolf, E., 1970, "Principles of Optics,"
    Pergamon.

Born, M., (tr. by Fisher, J. W.) "Mechanics of the Atom,"
    Ungar.

Lie, S. and Scheffers, G., "Beruhrungstransformationen."

Whittaker, E. T., 1944, "Analytical Dynamics," Dover
    (New York).

Sturrock, P. A., 1955, "Static and Dynamic Electron Optics,"
    Cambridge University Press.

Sygne, J. L., 1960, "Handbuch der Physik III/1, 1-225,
    "Classical Dynamics," Springer.

Landau, L. S. and Lifshitz, E. M., 1960, "Mechanics,"
    Addison-Wesley.

Langer, R. E., 1934, 1937, "WKB Connection Formulas," Bull.
    Amer. Math. Soc. 40,574; Phys. Rev. 51,669.

Chapter IX

Minorsky, N., 1962, "Nonlinear Oscillations," Van Nostrand.

Van der Pol, B., 1926, "On Relaxation Oscillations," Phil. Mag. 2, 978.

Van der Pol, B., 1934, "The Nonlinear Theory of Electric Oscillations," Proc. Inst. Radio Eng., 22, 1051-1086.

Liapunov, A., 1907, tr. by Davaux, E., Annales de Toulouse (2), 9, 203.

Chetayev, N. G., 1961, "Stability of Motion," Pergamon.

Lefschetz, S., 1963, "Differential Equations: Geometric Theory," (Van der Pol Equation), Interscience.

LeCorbeiller, Ph. 1931, "Systemes auto-entretenus," Herman, Paris.

Chapter X

Ince, E. L., 1926, "Ordinary Differential Equations," pp. 381-384, (Floquet Theory), Dover.

Kronig, R. deL. and Penney, W. G., 1930, Proc. Roy. Soc., London A130, 499.

Courant, E. D., Livingston, M. S., and Snyder, H. S., 1952, "The Strong-focusing Synchrotron," Phys. Rev. 88, 1190.

Birkhoff, G. D., 1913, "Proof of Poincare's Geometric Theorem," (Fixed Points), Trans. Amer. Math. Soc. 14-22.

Siegel, C. L., 1956, "Vorlesungen uber Himmelsmechanik," pp. 148-152 (Fixed Points), Springer.

Stability

Poincaré, H., 1892, 1893, 1899, "Les Methodes Nouvelles de la Mecanique Celeste," Vol. 1-3, Gauthier-Villars, Paris.

Arnold, V. I., 1963, Uspekhi Mat. Nauk., No. 5 (113) 13, 91.

Bartlett, J. H., 1961, Mathematical Congress, Leningrad.

Bartlett, J. H., 1968, "Motion Under a Periodic Cubic Force," Mat. Fys. Medd. Dan. Vid. Selsk 36, No. 11.

Bartlett, J. H. and Wagner, C. A., 1970, "Stability of Motion in a Periodic Cubic Force Field," Celest. Mech. 2, 228.

Jenkins, B. Z. and Bartlett, J. H., 1972, "Stability of an Area-Preserving Mapping," Celest. Mech. 5, No. 3.

Brahic, A., 1970, "Particle in Oscillating Box," Astron, & Astrophys. 12, 98.

Contopoulos, G., 1967, "Third Integral," Bull. Astron. Ser. 3, v.2, 223.

Contopoulos, G., 1970, "Periodic Orbits, Stability and Resonances," p. 326, Reidel.

Deprit, A. and Henrard, J., 1970, "Third Integral," Astrophys, & Space Sci. 7, 54.

Ford, J., 1973, "Integrability, AP Mappings," International Conf. on Point Mappings, Toulouse, France.

Hénon, M., 1965, "Stabilité, $\gamma = 0$," Ann. Astrophys. 28, 992.

Hénon, M., 1969, Quart. Appl. Math. 27, 291.

Hénon, M. and Heiles, C., 1964, "Third Integral," Astron. J. 69, 73.

Laslett, L. J. et al., 1968, "Long-term Stability," Courant Institute R & D Report NYO-1480-101, New York.

Lunsford, G. H. and Ford, J., 1972, "Stability in Stochastic Regions," J. Math. Phys. 13, 700.

Mel'nikov, V. K., 1962, Dokl, Akad. Nauk. 144, 747.

Moser, J., 1968, "Lectures on Hamiltonian Systems," Memoits Amer. Math. Soc. No. 81, Providence, R. I.

Poincare, H., 1899, Les Methodes Nouvelles de la Mecanique Celeste, Vol. III.

Roels, J. and Henon, M., 1967, Bull. Astron. Serie 3, 267, "Invariant Curves of AP Mapping."

Chapter XI

Cosmic Ray Trajectories

Lemaitre, G. and Vallarta, M. S., 1936, Phys. Rev. 50, 493.

Schremp, E. J., 1938, Phys. Rev. 54, 158

Godart, O., 1938, Ann. Soc. Sci. de Bruxelles 58, 27-41.

Vallarta, M. S., 1938, "Allowed Cone of Cosmic Radiation," University of Toronto Studies, Applied Math Series No. 3, U. of Toronto Press.

Banos, A., Uribe, H., and Lifshitz, J., 1939, Rev. Mod. Phys. 11, 137.

Störmer, C., 1955, "The Polar Aurora," Oxford.

DeVogelaere, R., 1958, "Symmetric Periodic Solutions" in "Contributions to the Theory of Nonlinear Oscillations," Vol. 4, pp. 53-84, Lefschetz, S., Ed., Princeton University Press. (See p. 80.)

Chapter XII

General Reference Books

Hill, G. W., 1905-1907, "Collected Mathematical Works," Carnegie Inst. of Washington, (Washington, D. C.).

Moulton, F. R., 1920, "Periodic Orbits," Carnegie Inst. of Washington.

Szebehely, V., 1967, "Theory of Orbits," Academic Press.

Special References

Alfriend, K. T., 1970, $L_4(1/2)$, Cel. Mech. 1, 351.

Bartlett, J. H., 1964, Mat. Fys. Skr, Dan. Vid. Selsk, 2, No. 7.

Bartlett, J. H., and Wagner, C. A., 1965, Mat. Fys. Skr. Dan. Vid. Selsk, 3, No. 1.

Broucke, R. A., 1968, "Periodic Orbits, Earth-Moon Masses," Jet Propulsion Lab. Tech. Report 32-1168.

Carpenter, L. H. and Stumpff, K., 1968, "Periodic Orbits, (k+1)/k," Astron. Nachr. 291, 25.

Deprit, A. et al., 1967, "Trojan Orbits, Earth-Moon," M.N.R.A.S. 137, 311.

Deprit, A. and Henrard, J., 1965, "Asymptotic Orbits, L_1, L_2, L_3," Astron. J. 70, 271.

Deprit, A. and Price, J. F., 1969, Astron. & Astrophys. 1, 427; 3,88, "$L_4(1/3$ and $1/4)$."

Deprit, A. and Rabe, E., 1969, "L_4 for 1/12," Astron. J. 74, 317.

Deprit, A. and Henrard, J., 1967, "L_4 Short and Long Periods," Boeing S.R.L. D1-82-0622.

Deprit, A., 1968, "Hecuba Gap and the Hilda Group," Astron. J. 73, 730.

Hénon, M., 1965a, "Orbites, $\gamma = 0$," Ann. Astrophys. 28, 499.

Hénon, M., 1966, Bull. Astron. Ser. 3, I, 1, 57.

Hénon, M., 1966, Bull. Astron. Ser. 3, I, 2, 49.

Hénon, M., 1969, Astron. & Astrophys. 1, 223.

Hénon, M., 1970, Astron. & Astrophys, 9, 24.

Hénon, M., 1973, "Vertical Stability, Hill's Case," Astron.
      & Astrophysics 30, 317.

Henrard, J., 1969, "L_4 Long Periods," Boeing S.R.L.
      D1-82-0910.

Jeffreys, W. H., 1970, "Periodic Orbits, etc.," (Sao Paulo
      Symposium) p. 397, Reidel; Symp. Mat. Bologna 3, 13.

Mullins, L. D. and Bartlett, J. H., NSF Technical Report
      No. 2, June 1, 1972; 1973, Cel. Mech. 7, 421.

Epilogue

Heisenberg, W., 1968, in "Topics in Nonlinear Physics,"
      Zabusky, N. J. Ed., Springer.

Galactic Orbits and Stability

Lindblad, B., 1959, "Galactic Dynamics," Handbuck d. Physik,
      v.53, pp. 21-99, Springer.

V. Hoerner, S., 1960, 1963, Z. Astrophysik 50, 184; 54,47
      ($n < 25$).

Ollongren, A., 1962, "Three-Dimensional Galactic Stellar
      Orbits," Bull. Astr. Inst. Neth. 16, 241.

Aarseth, S. J., 1966, M. N. Roy Astr. Soc. 132, 35 ($n \leq 100$).

Barbanis, B., 1968, "Spiral Field Orbits," Astron. J. 73,
      784.

Miller, R. H. and Prendergast, K. H., 1968, Astrophys. J.,
      151, 699.

Bok, B. J., 1970, "Structure and Dynamics of our Galaxy,"
      in Vol. 1 of "Galactic Astronomy," Chiu, H. Y. and
      Muriel, A., Gordon & Breach.

Contopoulos, G., 1970, "Periodic Orbits, etc.," (Sao Paulo
      Symposium) p. 322, Reidel.

Hohl, F., 1970, I. A. U. Symposium No. 38, p. 368, Reidel.

Lin, C. C., 1970, "Theory of Spiral Structure," in Vol. 2 of "Galactic Astronomy," Chiu, H. Y. and Muriel, A., Gordon & Breach.

Miller, R. H., and Prendergast, K. H., 1970, Astrophys. J., 161, 903.

Miller, R. H., Prendergast, K. H., and Quirk, W. J., 1970, I.A.U. Symposium No. 38, p. 365, Reidel.

Martinet, L. and Hayli, A., 1971, "High-Velocity Stars," Astron. & Astrophys. 14, 103.

Quirk, W. J., 1971, Astrophys. J., 167, 7.

Saslaw, W. C., Valtonem, M. J., and Aarseth, S. J., 1973, "Gravitational Slingshot and the Structure of Extragalactic Radio Sources," Astrophys. J. 190,253.

Index